Asebe Regassa Debelo

Wilderness or Home?

// Beiträge zur Afrikaforschung
ISSN 0938–7285

Herausgegeben vom Institut für Afrikastudien

in Verbindung mit der
Bayreuth International Graduate School of African Studies, BIGSAS
der Universität Bayreuth

Band

66

LIT

Asebe Regassa Debelo

Wilderness or Home?

Conflicts, Competing Perspectives and Claims of Entitlement over Nech Sar National Park, Ethiopia

LIT

Printed with the support of the German Research Foundation
(Deutsche Forschungsgemeinschaft), DFG

Bibliographic information published by the Deutsche Nationalbibliothek
The Deutsche Nationalbibliothek lists this publication in the Deutsche Nationalbibliografie; detailed bibliographic data are available in the Internet at http://dnb.d-nb.de.

ISBN 978-3-643-90709-7
Zugl.: Bayreuth, Univ., BIGSAS, Diss., 2014

A catalogue record for this book is available from the British Library

Klosbachstr. 107
CH-8032 Zürich
Tel. +41 (0) 44-251 75 05   Fax +41 (0) 44-251 75 06
E-Mail: zuerich@lit-verlag.ch   http://www.lit-verlag.ch

Distribution:

In the UK: Global Book Marketing, e-mail: mo@centralbooks.com
In North America: International Specialized Book Services, e-mail: orders@isbs.com
In Germany: LIT Verlag Fresnostr. 2, D-48159 Münster
Tel. +49 (0) 2 51-620 32 22, Fax +49 (0) 2 51-922 60 99, E-mail: vertrieb@lit-verlag.de

In Austria: Medienlogistik Pichler-ÖBZ, e-mail: mlo@medien-logistik.at
e-books are available at www.litwebshop.de

Table of Contents

Acknowledgements **v**

List of Illustrations **ix**

Acronyms **xi**

Part One: The Research Settings, Methodology and Theoretical Frameworks

Introduction 1

Chapter One: Setting the Context and Methodological Frameworks **3**

1.1 The Problematic of Protected Areas Conservation 3
1.1.1 Some Conceptual Debates on the Politics of Nature Conservation 3
1.1.2 Problematizing Conservation Enclosure in Ethiopia 9
1.1.3 Situating the Debate of Enclosure within the Post-1991 Political Reconfiguration 16

A Point of Departure 20

1.2 Research Questions 25

1.3 Methodological Frameworks 26
1.3.1 The Research Framework and the Choice of Methodological Perspectives 26
1.3.2 Methodological Approaches and Methods 32
1.3.3 The Challenge of Encountering Competing Actors in the Field 39

1.4 Actor-Oriented Approach as Analytical Concept 46

1.5 Ethical Considerations 48

1.6 Structure of the Work 50

Chapter Two: Conceptual and Theoretical Frameworks **53**

2.1 Conceptual Frameworks 53

2.2 Theoretical Frameworks 56
2.2.1 Notions of Nature-Culture Debates: Dualism vs. Mutualism 56
2.2.2 Political Ecology of Conservation 75

Chapter Three: The Study Area, the People and their Relations to the Land **95**

3.1 The Study Area 95

3.2 The People and their Land 98

3.2.1 The Guji: the People and their Economy, Settlement History and their Conception of Nature 99

3.2.2 The Koore and their Land 123

3.3 The Establishment of State Farm and Arba Minch Town: State expansion to the Periphery 134

Part Two: Political Ecology of Nature Conservation

Introduction 139

Chapter Four: The Exportation of National Park Ideals to Africa **141**

Introduction 141

4.1 The Countryside Ideal and 'Wilderness' Conservation 143

4.1.2 From Countryside Ideal to National Park Model in Africa 148

4.2 Colonial Legacies and Post-colonial Dynamics in Conservation Discourses 162

4.2.1 National Parks as Colonial Legacies 163

4.2.2 Global Influences 165

4.2.3 Changes in Value Systems among African Elites 167

4.3 Center-Periphery Relations in Ethiopia and the Establishment of Nech Sar National Park 170

4.3.1 The Center-Periphery Relations in Ethiopia 170

4.3.2 The 'Natural' South and the Establishment of Nech Sar National Park 176

Chapter Five: Ethnic Federalism and Contested Spaces in Ethiopia: Situating the Nech Sar National Park in the Political Discourse **191**

Introduction 191

5.1 The Reconfiguration of the Periphery within the New Political Order 192

5.1.1 Recognition of Cultural Pluralism 196

5.1.2 Decentralization of State Power 198

5.2 The Lived Reality of the New Political Order (mixed results) 201

5.3 Ethnic Federalism and Community Voices in the Periphery 209

5.4 The Experience from the Nech Sar National Park 214
5.4.1 Ethnic Federalism, Decentralization and the National Park 215
5.4.2 Privatization of Nature Conservation–the African Parks Foundation 222
5.5 Extending the Nech Sar Case beyond Local Contexts 231

Part Three: The Nech Sar National Park: An Arena of Contestation/Negotiation on Nature-Culture Relations

Introduction 237

Chapter Six: The Ethiopian State: Conservationist, 'Developmental' or Encroacher? 241

6.1 Situating the Conservation-Development Debate within the Identity of the Ethiopian 'State' 241
6.1.1 The State and Peripheral Representations 244
6.1.2 Development and Conservation in Ethiopia–Perspectives from the State 250
6.1.3 Towards the 'Making of Environmental Subjects'? 254
6.1.4 The National Park as a Field of Contention between the Two Regional States 263

Chapter Seven: Grassroots Actors, a Business/Conservation Organization and Human Rights Advocacy Groups 273

7.1 Grassroots Actors 273
7.1.1 The Guji Oromo 274
7.1.2 The Koore 294
7.1.3 Other Local Actors/Groups 302
7.2 African Parks Foundation and International Human Rights Advocacy Organizations 310

Chapter Eight: State Intervention and Local Responses 317

Introduction 317
8.1 Forms of State Intervention 318
8.2 Forms of Local Responses to State Intervention 330
8.2.1 The Guji and their Responses to Conservation Enclosure 332
8.2.2 The Koore and their Responses to Conservation 350
8.3 Ethnic Politics, Conservation Enclosure and Local Resistance Nexus 357

8.4 Situating the Contestation in Domination/Resistance Relations 364

Chapter Nine: Conclusions: Situating Conservation Enclosures beyond 'Environmental Degradation Thesis' 373

9.1 Situating the Establishment of National Parks in Africa beyond the 'Environmental Degradation Thesis' 376

9.2 Understanding the Nech Sar National Park from the Center-Periphery Relations in Ethiopia 381

9.3 Problematizing Participation 386

9.4 Theoretical and Policy Considerations 392
- 9.4.1 Epistemological/Theoretical Considerations 393
- 9.4.2 Policy Dimensions 395
- 9.4.3 Issues for Future Research 401

Bibliography 405

List of Informants 423

Appendices 429

ACKNOWLEDGEMENTS

The successful completion of this study is the result of direct and indirect contributions from institutions and many individuals. Above all, institutional accommodation and support I received from Bayreuth International Graduate School of African Studies (BIGSAS) has played a paramount role in the overall PhD study. Thanks to BIGSAS for providing me suitable work environment, and fieldwork and conference grants. Likewise, I am indebted to the German Academic Exchange Service (DAAD) for the scholarship it provided me all throughout my study. My stay in Germany would have been much difficult had it was not been for DAAD's financial support. In addition to these basic building blocks in the study process, the academic dimension of the support deserves significant recognition. I would like to express my heartfelt appreciation to my supervisor, Professor Dr. Beate Lohnert who has consistently been encouraging me to pursue independent work while at the same time providing me with the necessary supervision. I am also grateful to my first Mentor, Professor Dr. Georg Klute whose critical comments on some chapters of my research made me consider various perspectives in the work. My second Mentor, Professor Dr. Kocra Assoua also deserves my appreciation for being interested to read my research at the initial stages.

I am especially indebted to my wife, Chaltu Feyisa, whose patience, care, love and encouragement has been my strength whenever I felt exhausted, bored and challenged. Chaltu has been central in my academic success by shouldering several responsibilities. I was able to focus on my research only because she has never complained whenever she passed sleepless nights and restless days taking care of our children. My two daughters, Telile and Meeti, have always been refreshing my exhausted mind when I go home late in the evening. Telile once said, "Papa! I came to Germany to support you." This stunning word of encouragement from a three years old kid has been clicking my mind whenever I was in need of support. On top of these all, my mother, Dure Roro, played an extraordinary role in my academic success beginning from my childhood to this level. Her sense of pride, confidence and happiness in my achievements were part of my success.

Many individuals have been persistently encouraging me to pursue the long journey with determination. My special heartfelt thanks go to my friends Dr. Meron Zeleke, Dr. Dereje Feyisa, Megerssa Tolessa and Asefa Melka. Dr. Meron's unreserved moral support during her stay in Bayreuth and her uninterrupted encouragements since then have been very crucial in my successful completion of the study. Moreover, Dr. Meron deserves special thanks for her thought provoking comments and reflections on part of my chapters. I am also indebted to her mother, Mrs. Roman Tefera who rendered me moral supports. The initial stage of my PhD program, which would have been a period of confusion and frustration, went smoothly thanks to Dr. Dereje's incredible scholarly comments. My close friends, Megerssa and Asefa, have always been encouraging me all throughout my study periods.

I am also grateful to many other friends and colleagues for their supports. My gratitude goes to Dr. Ahmed Hussen, Berhanu Belayneh, Gashaw Hunde, Abiyot Legesse, Daniel Fekadu, Zelalem Tesfaye, Samson Seid, Arega Degife, Dr. Senbetie Toma, Roman Gobena, Dr. Tadesse Jaleta, Addisu Zegeye, Bayisa Faye and Temechegn Gutu. Moreover, Professor Hizekiel Gebisa, Dr. Ongaye Oda, Dawit Tilahun and Julian Sommerschuh helped me in reading part of my work for language and technical corrections. Thank you all! Furthermore, I learned a lot from all my colleagues in BIGSAS through formal and informal academic discussions. Most importantly, we had a vibrant discussion group with my colleagues Tamer Abd El Kereem, Jimam Lar, Peter Narh and Abdallah Mkumbukwa. In addition, I would like to extend my heartfelt gratitude to Mrs. Anette Volk, BIGSAS secretary, for her wonderful technical supports in setting the book formats. Thank you all for your critical comments that sometimes prompted me think twice before writing a word.

Furthermore, the fieldwork and the entire knowledge production process would have not been easy without kind cooperation on the side of my informants, field assistants and government officials at different levels. I am particularly indebted to my informants who shared with me their invaluable time and knowledge during my fieldwork. I am also very grateful to my field assistants who paid much sacrifice during the arduous fieldwork periods among the Guji and Koore communities. Thanks also to my brother in-law, Demisse Feyisa, for accompanying me during some of my fieldwork trips and for being part of my strength particularly during the moments of climbing and descending the Amaro Mountains that took us a total of eight hours. Thanks also to the government officials whose research permission and information were vital in opening the gate to the sources of knowledge. In this regard, I owe sincere appreciation to Obbo Kebede Jilo, an official in Borana Zone, who clearly understood my research objectives and connected me with other officials without reservation. I am also grateful to Dilla University for giving me study leave for the current program. I feel at home with Dilla University because it is where I began my academic career and gained many experiences in teaching, research and administration.

Finally, thanks to the Almighty God for giving me strength during hard times and hope at times of hopelessness all throughout my life.

Asebe Regassa, March 2014
Bayreuth, Germany

LIST OF ILLUSTRATIONS

Figure 1: Location Map of Nech Sar National Park 11
Figure 2: Map of the Federal Democratic Republic of Ethiopia 17
Figure 3: Photo Showing our Boat Trip on Lake Abaya 42
Figure 4: "Map of the Nech Sar National Park" as depicted by SNNP Regional State 96
Figure 5: "Nech Sar" (White Grass) 97
Figure 6: Guji's Pastureland Burning Practices 109
Figure 7: Mango and Avocado Trees in Golbo/Tsalke 113
Figure 8: Guji and Koore women exchanging crop and animal products in Golbo market 115
Figure 9: Amaro District Map 123
Figure 10: Vegetable Market in Golbo/Tsalke 132
Chart 1: From Countryside Ideal to National Park Model in Africa 143
Figure 11: A Few Areas of Large-Scale Agri-business in Ethiopia 207
Figure 12: Actors and their Interactions 240
Figure 13: Guji Oromo slaughtering a Calve during a rite of transition in the *Gadaa* System 278
Figure 14: Guji Resource Mapping done through Participatory Mapping Method 288
Figure 15: Horny trees grown in parts of the national park after the banning of burning 289
Figures 16 &17: Farm Land in Amaro Highland; and Coffee Plants in Tsalke/Golbo Lowland 296
Figure 18: Resource Sites in/around the Park Important for the Koore 298
Figure 19: Guji and Koore Women in Golbo/Tsalke Market 327
Figure 20: Guji individuals traveling on local boat (Jatara) 328
Figure 21: A Guji boy with his AK47 gun 340
Figure 22: Guji children in classroom attending lesson, Arda Guddina Primary School 349

ACRONYMS

APF	African Parks Foundation
EPRDF	Ethiopian Peoples' Revolutionary Democratic Front
EWCA	Ethiopian Wildlife Conservation Authority
FDRE	Federal Democratic Republic of Ethiopia
GTP	Growth and Transformation Plan
IUCN	International Union for Conservation of Nature
OPDO	Oromo People Democratic Organization
PA	Protected Area
SNNP	Southern Nations, Nationalities and Peoples
TPLF	Tigray People Liberation Front
UNESCO	United Nations Educational, Scientific and Cultural Organization

PART ONE: THE RESEARCH SETTINGS, METHODOLOGY AND THEORETICAL FRAMEWORKS

Introduction

Since the onset of the colonial era in the late nineteenth century, Africa has become one of the most glamorized places on earth for the conservation of nature. Despite a long history of human habitation, the continent has been represented within mainstream Western media and academic works as 'natural', 'wild' and 'unspoiled' by human influences (Adams and McShane 1996: xii). While it was part of the colonial discourses of 'othering' non-Western traditions, ways of life and territories, such representation of Africa eventually gave rise to the argument that unless they were 'rationally' managed, natural resources, particularly wildlife in Africa would be 'threatened' to extinction by 'unwise' management and utilization on the side of Africans. Eventually, the colonial powers imposed hegemonic colonial rule, knowledge and worldviews upon colonial subjects through the territorialization of African customary land into protected areas, whereby they criminalized customary rights and practices of resource use, ownership and management. The situation becomes more complex when we incorporate issues of environmental degradation and biodiversity debility into our debates on the human dimension of environmental conservation. Moreover, the involvement of multilayer actors[1] with competing interests, actions, perspectives and narratives in the process of nature conservation necessitates a comprehensive investigation into the political, economic, cultural, ecological and social dimensions of environmental conservation, particularly in the form of national parks.

Despite its history as a country never being colonized by external forces, Ethiopia has much in common with other African countries in importing the dominant Western conservation discourses that consider the separation of humans from non-humans as an appropriate approach to nature conservation. Following its involvement in the international environmental conservation initia-

[1] I use the concept, "multilayer actors" in this study to indicate actors at different levels that include local, regional, national and international actors who interact, cooperate, contest, negotiate or at times enter into conflict in the struggle over environmental resources such as the case in the Nech Sar National Park in Ethiopia.

tives in the early-1960s, Ethiopia introduced the protectionist conservation principles upon its establishment of national parks in the 1960s and 1970s. Taking the case of Nech Sar National Park in southern Ethiopia, which was established in 1974, this work investigates how different actor groups conceptualize, define, negotiate and contest nature conservation in the country.

This first part of the study encompasses the first three chapters. Chapter one begins the investigation by problematizing notions of nature conservation in the form of protected areas in Africa in general and the Nech Sar National Park in particular. In doing so, it hints at how the colonial representation of the African landscapes as 'natural' and 'wild' informed the establishment of national parks based upon ontological orientations of human-nature dichotomy. Taking the case of the Nech Sar National Park, the chapter frames the problems, research questions and methodological frameworks that are central to the entire research project. Chapter two articulates the conceptual and theoretical frameworks drawing upon the notions of nature-culture relations and the political ecology of conservation, providing theoretical debates that are important platforms for analytical discussions. The third chapter deals with human-environmental interactions in the study areas, thus grounding the theoretical discussions in empirical realities. Under this chapter, I will present the settlement history of the people in and around the area that later became the Nech Sar National Park and thereby discuss some empirical data on human-environmental interactions from multidimensional angles.

CHAPTER ONE: SETTING THE CONTEXT AND METHODOLOGICAL FRAMEWORKS

1.1 The Problematic of Protected Areas Conservation

1.1.1 Some Conceptual Debates on the Politics of Nature Conservation

The practice of setting aside nature as a protected area (game reserves, sanctuaries and national parks) is premised on the perspectives of delineating a boundary between humans and non-humans (Cronon 1996; Merchant 2003). Debates concerning nature conservation among scholars, environmentalists, human rights activists, nation states and local communities in different parts of the globe have been contentious, given that there are different ontological and epistemological foundations in the understanding of nature-culture relations (Pálsson 1996). The history of Protected Areas (PAs) in the form of national parks in the modern era can be traced back to the Yellowstone National Park in the United States, which was established in 1872 (Brockington et al. 2008). In the context of this study, I use the concept Protected Areas to refer to a geographically defined area that is controlled, designated, managed and utilized by powerful actors for the sake of political, economic, social, cultural and environmental/conservation objectives. Although there are conventional definitions of PAs, such as those in Chape et al. (2008: 9), defining the concept as, "a geographically defined area which is designated or regulated and managed to achieve specific conservation objectives", I argue that such a definition does not fully capture the complex motives and involvement of various actors in the establishment of PAs. Indeed, the above definition does not thoroughly address inter-actor relations and issues of power positions among competing actor groups.

In the African context, different societies in the continent have multidimensional indigenous approaches to nature conservation, as well as specific norms, values and beliefs that shape their interaction with their surroundings. In these indigenous approaches and practices, humans are perceived as part of the natural environment. During the colonial period, Europeans introduced a dualistic view that separates humans from non-humans (Adams and McShane 1996). Within this perspective, Europeans' representation of Africa as pristine territory came to underpin ideologies that designated Africa as a continent deserving of preservation. As colonial rule progressed, colonial officials, conservationists and private investors continued to translate Western views of wilderness into practice by establishing national parks from which native Africans were either evicted or they were denied their customary rights to utilize the resources (Neumann 1998). This conception of nature in Western thought has brought attempts to dissect human-nature relations to areas of conservation. This eventually led to the introduction of a protectionist conservation paradigm that often excludes local people from their customary lands (Campbell 2005). Despite differences in the objectives and approaches in the establishment of national parks, most of such projects resulted in the exclusion of Africans from 'nature' and the restriction over access to their customary resources.

However, the colonial discourse and practices did not end with the passing of the colonial period. Both the colonial and postcolonial states in Africa adopted various forms of power, such as the eviction of local people and enclosure of formerly common-access resources in the process of making these resources legible for governance and commoditization. In a similar manner, as Scott (2009) rightly argues in the case of South East Asia, nation-states' attempts to convert common property land into privately owned or state controlled enclosures were mainly processes of controlling the society. Scott (2009: 5) further contends that such state projects are designed "to ensure that their economic activity was legible, taxable, assessable, and confiscatable or, failing that, to replace it with forms of production that were [replacing for example pastoralism by sedentary agriculture]." Common-access pasturelands, forests, fishing areas and hunting grounds were enclosed as national parks or game reserves in many parts of Africa, perpetuating the "tragedy of enclosure" (Bryant and Bailey 1997: 154). By tragedy of enclosure, following Bryant and Bailey, I refer to the social, cultural and economic crisis enclosure brings to the people living in or adjacent to protected areas, and to the ecological and biodiversity implications of such practices that often threaten the livelihood of local people. Against this background, I argue that the establishment of protected areas or enclosures and the subsequent introduction of modern conservation discourses and practices have brought to the scene a power-laden relationship between states and other external actors on the one hand, and local communities on the other hand. In this regard, it is essential to position competing perspectives and practices such as questions of scientific and indigenous knowledge, state control and customary rights as well as state intervention and local responses within different power relations that embody domination and resistance as mutually constitutive aspects of power.

Despite rich literature on conservation enclosure in Africa, only a few studies have tried to comprehensively address the complex relationships between the practices of exclusionist conservation approaches, colonial discourses and European images of Africa. A few have proceeded beyond the 'taken-for-granted' discussions of the social and economic impacts of protected areas on the 'poor' local people (Garland 2008). Some scholars, such as Brockington (2002) and Igoe (2004), have emphasized the social impacts of colonial and postcolonial conservation enclosures in the form of evictions and restriction of local communities from access to resources that eventually led to impoverishment of indigenous communities. Others have delved into discussing how the colonial imposition of the 'wilderness' conservation disempowered the African local communities through asymmetrical power relations (Anderson and Gove 1987; Mackenzie 1990; Neumann 1998). Although the scholarship on conservation projects in Africa has attempted to establish a nexus between colonialism and the introduction of the national park ideals, works in this category do not adequately interrogate colonial rule and its conservation discourses. Such works take colonial imposition as inevitable without probing the motivations and ideologies behind the colonialists to introduce and promote the establishment of national parks based upon the nature-culture dichotomy. Moreover, they also fail to pay due attention to the outcomes of interactions between competing forms of environmental knowledge (Adams and McShane 1996; Garland 2008).

Neumann's (1998) seminal book, *Imposing Wilderness: Struggles over Livelihood and Nature Preservation in Africa*, was one of the pioneer works linking Western perspectives of nature and the exportation of the national park ideals to Africa. Neumann himself emphasizes the aesthetic or pristine dimension of nature in Africa as a major driving motive behind the colonial powers' establishment of national parks in the continent. He argues that "it is principally the vision of Africa as earthly Eden–a romanticized wilderness in opposition to the decadent metropole–that underpins the historical development of the national park ideal in colonial Africa" (1998: 18). In this way, Neumann underlines the glamorized and idealized landscape visions of Africa, which was imagined to be in sharp contrast to the despoiled metropolitan cities of Europe as the fundamental historical and discursive phenomenon behind the establishment of national parks in the region. While this line of argument is valid to a point, it glosses over complex political, ideological, economic, social and environmental/ecological motivations within the general colonial political and economic discourses of the period.

Many scholars wrote much concerning issues of colonialism, environmental degradation, conservation discourses and the establishment of protected areas in Africa (Anderson and Grove 1987; Grove 1990; Neumann 1998; Igoe 2004; Brockington et al. 2008). Nonetheless, we have limited information on how the nexus between pre-colonial European imagination of Africa, colonialism and the introduction of the national park ideals to Africa outlived the colonial era. For instance, unanswered questions include the following: how did the countryside ideal in Europe and the national park ideal in Africa resemble each other? If national parks and other forms of protected area conservation discourses and practices were attributed to colonialism, why did the postcolonial African states fail to dismantle this colonial institution, instead of building, strengthening and expanding it? What forms of knowledge, values and discourses did the interplay between colonial forms of conservation and African customary practices reproduce? Alternatively, what kind of discourses on and practices of nature conservation did postcolonial African elites and states inherit from the colonial powers?

All these questions call for broader investigation into the historical trajectories of the ideas and practices regarding the introduction of national park to Africa and into how the changes and continuities regarding conservation discourses and practices unfolded. Academic attempts to find answers to these questions would enable us understand changes and continuities in the politics of conservation enclosures during the colonial and postcolonial epochs in Africa. Garland (2008) problematizes essentializing colonial impositions of wildlife conservation in Africa because, according to the author, wildlife conservation trajectories in Africa were more informed by changes in perspectives and value systems that occurred as a result of interactions between competing discourses and perceptions among Africans and European colonial powers. Thus, the post-colonial conservation practices are products of such interactions, rather than being imposed by external actors. Along these lines, Mitchell (2012) illuminates how conflicts between the European settlers and the Khoisan indigenous peoples in southern Africa were more of ontological contradictions between the two groups (settlers and Khoisans) in terms of their representations of nature and the place of humans in it. Mitchell (2012) goes further in arguing that contestations between competing social groups over environmental resources that often emanated from competing perceptions of human-nature relations eventually compelled the colonial rulers to negotiate their understandings of human-environmental relations and livelihood practices, as in the case of settlers' appropriation of transhumant nomadism.

Most research works on conservation enclosures in Africa have rarely addressed two fundamental areas of interrogation. These are the issues of unmasking the political ecology of colonial conservation discourses and the questions of treating postcolonial conservation practices not only as inheritances of colonial impositions but also as modes of production and a product of contestations, negotiations, appropriations and changes in value systems among competing perspectives and actors (Garland 2008). This work addresses these issues by exploring beyond the common understandings of considering national parks as mere colonial impositions. It is the central argument of this study that struggles over environmental resources such as the case of the Nech Sar National Park have partly emanated from competing perspectives on human-nature relations, while partly reflecting forms of exercise of power/knowledge among competing social actors.

In discussions concerning issues of environmental conservation in the form of protected areas, it is important to contextualize the debate with regard to pervasive competing ontological positions on nature-society relations. It has been evident that the general popular imagination and academic tradition in the West has established an ontological dualism between nature and society in which the former was presumably alienated from the latter (Descola 1996; Merchant 2003). In contrast, most non-Western and non-industrialized societies have a holistic understanding of nature and culture, believing in the conjoint constitution of the human and non-human worlds (Braun 1997; Escobar 1999; Ingold 2000, 2005). While the dominant Western notions of nature-society relations distinguish the biophysical, human and supernatural power, local models in non-Western contexts are often based on the tripartite relations between these three domains, conceptualizing the living, non-living, and supernatural beings as conjointly constituted (Braun 1997; Ingold 2000). Although studies reveal the embodiment of the social and the physical, which in turn implies that different conceptions of nature are social and cultural constructions (Freudenburg et al. 1995), conservation practices in Africa are expressions of the dualist notion of nature-society relations. This is why most national parks in the continent still subscribe to policies of delineating humans and the wildlife.

For instance, in debates about nature conservation, the notion of dualism entails the separation of people from the natural environment, in presuming that human activities disrupt the 'natural beauty' or 'wilderness' of nature (Cronon 1996). This is why we should critically scrutinize these notions and the practices that have been translated from the discourses. As dichotomic thinking is the basis of modern science in dividing between object and subject, body and mind, nature and culture, the notion of dividing humans from non-humans in the conservation practices should also be understood within this broader way of thinking. In this regard, the exportation of different ideas, policies, ideologies and perspectives such as development paradigms, conservation discourses, 'good governance', democratization, Structural Adjustment Programs and, very recently, climate change protocols and policies on the fight against terrorism all reveal some aspects of asymmetrical power relations between various actors at different levels and are subject to different forms of contextual interpretations. Nonetheless, this does not rule out the agency of postcolonial African states in appropriating and re-imposing foreign discourses for their local political or economic interests.

In Ethiopia, successive regimes have adopted and used different foreign development models and conservation approaches in the process of empire-building cementing the control of the people, resources and territories in the southern part of the country since the late nineteenth century. The state also used such models as mechanisms of building legitimacy by presenting the people with development programs that seemed to 'improve' their lives. In this regard, different governments attempted to emulate development models from different parts of the world. However, the adopted socio-political and economic models were in most cases incompatible with local contexts and proved failure stories (Clapham 2006a). They have also been practicing similar approaches of imitating the West in conservation schemes, and particularly in the southern parts of the country, where the center has been continuously engaged in the processes of engulfing the periphery by establishing state controlled enclosures. As will be discussed in the next chapters, most of the national parks in Ethiopia were established in the southern half of the country, not because of mere ecological necessity but rather as part of the political and cultural representation of the region as 'natural' and 'wild' within the north-south dichotomy and within center-periphery asymmetrical power relations.

There has been a tradition of drawing boundaries between the north and the south in Ethiopia, whereby the north has been represented in mainstream society and academia as 'historical', whereas the south is portrayed as 'natural'. For instance, the Ethiopia saying that, 'historians go to the north whereas anthropologists go to the south' is more than a mere territorial representation of historical and anthropological research. Such a saying implies that people in southern Ethiopia have 'intact' culture to be studied by anthropologists, watched by tourists and preserved by conservationists. When it comes to foreign researchers in Ethiopia too, anthropologists mostly researched the south, whereas historians and archaeologists mainly focused on the north that represented the core of the Abyssinian state. The Ethiopian historiography, built on the myth of 'the Great Tradition', depicts the north as the source of Ethiopian 'civilization' and the south as 'unspoiled' natural environment, which later seems to have contributed to the representation of the territories in the south as 'wilderness' to be watched, protected, conserved and commoditized through tourism economy (Ullendorff 1965; Donham 1986; Turton 2011). This is quite analogous to the Western conception of nature conservation in Africa during the colonial period. Therefore, the establishment of many national parks and protected areas, including the Nech Sar National Park in the southern part of the country, should be understood within the macro-historical trajectories that have shaped modern Ethiopia since the late nineteenth century.

Contrary to the dominant/mainstream perspectives about the people and territories in the south, the peoples in southern Ethiopia have had their own conceptions of nature. For example, the Guji people–one of the research communities of this study–understand human, non-human and divine beings as mutually reinforcing constituents of nature. The Guji believe that any disruption in the relationship between these three constituents of their surrounding affects the overall dynamics (Baxter 1991; Asebe 2012c). Concisely, competing views of nature and the place of humans in it prevail between the state, local communities and other actors in the Nech Sar National Park case. The space now designated as the Nech Sar National Park represents different meanings and purposes attached to economic, cultural, religious, political and symbolic/identity issues for different actors involved in the competition over the territory. Therefore, understanding the multiplicity of views on human-environmental interactions from epistemological and ontological dimensions, political and economic motives and socio-cultural representations is essential before attempting to address issues of conservation interventions and local responses.

1.1.2 Problematizing Conservation Enclosure in Ethiopia

The plan to establish the Nech Sar National Park started in the 1960s, following the request from the then imperial government to UNESCO seeking technical support and expertise for the process of surveying potential territories for the establishment of protected areas. Based on the request from the Ethiopian government, UNESCO commissioned a team of five renowned experts in the field of conservation and national parks management, including Sir Julian Huxley, a former director-general of UNESCO, to survey potential areas to be designated as protected areas (national parks, game reserves and sanctuaries) (Huxley et al. 1963; Abiyot 2009). In 1963, the team submitted a forty-page report to the Wildlife Conservation Board of Ethiopia, which included Nech Sar among other seven proposed protected areas. These include Managasha National Park (located at about 20km to the west of Addis Ababa), Mathara National Park (in the Awash valley), Abijata Lake National Park (in the Rift Valley), Nech Sar National Park and a scenic National Park at the source of the Nile. it also included controlled wildlife areas (in the south-west and southern peripheries, including Lake Rudolf, Lake Stefanie and portions of Omo Valley, further north the

Dinder tributaries along Ethiopia-Sudan border) and, finally sanctuaries for very rare animals (this category included areas such as Semein Mountain, Arsi-Bale Mountains, and Afar and Somali Lowlands) (Huxley et al. 1963).

In these areas, the UNESCO team recommended the relocation of the local people, as in the case of pastoralist communities in the Awash Plains. The team came up with the exclusionist notion of nature conservation, which had been the guiding principle of conservation discourses in the West recommending, for example:

> The question of the nomadic Matahara people within a future National Park is a difficult one. At present, they have a reputation for keeping other people away and this may be of some advantage to the surviving wild animals. Experience from many other parts of Africa indicates, however, that the chances of the present small numbers of wild life multiplying and becoming sufficiently tame to be a major attraction will be remote unless these people and their stock can be induced to leave the National Park area, and also unless the cultivation which is taking place in certain places is stopped (Huxley et al. 1963: 22).

Accordingly, the government established a number of national parks in the 1960s and 1970s, namely Abijata Shala Lakes, Awash, Bale Mountains, Omo, Mago, Nech Sar and Semien Mountains national parks. By adopting the recommendations from the USESCO team (external 'experts'), who advocated strict separation between humans and wildlife in the conservation practices, the Ethiopian government exercised full control of the territories through territorialization and the enactment of administrative measures on people who were deemed to have 'violated' national park rules. Enclosure entailed the eviction and resettlement of local communities, restricted access to their customary resource areas and the criminalization of their ways of life practices, which involved ritual practices and extraction of resources from the protected areas. Another striking point that warrants mention is the question of compensation for people who were relocated from their customary land. Issues of compensation and citizens' rights to it go in line with the broader political context, which encompasses aspects of land rights and government's accountability in responding in accordance with the law pertaining to such rights (if any). In Ethiopia, land effectively remained under the control of the state across several successive regimes and particularly in the subjugated regions of the south. In this regard, the late nineteenth century conquest stripped the right of land ownership from the people in this region. Under such circumstances, issues of compensation would not have legal or political grounds because the state considers itself as the rightful owner of all land and its resources.

Although it is not yet being gazetted due to multiple competing claims and conflicts, the Nech Sar National Park was formally designated with the status of a national park in 1974, a few months prior to the downfall of the imperial regime. In fact, the government had already started to implement some forms of control over forest utilization, hunting and grazing in the area since the mid-1960s, strengthening its structured arrangement and administration following the designation of the territory as a national park in 1974. The major objective of the establishment of the park was initially for biodiversity conservation, and particularly concerning an endemic Swayne's hartebeest. However, the economic dimension of generating economic benefits through tourism projects was also central to the conservation objectives (Dessalegn 2008).

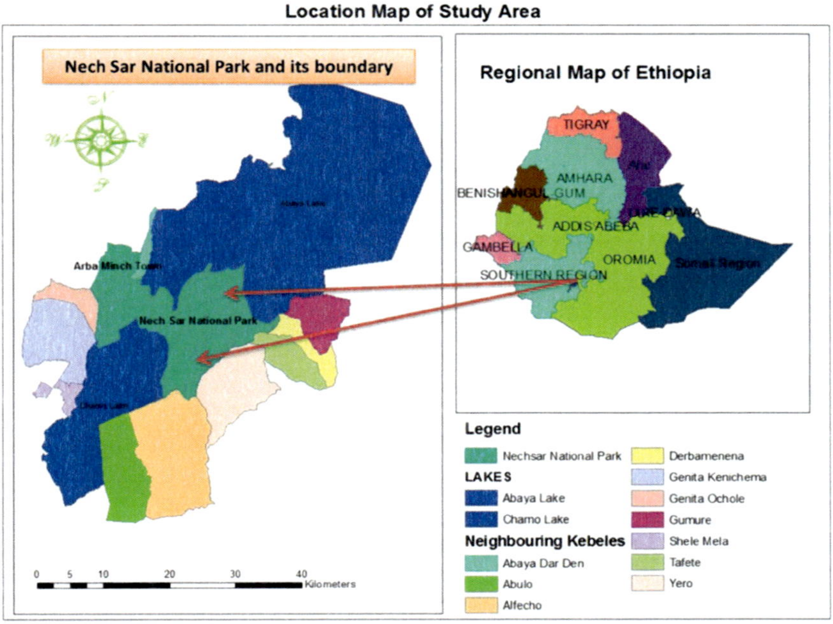

Figure 1: Location Map of Nech Sar National Park[2]
(Source: Zenebe Hailu, Arba Minch University, March 2014)

The park harbors over 800 species of plants, 91 species of mammals, 351 species of birds, and others such as insects. It also features a vast diversity of animal population, predominantly including Burchell's zebra, Grant's gazelle, the endemic Swayne's hartebeest, Nile crocodile in Lake Chamo, Lesser kudu, lion, wild dog and other animals (Dessalegn 2004; APF Annual Report 2007). More-

[2] Although this map of the study area shows the Nech Sar National Park in reference some adjacent villages, it does not show the Guji villages on the eastern side that were established in 2008.

over, the forest and the 'Forty Springs' adjacent to the town of Arba Minch on the western side of the park accentuate its scenic beauty (Briggs 2006). It was with dual aim of preserving immense natural resources and generating economic benefit from tourism for the country that the government established the national park. By the time of its establishment, the park covered an area of 514 square kilometers, although the exact size of the park remains contentious. Indeed, given that it has undergone different forms of expansion under the three regimes, some authors estimated its size to be greater than the above figure (Taddesse 2004).

Despite the government's claims that the area was uninhabited during and before being designated as a national park, Huxley et al. (1963) confirm the presence of pastoral and agrarian communities in the area. Local accounts and some historical sources also attest the presence of the Guji Oromo in the area to the west of Lake Abaya since the sixteenth century (Dhadacha 2006; Getachew 2007). Such competing narratives concerning the history of human settlement in the area have persisted as points of contestation between the state and local communities. The central issues in the study of protected areas in Ethiopia are related to two fundamentally interconnected situations. First, the country adopted the politics of nature conservation in the form of national parks based on the dominant and hegemonic nature-culture dichotomy that nullified the history and rights of local people from their land. Second, most national parks in the country, particularly those in the south were established on territories that were represented as 'natural', 'wild' and 'uninhabited'. Such representation of the territories delegitimized local communities' rights over their land. Therefore, the convergence between the global discourses of nature-culture dichotomy and the exclusionist approach of conservation practices, and the north-south dichotomy in Ethiopia has been part of the complex scenario in the conservation practices and contestations over the Nech Sar National Park.

As discussed at the beginning of this chapter, conservation enclosures or protected areas in Africa entail the engagement of various actors. Such inter-actor interaction includes aspects of contestations, negotiations and appropriations of different narratives, discourses and practices. Thus, studying protected areas without paying due attention to the voices of local communities is a futile exercise for a researcher or practitioner attempting to understand the complex human-environmental interactions. Hence, this study takes the Guji Oromo and Koore communities who live in and around the Nech Sar National Park as its core research subjects.

The Guji Oromo traditionally inhabited and used the park for pastoral activities and to some extent practiced farming along the eastern side of the park

(Getachew 2007). The Guji belong to the Oromo, which form the largest ethnic group in Ethiopia (Taddesse 2009). While the majority of the Guji live in Guji and Borana Zones of Oromia Regional State, according to the post-1991 political and administrative reconfiguration, a significant number of the people have inhabited different zones and districts in the Southern Nations, Nationalities and Peoples (hereafter SNNP) Regional State. However, over 5,000 Guji individuals who lived in and adjacent to the Nech Sar National Park remained outside of any formal government administrative structures from 1991-2008, because they were in neither the SNNP nor Oromia regional states.

The Koore ethnic group belongs to the speakers of the Omotic language group and lives in the SNNP regional state. The Koore were predominantly engaged in agricultural activities and strongly depended on the territory along the eastern and southeastern confines of the park for their livelihood. According to Koore oral information, the people moved downslope from the Amaro highlands for seasonal cultivation of the Tsalke Valley along the eastern escarpments of the national park around the mid twentieth century (Awoke 1985). Although the Koore did not inhabit the central part of the national park, they still claim that it belongs to their district because the former Amaro sub-province partly administered the area during the imperial and military regimes.

On the western side of the park, most of the low-income households of Arba Minch town established their livelihood by extracting forest resources from the park. As the town is constituted by members of diverse groups, it is difficult to distinguish this category in either purely ethnic categorization or in terms of specific livelihood practices regarding their interaction with the national park. Low-income sections of the town dwellers, mainly women and youth, sustain their living through income that they generate from selling charcoal, firewood, timber, forage and construction materials obtained from the forest adjacent to the town. Moreover, others have been engaged in fishing from the two lakes, Abaya and Chamo. These three social groups–the Guji agro-pastoral community, Koore farmers and the town dwellers–who I consider in this research as grassroots actors–have different claims. In claiming their rights of entitlement, ownership and utilization of the resources in the territory, these groups of actors also refer to different narratives, histories and resources (Dessalegn 2008; Bayisa 2011).

The military regime evicted the Guji and the Koore communities from the park in 1982 and restricted them from the area until they returned during the transition period in the early-1990s (Dowie 2009). In the post-1991 period, the former evictees and some others from adjoining Guji and Koore villages moved back to the park territory until the next round of eviction and relocation in 2004.

By the time the government agreed to transfer the park management to a Dutch company called African Parks Foundation (APF) in 2004, "police and park rangers torched about 463 Guji houses situated in the park and forcefully expelled the people out of the park boundaries" (Dowie 2009: 227). After the eviction, over 5000 Guji were cornered to the eastern side of the park, where they encountered day-to-day pressure from the park authority. In the same year of the Gujis' eviction, around a thousand Koore families were relocated to an area called Abulo Alfacho, located to the south of the park (Dessalegn 2008; Dowie 2009). In addition to the challenges faced by the local people due to the enclosure of their traditional resources, there have also been reports of biodiversity depletion, particularly concerning the wildlife in the national park, although the cause of such environmental problems remains contentious.

Following the establishment of a federal government in 1991, the government formally placed the park administration under the SNNP Regional State until 2004, when it transferred to the park administration to the African Parks Foundation. However, as the park adjoins two regional states–Oromia and SNNP–inter-regional competition seems self-evident. Indeed, some even argue that the removal of the Guji Oromo from the park area itself was part of an ethno-federalist fixation that has been premised on establishing administrative units along ethno-linguistic lines (Dowie 2009). In this regard, many actors with different interests, narratives of entitlement and varying degrees of power relations involved in competition over the park, in direct or indirect ways. Some of these actors include the state (the federal state, the regional states of Oromia and SNNP and their respective local administrative units), the two communities (Guji and Koore), African Parks Foundation, park administration and tour operators, among others. However, it would be overly simplistic to presume the interests and claims of entitlement of these actors to be unified in alignment. Because of internal variations such as gender, age, political affiliation and economic activity exist within each category of actors, variations in perceptions and conflicting interests over the territory are likely to be part of the complexity of the situation. Inevitably, there will be winners and losers. Given that it pays attention to different grounds of claims and referents of entitlement, this study seeks to address internal variations within the actor groups.

The selection of the Nech Sar National Park has been based on some fundamental reasons. First, apart from having existing across the three regimes–unlike the majority of national parks in the country–the Nech Sar National Park has remained particularly contentious due to its location on the border of Oromia and SNNP regional states. It thus has posed a strong challenge to the federal government due to strong claims of entitlement on the part of the two regional states. Second, it has been the most visited national parks in southern Ethiopia and the second in terms of tourism income generate at the national level. Nonetheless, in contrast to other national parks in the region, this national park has been noted for its exclusionist approaches to conservation, due to which local communities have been displaced, restricted and denied their right of participation and benefit. Third, its location in the region that was conquered by the Ethiopian empire and was subsequently considered as 'natural' will give us with the opportunity to establish a nexus between the establishment of national parks in Ethiopia and the broader trajectories of exportation of national park ideals to colonial Africa. Fourth, this national park has been rarely researched and not at all within the broader debates on nature-culture relations. Its proximity to the town of Arba Minch and scenery of the landscape that could be gazed down from the town, its long history, being a home to endemic animals and highly contested among multilayer of actors all provide opportunities to investigate how different actors define, understand and claim various aspects of rights over the national park.

Although the social and environmental repercussions of colonial forms of protected areas management are also common in other African countries, the Ethiopian case is a paradigmatic one to understand the colonial discourses of conservation, because this work deals with an African nation that does not have the colonial past, as opposed to its counterparts in other African countries. Unlike other scholars who are problematizing the colonial discourse of conservation in a given context where the colonial history and its legacies are quite visible and active, this inquiry takes another trajectory of analysis. It probes into two key areas: the dynamics of importation of nature conservation discourses and practices; and issues of power relations between the center and periphery within Ethiopia that have been created following the birth of the modern Ethiopian state in the late nineteenth century.

Therefore, by locating the establishment and management of the national park and different forms of contestations, negotiations and claims of entitlement to the territory within broader political and historical antecedents that shaped the center-periphery relations in Ethiopia, the study sheds some light on issues of state control over the people and their territories. For example, while discussing how the incumbent Ethiopian People's Revolutionary Democratic Front (EPRDF) government articulated rhetoric of conservation and development with reference to this particular national park, the study interrogates the state's strategies of channeling down its policies in its quest to exercise effective control over the periphery in which its predecessors significantly failed to achieve. Moreover, by paying attention to different actor groups who compete over the territory for different reasons, I also examine how the actors position their claims over the territory. Thus, the study investigates the issues from ontological representations of human-environmental interactions and day-to-day practices such as dependence on the territory for livelihood, conservation practices and tourism interests, as well as local responses to state intervention by employing an actor-oriented approach in the analysis. This work locates this specific investigation within broader debates on nature-culture relations by focusing on how notions behind the establishment of national parks were introduced to colonial Africa and sustained across the postcolonial period including Ethiopia despite some differences.

1.1.3 Situating the Debate of Enclosure within the Post-1991 Political Reconfiguration

In 1991, Ethiopia embarked on a new political experiment following the overthrow of the military regime by the Ethiopian People's Revolutionary Democratic Front (EPRDF)[3], Oromo People Liberation Front (OLF) and Eritrean People Liberation Front (EPLF). Following a period of transition (1991-1995), the EPRDF took control of state power and introduced a new political model and administrative structure based on ethnicity (Clapham 2002; Turton 2006; Dereje 2010). The formal recognition of ethnic questions and the institutionalization of ethnicity as a foundation for political, economic, cultural and social entitlements marked a new political order in the country. It was a fundamental departure from the past, both in deconstructing the ambitious homogenization policies of the state building project, as well as in presumably reversing the century-old hegemony of the center (Markakis 2011). In this regard, the 1995 Constitution of the Federal Democratic Republic of Ethiopia recognizes, above all, the right to self-determination of nations, nationalities and peoples, including the right to secession (FDRE Constitution 1995, Article 39). This implies the devolution of state power to lower administrative units through principles of ethnic federalism and policies of decentralization. In economic terms, the regime has tested different development paths from market liberalism to the recently introduced developmental state paradigm. In contrast to the neoliberal political and economic discourse that advocates limited state involvement in development schemes, the developmental state paradigm calls for strong leader-

[3] The EPRDF is a coalition party of four ethnic-based parties that include Tigray People Liberation Front (TPLF), Amhara National Democratic Movement (ANDM), Oromo People Democratic Organization (OPDO) and Southern Ethiopian Peoples' Democratic Movement (SEPDM). The EPRDF was formed in 1989 by the initiative of the TPLF as a strategy to frame its struggle within the broader questions of nations and nationalities in the country. Since 1991, EPRDF continued to be the only ruling party in the country's political system.

ship in the political and economic spheres and endorses the use of authoritarian means to protect the state's developmental goals (Meles 2012).

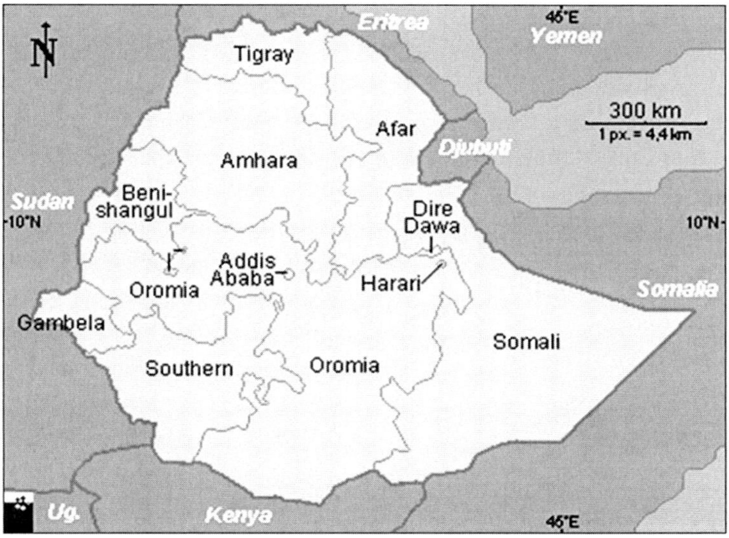

Figure 2: Map of the Federal Democratic Republic of Ethiopia
(Source: http://fotw.net/flags/et(.html)

Ethnic federalism was expected to address questions of local ownership of development programs through promoting decentralized governance and the participation of local communities in decision-making. The discourse presumably seeks to redress the past asymmetrical power relationship between the state and society, thus ensuring equal opportunities for citizens (Mohamed and Markakis 1998). Mohammed and Markakis (1998: 8) further argue that:

> Decentralization in Ethiopia is not seen merely as a device for the satisfaction of ethnic political demands, but also as the path leading to democratization through devolution of decision making in a manner that enables more people to influence the political process. Furthermore, since decentralization and democratization are regarded as requisite to development, the empowerment of ethnicity is intended to harness ethnicity to the purposes of development.

While historical antecedents in the country dating back to the late nineteenth century had significant influences in shaping center-periphery relations in Ethiopia, the post-1991 political reconfiguration has significantly influenced developments regarding the national park and its relations with the surrounding communities. The investigation of this political order is important because, while the new system has supposedly deconstructed the old political culture and identity of the Ethiopian state that previously marginalized the diverse nations and nationalities in the periphery, the government did not devolve power to the regional states in key areas such as land, budget, taxation, investment and privatization. Indeed, the center still maintains effective control and inspection of the activities in the periphery. Moreover, this Ethiopian oxymoron of a political project has also created another pattern of internal heterogeneity within the state, whereby claims of entitlement to environmental resources have now adopted ethnic and regional dimensions as the case of the Nech Sar National Park clearly illuminates.

Although the institutionalization of ethnic federalism was supposed to ensure the self-determination of the nations and nationalities in Ethiopia, many scholars have criticized the new political arrangement, particularly with regard to its practical implementation. For instance, as Dereje (2006) contends in his study of the Gambela case, despite a promising start that ensured formal and symbolic empowerment of the historically marginalized periphery, the new political order has proven to be a curse for the majority of the people in this region, having escalated the existing conflicts following the new political experiment. Detractors and supporters of ethnic federalism alike have questioned whether the political rhetoric has been effectively translated into practice (Assefa 2006; Clapham 2006b; Abbink 2009; Dereje 2010).

Despite claims of using participatory approaches to development, hierarchical structures persist under the new political landscape in the state-society relationship in the process of implementing development programs. Some scholars argue that the concept of 'development' has become a rhetorical device through which ethnic entrepreneurs exercise hegemonic control over the people rather than serving as a grassroots level platforms upon which all actors can negotiate based on their understandings (Zerihun 2004). In line with this concern, Mohammed and Markakis (1998: 8) cautiously note that the success of this unfinished political project depends on "whether the formal i.e. constitutional provisions of decentralization and democratization are realized in practice." Turton (2002) also invites us to further scrutinize the rhetoric concepts such as participation, decentralization and empowerment, which the EPRDF frequently uses in the process of broadcasting its power, political programs, development policies and conservation schemes. According to Turton (2002: 110), "the essential feature [of participation] is that local people are involved, from the start in design and implementation. It must surely be accepted by now that, unless people have real power to influence the way a project is designed and managed, they will not feel that it is 'theirs', whatever (often temporary) benefits they derive from it." As Jennings (2000) also discusses, participation is conceptually understood as:

> [I]nvolvement by a local population and, at times additional stakeholders in the

> creation, content and conduct of a program or policy designed to change their lives. Built on a belief that citizens can be trusted to shape their own future, participatory development uses local decision making and capabilities to steer and define the nature of an intervention (Jennings 2000: 1).

Participation is far beyond consulting a particular society about development programs or any other project; rather, it is intertwined with a political ecology approach that recognizes the power and capacity of local communities legitimate at local and national levels in decision-making processes. Put differently, believing in and promoting participatory development and conservation is to believe in the fundamental importance of self-determination of the subjects. Although participation as a methodological and theoretical instrument has presumably been a focus of many development and conservation practices over recent decades, it has failed to bring meaningful social changes, mainly due to its failure to engage with issues of power and politics (Mohan and Hickey 2004). According to Mohan and Hickey (2004: 66), participation should be situated within a wider range of citizenship rights through which "people extend their status, agency and rights as members of particular political communities thereby increasing their control over socio-economic resources." Issues of citizens' right to participation are also important in areas of nature conservation, because most conservation projects such as national parks often ignore the voices, knowledge and perspectives of local communities and rather impose scientific and expert-based approaches to conservation. It will be under such context that this study interrogates whether the Ethiopian political model under ethnic federalism enhances or hinders the customary rights of local communities in and around the Nech Sar National Park.

Critics of management approaches to protected areas point out that states and non-state actors often use these schemes as instruments of control over people and territories. Building his arguments on experiences from the Lower Omo Valley in Ethiopia, Turton (2011) argues that conservation and development discourses advocated by the Ethiopian state are being used as instruments of control and mechanisms of co-opting people under the guise of conserving the environment and improving people's living standards. Since the late nineteenth century conquest and the subsequent incorporation of territories in the south into the Abyssinian Empire, the state has employed different strategies to maintain its control over diverse groups in the subjected regions. Among these strategies, state control over the territories through state enclosures of indigenous territories under the guise of development and conservation were cases in point (Turton 2011). Peet et al. (2011: 28) substantiate this argument with a more general statement, suggesting, "Any effort to environmental conservation in its many forms, for better or worse is basically a form of environmental control." I would extend the above statement to how control over the environment entails control over the people, because any control over people's environmental resources is tantamount to putting them under the control of the state or non-state actors, particularly in developing countries where most people depend on the environment for different purposes.

In the case of the Nech Sar National Park, there are contestations between different actors over the management, ownership and utilization of the territory and its resources. Local communities claim that the state has relegated their customary rights of resource ownership and indigenous knowledge of resource management through its version of development and its protectionist conservation approach. They accuse the state of interfering in their livelihood and imposing upon them what to do and what not to do in conservation and development questions. This brings us to the fundamental themes of this study, namely conflicts, narratives of entitlement, competing views of nature-culture relations and power positions. While addressing local struggle between different actors, the study simultaneously probes into interplays between the national political and economic discourses and local conceptions of these issues.

A Point of Departure

The initiative for this project was initially conceived from my personal encounter with competing views of government authorities, local Guji and Koore communities and an international conservation company on people-animal relations in the Nech Sar National Park. The following narrative elucidates the intricate and multilayer competing interests, perceptions of nature and stories produced by different actors who compete for/argue about the park. In 2005, a large number of zebras moved out of the park and flocked to Guji villages close to Dilla University where I was teaching at the time. Apart from being a resident in a nearby town to the village, I had long research experience among the Guji since 2003, which provided me the opportunity to take a closer look at the dynamics within the community. The incident of outmigration of the zebras happened a few months after the local Guji and Koore communities were moved out of the park in the process of transferring the management of the park to the African Parks Foundation. Explanations of why the animals moved out of the park differed. The incident was intriguing because it contradicted the claims made by the government and the conservation company that the resettlement was carried out to enable wildlife to roam freely in the park territory and to enhance tourism activities. Government officials linked to the park administration listed apolitical, ecological and demographic factors such as hunting threats, fire and shortage of pasture as driving forces behind the withdrawal of the animals. In 2005, I received a peculiar explanation from a park biologist concerning the incident:

> We came to understand that the presence of people in and close to the park territory has been a major threat to the wildlife. The people encroached into the park territory through poaching, cattle trespass, setting fire, clearing the land for farming and so on. All these threatened the animals. Moreover, the contact between livestock and the wildlife enhanced transmission of contagious diseases from livestock to wildlife. On the other hand, tourists often complained when they saw wildlife mixed with livestock because the major interest of tourists for traveling such long distances and paying their money was not to see cattle. They could see cattle from everywhere. This, no doubt could affect our tourism sector and the image of the park. However, because of overgrazing that occurred actually before the resettlement, the animals migrated in search of pasture and water. The people who were relocated outside the park also continued hunting practices. Therefore, the animals moved out of the park territory to escape such threats (informant Girma, April 2005).

In contrast, local Guji elders saw the phenomena from a different perspective. A Guji elder whom I interviewed in 2006 during my fieldwork among the Guji for my Master's thesis commented that:

> The animals followed the people who were evicted out of the park. In the past, they lived together with our people [the Guji] and their livestock. Now, after the government removed the people, the animals were left alone in the hands of their enemies [predators]. In the past, the animals were able to stay close to the kraals and the people rescued them from predators. In our culture, we care for all animals. They are harmless just like cattle. We treat those animals like our cattle because our ancestors had oath with *Waaqa* [God] to take care of those animals. We know that *Waaqa* is pleased and reciprocates us by blessing our people and livestock when we take care of the animals. If people kill these innocent animals of *Waaqa*, the action would provoke retribution from *Waaqa*. It would cause punishment like infertility of people and livestock. Sometimes drought and flooding destroy Gujiland when people are involved in such activities. The animals migrated also for food. The government banned burning grass in the park. These animals could not get fresh grass in the park. No fresh grass regenerates for the animals. These people [the state/park officials] dismantled both the people and the animals. I am afraid that *Waaqa* may punish them with something bad (discussion with Boru Galgalo, July 2006).

The Koore had mixed views about the incident. While some of them shared the

Guji's arguments concerning the ban on burning and the subsequent shortage of fresh grass in the park as a cause for the animals' outmigration, others still share the park authority's views and cast the blame on their rival Guji community for threatening the animals through hunting. According to Gudo Mamacha, a Koore elder who was among those relocated to Abulo and Alfacho villages in 2004, "the outmigration of zebras was related to Guji's hunting practices in the park. The Guji retaliated against the animals following their eviction from the park core areas. The animals moved out of the park to escape Guji hunters" (Informant: Gudo Mamacha, April 2008). Although the arguments of the park officials and the Koore elders do not fully explain why the animals moved out of the park and sought shelter close to the Guji villages, given that it contradicts the alleged hunting practices of the Guji, it is nevertheless an equally important argument in revealing the divergence of perspectives among different actors. That is why it is now important to probe into the depth of such competing perspectives and narratives of each actor group.

The displacement of the local people from the park territory by the EPRDF government, which ostensibly advocates decentralization and self-government, has sparked criticism from international human rights organizations (Dowie 2009). The EPRDF government's actions attracted the attention of many observers of the Ethiopian political order, not because it was the first to take such actions of displacing local communities under the guise of conservation or 'development', but rather due to the paradox between the EPRDF's official narrative of local autonomy and its practical actions on the ground. Some of the groups, particularly the Guji Oromo, claim the area as their ancestral land with cultural, historical, livelihood and religious significances. Therefore, the displacement contradicted the EPRDF's constitutional provision, which gives ethnic groups the right to self-determination on territories that they historically inhabited, as well as deciding on development matters that affect their communities.

In the context of this narrative, it is worth mentioning that the specific territory now designated as a national park has different meanings for the actors based on their worldviews. Regarding contestation between social groups over space, Sharp et al. (2000: 26) once indicated that space is inscribed with different meanings suggesting that, "different social groups endow space, of course place, with amalgams of different meanings, uses and values. Such differences can give rise to various tensions and conflicts within a society over space for individual and group purposes, and over domination of space by the state and other forms of dominating social (and class) power." This hints at contestation between social actors, not only due to conflicts of interests (material, economic, political), but also owing to disparities in ontological orientations through which they represent space. In other words, if competing groups have conflicts of interests and differences in meanings ascribed to a particular space or territory, this reinforces tensions over environmental resources. It has also become evident that notions of establishing protected areas that mediates human-environmental interaction through categorization of 'appropriate' and 'inappropriate' activities have resulted in the exclusion of indigenous peoples from their customary homeland and have impinged on their access rights to their resources (Adams and Hutton 2007; Campbell 2005).

In this study, I will discuss the prevailing competing perceptions of nature, claims of entitlement and contestations over territories between and among actors, based upon the case of the Nech Sar National Park. Through this case, the project investigates the nexus between the official narratives of ethnic federalism, which presumably provides opportunities of self-government, local development, decentralization, participation and empowerment to local communities, and their lived realities. In this regard, it attempts to draw contours of connection/disconnection between the state's official narratives of development and conservation and local contexts. In doing so, the project probes into different constellations of actors, investigating how their conceptions and definitions of nature, conservation and development are contested and negotiated. By relocating the establishment of the national park within conservation and development discourses across three regimes in Ethiopia, the research project investigates how the state addressed customary rights, local knowledge of resource management, the livelihood dependence of local communities and the people-environment relationships, with particular reference to their relations to the wildlife in and around the park territory. The claims and struggles over the territory are signals to questions of entitlement and ownership rights, issues of legitimacy, cosmological schemes, and knowledge/power relations between different actors. The contestations and negotiations over the Nech Sar National Park embody complex interactions among local communities, the state and non-state actors. There have been elements of interactions and engagement in the struggle which take the form of competition between local communities and the government, as well as within local communities. Moreover, there have also been cases where local communities have struggled against intervention from non-state actors, including business organizations and conservationists.

Therefore, this study aims to understand multilayer contestations over the Nech Sar National Park by synchronizing the micro level claims and competitions with the macro level political, economic and environmental discourses. By doing so, it aims to unveil how competing notions of nature and conceptualizations of human-environmental interactions affect conservation discourses and practices in Ethiopia, and more importantly how local communities articulate, contest, negotiate and appropriate different perspectives of conservation and development.

1.2 Research Questions

From the short ethnographic description above, it is evident that there have been competing perspectives on notions of human-environmental interactions in the Nech Sar National Park. The territorialization of the Nech Sar National Park and the subsequent control over the multidimensional spaces of the local Guji and Koore communities and low-income town dwellers near the park have unequivocally threatened their livelihood conditions and disrupted their historical coexistence with the wildlife (Dessalegn 2004; Getachew 2007). This phenomenon shares much in common with many national parks in other parts of Africa that contributed to impoverishment and marginalization of local communities during colonial and postcolonial periods (Neumann 1998; Igoe 2004). Despite the government's rhetoric of creating a suitable environment for a thriving and vibrant tourism industry by removing the settled inhabitants from their longstanding homesteads, the biodiversity in the national park has not prospered in the last twenty years (APF Annual Report 2007; Freeman 2006). Animal population dramatically decreased since 1990s and particularly after the withdrawal of the African Parks Foundation in 2007. Thus, the fundamental problems related to the state projects of nature conservation should be addressed within the broader debates on nature-society relations.

Another complex phenomenon that I interjected in the previous section, the interrogation of which I postpone for the upcoming chapters, is the contradiction between ethnic federalism's promises of local autonomy on the one hand, and its lived realities that enhanced strong state intrusion into the periphery on the other. In the context of existing struggles over the national park, continuing state interventions and local responses to different forms of domination, this study will address four major research themes, each involving specific questions:

Theme One: Introduction of Notions of "Wilderness" Conservation to Ethiopia

1. What kind of link can be drawn between the Western notions of wilderness conservation, colonial conquest in Africa and the establishment of national parks in Ethiopia?
2. How can the establishment of the Nech Sar National Park be understood within the context of the empire-building process in Ethiopia?

Theme Two: Competing Actors over the Nech Sar National Park

1. Who are the major actors involved in the contestation/negotiation over the Nech Sar National Park, and what are their interests over the territory claims?

2. How do the actors understand nature-society relations? More specifically, how do they define the territory and the place of local communities with regard to resource conservation and ownership rights? Moreover, whose views have been hegemonic, and when and how did they prevail?
3. How was the relationship between different actors established? Who is allied with whom and why?

Theme Three: Conservation Enclosure, Ethnic Federalism and Community Voices/Participation:

1. How is the designation of the Nech Sar National Park understood within the context of ethnic federalism?
2. What forms of inter-group relationship has the new political reconfiguration brought into the contestation between different actors?
3. How were issues of local community voices, participation, customary rights and self-government addressed within the context of ethnic federalism on the one hand, and the government's efforts to promote conservation and tourism on the other?

Theme Four: Domination/Resistance Interface in Contestations over the Nech Sar National Park:

1. What were the patterns of state intervention regarding conservation practices in the Nech Sar National Park?
2. How have the local communities responded to external interventions?
3. What forms of power relations have prevailed in the interaction between the state and grassroots actors?
4. How have the actors been contesting, negotiating and coming to terms with different perspectives and power positions?

1.3 Methodological Frameworks

1.3.1 The Research Framework and the Choice of Methodological Perspectives

Before embarking on the murky debates on methodological choices, I will explain here how I associate a research process with a journey that one follows along different paths from one point to another. Let us imagine that a person travels to a certain destination, which might be a Western metropolitan city, an Amazonian forest, the African Kalahari Desert, an Australian Seashore or Alaskan Polar Iceberg. The knowledge that the traveler would report to colleagues upon completion of the journey would be highly influenced by how the traveler conceived the spaces, what he/she wanted to know and how he/she went about collecting data. One might travel to the Amazonian forests to either 'discover' the presumed 'savage' and 'wilderness' creatures of the Amazon or produce knowledge in collaboration with the indigenous peoples about how the people understand, conceive, and interact with the environment and thereby give meaning to the space. This forms a foundation for the traveler to make important decisions about what equipment to take with him or her, where to go and what to do before beginning the journey.

Likewise, a social science researcher should take decisions at every step of the research process, given that such decisions influence the outcome of knowledge production (Crotty 1998). The major purpose of this section is not to delve into the bulk of discussions on social science research paradigms, but rather to pinpoint specific research paradigms that have informed and shaped my methodological framework in this particular work. Four research paradigms are widely discussed in academic works, namely positivism, post-positivism, critical theory and constructionism (see Denzin and Lincoln 2005; Guba 1990; Crotty 1998). After outlining each research paradigm, I will explain why I selected one of these approaches: the social constructivist.

A positivist thinking that presupposes a social world as a fixed, predictable and external object informed traditional social science research. Positivist thinking assumes the existence of an objective 'reality' 'out there' to be discovered through scientific investigation (Phillmore and Goodson 2004). Traditionally, this paradigm was the focus of natural science researchers although it was subsequently introduced into social science research during the fifteenth and sixteenth centuries' colonial conquests. During this particular phase in social science research, early colonial researchers used evolutionary ethnography particularly to collect data asserting the assumption that natives were at the bottom of human evolutionary progress, with the West at the top of the pyramid. With its assumption of the social world as something fixed, predictable, controllable and measurable through an independent investigation of the inquirer, it served as an instrument to justify the predetermined conceptions of European travelers, explorers, missionaries and colonial administrators about native people in the 'New Worlds' (Phillmore and Goodson 2004). For instance, explorers and missionaries used positivist ethnography to collect data about the culture, religion, economy and governance systems of indigenous peoples. This research approach went along with colonialism, given that researchers who co-operated with the nineteenth century colonial governments in the process of suppression and exploitation of the colonial subjects mainly used this approach. In this regard, researchers' interpretations of the culture, governance systems, knowledge and ways of life of the natives were taken for granted as ultimate truth, which in turn informed colonial policies. The basic premises in positivist modes of thinking are based on its assumption of 'reality' as a fixed object independent of human knowledge, its views of the social world as predictable and the assumption that social research attempts to provide predictive, analytical and explanatory tools based on generalizations, which are then applied to broader populations (Wagner and Okeke 2009).

Some scholars criticized this perspective fundamentally for its rejection of multiple interpretations of 'reality'. It silences multiple voices, views and interpretations of the subjects, while representing a researcher as an 'expert' whose findings are taken as ultimate truth and as generalizable. Particularly postcolonial researchers have challenged the perspectives of positivism for its simplistic view of representing the 'Other' without making visible the multiple voices and views of the subjects. More importantly, positivism fails to recognize the diversity of individual experiences among different dimensions such as age, gender, race, class, religion, ethnicity, caste, education and other forms of internal categorizations. These aspects of a person's social position profoundly shape the perceptions, emotions, views and experiences he/she has with regard to a given social phenomenon. This reduces the credibility of generalizing research findings without due consideration of specific contexts (Crook and Garratt 2005).

A slight modification of positivism entered the social science research discourse in the 1970s, not as a complete rejection of the fundamental assumptions of positivism, but rather in the form of improvising on the existing paradigm, termed "post-positivist paradigm". While the ontological thought of positivists continued as the basic framework of social science research within the post-positivist paradigm, the approach to understanding the social world was changed. It recognizes that "although real world driven by real natural causes exists, it is impossible for humans truly to perceive it with their imperfect sensory and intellective mechanisms" (Guba 1990: 20). As a result, it urges inquirers to be 'critical' in their inquiry process by reflecting on why things become what they happened to be. Nonetheless, positivist rigor remained, as rule-bounded methodologies continued to be part of the inquiry process (Crook and Garratt 2005).

Another paradigm that shares some similarities with the positivist and post-positivist paradigms is the critical theory. This paradigm emerged neither as a counter argument against those discussed before nor an attempt to improvise on the former ones. It is not necessarily an immediate successor of the first two either; rather, the similarity concerns their assumption of reality. Critical theorists consider the social world from the perspective of inequality and asymmetrical power relations. They believe that people are naturally capable and have the creative and substantial potential to engineer social change but that some people block others from exercising their potential for change through different forms of oppression (Smith 2010). According to Kincheloe and McLaren (2002), aspects of hegemony and domination that prevail in the contexts of oppressive power can only be understood through critical approaches of investigating the underlying political, cultural, and socio-economic practices in a given context. According to this line of argument, critical theorists investigate how different forms of domination are expressed in popular culture, rituals, media, education and politics, as well as how people who are made subordinates to the dominating power respond to it. Driven by its presupposition of a social world as an arena of power struggle and inequality, critical theory positions the inquirer as an agent of social change and action through a value-oriented methodological approach in which he/she is supposed to reveal the 'truth' of oppression (Smith 2010). Like the positivist variants, this paradigm also blocks the mirror of unveiling multiple interpretations of reality. Another drawback of critical theory relates to its consideration of social science research as an instrument to liberate society from a certain social crisis, which in turn reduces the agency of research subjects in their struggle to position themselves within different social contexts. Although they criticize social inequality and asymmetrical power relations between different social actors, they also advocate for another hierarchical ordering of values and discourses by considering the researcher(s) as the main actor(s) of social change.

Since the 1980s, social science research has moved away from the notion that reality is 'out there' in the social world to the consideration of multiple realities held by inhabitants of the social world, which are deemed to be reachable through qualitative research (Denzin and Lincoln 2005; Phillmore and Goodson 2004). This entails a paradigmatic shift in many social science disciplines, from positivist notions of perceiving the social world as an objective reality that can be discovered, predicted, measured and quantified, to a perspective that accepts the social world as multiple and context-oriented. More importantly, it necessitates a way of looking at knowledge production is now considered a collaborative process between the inquirer and the subjects of knowledge (the inquired) (Phillmore and Goodson 2004). This approach is constructivist methodological perspective (Crotty 1998).

Crotty (1998) warns us that many of the fundamental challenges facing social science researchers are related to the core activity of decision-making and the justification of the strategy and method(s) adopted. Although methodology chapters in research works like this are often dominated by methods and field experiences, it is essential to make clear the form and nature of reality (ontology), what counts as knowledge (epistemology) and the way(s) of knowing (methodology) prior to embarking on the choice of research method (Jordan and Gibson 2004). Thus, our understanding of knowledge (ontological position) and our positions concerning the relationship between the inquirer and the knowable (epistemological position) frame our positions as social inquirers, including how we go about finding knowledge (methodological approaches) and the specific tools employed, namely the method(s). As I outlined at the beginning of this section, a social science researcher faces decision junctures at different stages of the research process, which profoundly influences not only the course of the process but also the outcome(s). Like a traveler who should decide on the type of travel equipment with which he/she has to be armed based on the nature of the journey and the purpose behind the journey, I had to make decisions on the methodological framework of my project. Accordingly, my methodological framework has been informed by a constructivist paradigm. In the next few paragraphs, I will discuss what informed this paradigmatic choice in respect of this particular research project. For this purpose, it is imperative to contextualize the methodological choice within the research project.

As I outlined earlier, this study addresses how different actors contextually define and redefine the contested territory, development projects, conservation programs and ownership rights over the land. It has been evident that the state and local communities have been internally differentiated and espoused different narratives of entitlement to the national park. To this end, I adopted a methodological paradigm that recognizes the persistence of multiple views and interpretations and helps me to gain comprehensive insights into the complex nature of the struggle over the territory.

With the involvement of human rights advocacy organizations, the competition over the Nech Sar National Park has international dimensions as well. As a space designated for conservation and 'ecotourism', it is a site where people from different backgrounds and perspectives have converged, including conservationists, tourists and human rights organizations that are allied with some of the local or national actors. In this regard, environmental concerns, economic aspects and human rights issues have become central agenda among the competing actors. For instance, while the involvement of the African Parks Foundation in the management of the park was a clear indication of how the 'marriage' between national and global economic and environmental discourses, reactions from international human rights organizations against the displacement of local communities in 2004 illuminate a different dimension of resistance against the hegemony of state and non-state actors. This hints at the actors' utilization of available resources (social capital, knowledge, power, agency and network) to achieve their claims or deterring others from gaining a certain advantage from the same territory.

Actors' production of different forms of knowledge in pursuing their claims over land, as well as the way in which conservation approaches, people-environmental interactions and development practices are defined, contested, articulated and negotiated, hint at multiple interpretations of 'reality'. For example, in the aforementioned story of migration of zebras out of the national park territory in 2005, competing epistemological positions are held by different actors regarding their understandings of environmental degradation, human-wildlife interactions and solutions to redress environmental problems. In this context, I had to ask myself whether I was going to produce a single interpretation of the incident or multiple interpretations representing different voices of the actors involved. The answer is that if I had attempted to take the first option, the knowledge production process would have silenced the voices of some actors while favoring others. To avoid such methodological bias, my inquiry process should embrace from the onset of the journey an understanding that various subjects produce multiple voices and interpretations about a particular social world. My role as a social science researcher is thus to bring these diverse narratives into a dialogue, not necessarily at the same practical level in the field, but through interpretive approaches while writing the interpretations of the research subjects.

The complexity extant in the various strands of the narratives cannot be thoroughly understood unless the research process recognizes multiple voices and multiple interpretations given by different subjects involved in the contestation. As a result, a constructionist perspective of knowledge production has been used as an epistemological basis, because the narratives held by different actors show how conceptions of certain 'realities' are socially constructed. As Crotty (1998: 42) maintains, "constructionism is the view that all knowledge, and therefore all meaningful reality as such, is contingent upon human practices, being constructed in and out of interaction between human beings and their world, and developed and transmitted within an essentially social context." This argument is in line with Long (2001: 3), who posits that, "In actor-oriented perspective, social constructionism is often used closely with the use of discourses and controversial analysis for understanding processes of social interaction and negotiation, constitution of power relations, and the co-production of knowledge."

According to the constructionist research paradigm, researchers do not necessarily attempt to discover meaning, but rather construct it from multiple ways of interpretations by the subjects. Thus, human beings construct meanings about a particular object, phenomenon or practice through their iterative engagement and interaction with the world they are interpreting. For example, "a tree is a tree in a 'commonsense' but a tree might possess different meanings or constructions in a logging town, an artists' settlement and a treeless slum" (Crotty 1998: 43). Among the Guji Oromo in the Nech Sar area, for instance, a tree is given different meanings from cultural, spiritual, and historical dimensions, as well as in terms of its relations to humans. In contrast to positivist views, "constructionism takes the world and objects in the world as partners to human beings in the construction of meaning" (Crotty 1998: 44). Rather than attempting to ascertain a universal reality about the contestations over the protected area, I posit that the complex dynamics in park-people relations, claims of entitlement, perspectives on environmental conservation and development discourses, and state intervention and local responses can be well understood from the perspective of those who operate within it. As Ingold (2005) also argues, individuals' environmental knowledge is built through iterative interaction between people and the environment and through people's expectation of environmental affordances. Thus, I agree with Phillmore and Goodson (2004: 36) that "research is undertaken in a collaborative fashion, with the researcher and the researched viewed as partners in the production of knowledge and the interaction between them being a key site for both research and understanding."

1.3.2 Methodological Approaches and Methods

Drawing upon the above theoretical foundation, I have developed methodological approaches for data collection. At this decision juncture, it should be repeated that no single comprehensive methodological approach in social science studies is used as a toolkit in addressing a certain topic. For instance, the existence of multilayer actors, processes and forces in the field entails the use of multidimensional approaches for balanced academic inquiries in my study. Thus, this research employs different methodological approaches from different fields of studies. I used an ethnographic method to collect data about local communities' lived-in experiences and their day-to-day encounters with the national park, their traditional eco-cosmologies, and engagements the state. In addition, I used the extended case method to supplement the ethnographic method to link the dynamic interactions between local or micro contexts and national or macro forces that shape the subject of this study.

I selected the ethnographic method due to its recognition of context-specific accounts of the subjects. The analysis of data in ethnographic work "involves interpretation of the meanings, functions, and consequences of human actions, practices and views and how these are implicated in local, and perhaps also wider, contexts" (Hammersley and Atkinson 2007: 3). Interpretive ethnography gives voice to people in their own local context, through "thick description" (Geertz 1973) of their everyday accounts, experiences, and encounters. Unlike empiricist methodologies, ethnography downplays the search for universal laws but rather considers meanings by unveiling the detailed experiences of the subjects within their particular culture. This is achieved by recognizing underlying social rules and beliefs that are used as resources within that particular context (Hammersley and Atkinson 2007).

Ethnography has undergone some paradigm shifts, with traditional ethnography having been characterized as a colonizing methodology, whereby ethnographers complemented colonial regimes by collecting data about the native people based on predetermined assumptions about them. By presenting their findings as the ultimate 'truth' without recognizing the multiple interpretations of natives' accounts, colonial ethnographers served as 'academic' agents in the process of oppression and exploitation of indigenous peoples (Phillmore and Goodson 2004). From the mid twentieth century onwards, ethnographic work moved away from its traditional positivist orientation to describing the accounts and experiences of local people in their natural contexts as per the meaning conferred upon the phenomena by the people holding the culture, belief, practice and social institutions. However, during the post-1980s period, ethnography reached another level, where, according to poststructuralist and postmodernist theorists, it is presented as an advocacy work in favor of a minority position rather than being solely used as a process knowledge production (Hammersley and Atkinson 2007). In this work, I have used ethnography as a methodological approach to study the research subjects' lived-in experiences and interpretations within the research settings.

The second methodological approach that complements ethnography is the extended case method. As Burawoy (2009) discusses, this approach is a methodology that stretches or extends the researcher's scope of investigation beyond specific contexts. It is defined by four extensions: the extension of observer into the lives of participants under study; the extension of observation over time and space; the extension from micro processes to macro forces; and finally, the extension of theory. I am mainly interested in the first three dimensions of the extended case method. The extended case method is not used as a substitution for ethnography, but rather as its complementary methodological approach that fills some of the gaps for which traditional ethnography has been criticized. For instance, ethnography has been criticized as ahistorical and location-specific due to its emphasis on on-the-spot and face-to-face interactions with the subjects as a way of recording their day-to-day experiences (Burawoy 1991).

Extending the investigation beyond the temporal and spatial limits of a particular phenomenon helps to grasp the social world from broader contexts and to create a nexus between local or micro processes and external or macro forces. This approach does not have its own unique features in terms of methods of data collection, but rather shares methods used by ethnographic approaches. However, unlike traditional ethnography, which encourages the concentration of the ethnographer on a natural setting of the research theme, the extended case method supports engagement and dialogue between multiple actors, processes and forces that directly or indirectly influence events in the field. This approach is important in understanding the interconnectedness between different actors and processes through interactive dialogue. According to this method, the dialogue "involves participant and observer (subjects and researcher), successive events in the field, micro and macro processes and forces and reconstructions of theory" (Burawoy 2009: xv). In its specific techniques of dialogue between various actors, the extended case method shares similarities with ethnographic methodology, employing observation, interviews and group discussion.

In this study, one cannot separately investigate local people's perceptions about the people-nature relationship, claims over territories, conservation and development programs and contestations over boundaries without linking them with external forces that shape the economic and political discourses of different actors. In this regard, I have made two major extensions from the present scenario to past trajectories and from micro processes to macro forces. In fact, I also analyzed the interpretations of my research subjects from their own points of views positing the research into their life worlds. Despite being difficult to conjure the past through observation, people's narratives of their experiences and encounters of the past help us to thread the past with the present. In this regard, I have made a historical reconstruction about the people's settlement history, economic practices, rituals and inter-group relations to gain an insight into current claims and conflicts over the territory. As Neumann (1998) argues, local people's resistance against state conservation discourses in East Africa is part of the broader peasant resistance against hegemonic states that spanned from the colonial period to the postcolonial era. In the Ethiopian context, it is important to contextualize the territorialization of indigenous lands as a continuation of the empire-building project that began in the late nineteenth century (Turton 2011).

The second aspect of the extended case method involves extending micro processes to macro forces. In this regard, there are many interrelated dynamics; for instance, I investigated aspects of local people's claim for customary rights over the land in the context of the official political discourses of ethnic federalism in Ethiopia that ostensibly grants rights of self-government to local ethnic-based administrative units. By tracing the historical trajectories that brought about the establishment of the Nech Sar National Park in 1974 and extending the observation and investigation beyond temporal and geographical limits of the current fieldwork site through the minds of my research subjects, the project positions the contestation within broader political trajectories at different levels. Issues related to inter-ethnic or inter-regional claims over resources and their management, national political discourses of ethnic federalism and its lived experiences, as well as interplays between local and external factors, were addressed by extending the investigation of local phenomena to extra-local forces that influence the micro contexts. Therefore, the extended case method reflects one of the appropriate methodological approaches to supplement the ethnographic approach.

1.3.2.1 Methods

The epistemological and methodological perspectives discussed above are not simply ends in themselves within the process of knowledge production; rather, they are used to guide the journey towards the end goal. For topics such as mine, which involve competing actors whose interests, narratives and claims are significantly different, any attempt to use a single method of data collection cannot be a fruitful exercise. Therefore, I employed in-depth and narrative interviews, semi-structured interviews, focused group discussions, participatory methods and transect walk observations, according to their suitability in specific contexts. I used each method or combination of methods depending on the types of issues and the type of research participants involved. I employed these methods during the data collection process in 2011 and 2012 among the Guji and Koore local communities who inhabited the eastern side of the park and government officials related to the national park.

In-depth/Narrative Interviews: In-depth interviews are used to investigate an issue from its source. Unlike structured or semi-structured interviews where informants or research participants are expected to answer questions with the interviewer's frequent interventions, an in-depth interview is open-ended and gives considerable time and freedom to the interviewees to narrate their experiences. In this study, I used in-depth interviews to unveil complex experiences of the communities under study. In most cases, these types of interviews share some similarities with the narrative method, which help the researcher to "organize a sequence of events into a whole so that the significance of each event can be understood through its relation to the whole. In this way, a narrative conveys the meaning of events" (Elliott 2005: 3). In this study, research participants' lived experiences related to the national park, their historical attachments to the land and their claims thereof, and narratives of displacement following the establishment of the national park are traced through narrative interviews. Remarking on the significance of narratives in qualitative social science research, Shkedi (2005: 13) states that, "human beings organize and manage their perception of the world through stories they construct and tell. These human narratives give a structured quality to experience and are the way people make sense of the world around them."

For in-depth/narrative interviews, I went on selecting informants based upon their knowledge about the research theme. To this end, I first contacted village chairpersons and schoolteachers in both Koore and Guji villages to obtain a list of a few local elders for this purpose. Once I met some individuals, the selection of knowledgeable elders who were locally known as 'culture men' or 'men of history', was conducted through the snowball method for the purpose of key informant interviews. Although men dominate public spaces in both the Guji and Koore communities, women were also included in the interview to understand the gender dimensions of the themes investigated. The inclusion of women into the research was not merely for the purpose of representativeness, but rather because the lived-in accounts of women are often different from their male counterparts regarding human-environmental interactions. Therefore, women's participation in the research has both ethical and paradigmatic imperatives. For instance, I asked elderly women to narrate their memories of state intervention and its implications on them. Therefore, interviews with the women were mainly important because the government's enclosure restricted the people from access to environmental resources (water and forest) and social services such as the route to the market. This is because in Africa in general, women play major roles in activities such as resource extraction for household consumption and engagement in local markets. Thus when these social and economic spaces are restricted, women are commonly prompted to travel long distances to find these resources or to identify market centers (Neumann 1998; Igoe 2004). Women were also the main actors in enhancing inter-societal relations, particularly during conflicts, due to the common understanding among Guji and their neighbors not to attack women, despite any sort of inter-ethnic conflict.

Semi-structured Interviews: I mainly used this method in interviews with selected government authorities, park staff members and tour operators, using some semi-structured interview questions, based upon which informants elaborated their views. From the government offices, people from different administrative structures of the government were included, including culture and tourism bureaus of Oromia and SNNP regional states, officials of Amaro and

Galana districts, former administrators of both districts, and chairpersons from villages adjoining the park and staff members of Ethiopian Wildlife Authority. In addition, former staff members of African Parks Foundation were also interviewed to get detailed information on the management approaches of the park from 2004-2007.

Participatory Mapping: This is a technique used in human geography, development studies and childhood research, which allows research participants the freedom to depict their views through concrete diagramming (e.g. mapping, time lines, cartoons and pie charts) and can also be used beyond cultural, language, and literacy barriers (Pain and Francis 2003). Participatory diagramming or mapping increases the degree of participants' involvement in the research process, enabling one to read their minds through what they do rather than through what they speak. Among possible participatory mapping techniques, my research subjects and I used historical and resource mapping in this research (see chapters 3 & 7). In historical mapping, participants drew 'maps' about their settlement in and around the park across time. Regarding resource mapping, the members of the Guji and Koore communities were asked to draw which resource type was/is located in a particular place. Their depictions of resources provided me with preliminary insights to the prevailing competing interests between the local communities and the park and within the local communities as well, because there were instances in which both the Guji and Koore people were engaged in exploiting the same ecological niche along the eastern border of the park.

Observation through Transect Walk: A transect walk is a guided walk in a village, farmland or pasture area, conducted by one or two members of a community to collect place-based information that might be difficult to uncover through other methods. In this particular project, transect walk has been used to gain information about some contested territories such as the boundary of the park, boundaries between Guji and Koore territories, as well as locating some sacred spaces such as ritual sites, burial spaces, a healing hot spring (*Xabala*) and the access route to a lake or a river (*malka*). By this method, two different transect walks were conducted by members of the Guji and Koore communities respectively. Throughout the walk, we made observations and discussions to elucidate the individuals' mental conceptualization of the territories. During the transect walk, we observed about cattle trespass, farming encroachment and some activities of park scouts, such as their response to cattle trespass.

During one of the transect walks, Danbala Edema, a Guji man who guided us (my field assistant and me) instantly changed his tone of conversation to one of distress, telling that the place was where three of his family members (his grandfather, uncle and nephew) were buried before the Guji were evicted in 2004. We continued our walk after Danbala dropped a leaf on the top of the graveyards of his relatives, as is the custom among the Guji. From conversations that we had afterwards, it became evident that members of the Guji who lived in the park territory and had experiences of eviction under successive regimes had developed a sort of negative attitudes towards the park because it detached them from their ancestral spaces. In this particular case, a transect walk was useful to observe the individuals' emotional attachments to their land or to particular spaces, such as sacred, ritual and burial sites.

Focused Group Discussion (FGD): The key qualities of this method are "related to the synergistic effect of group interaction and their potential to break down the researcher–researched power relationship" (Thomas 2004: 200). Group interaction partly replaces the interviewer–interviewee relationship found in individual interviewing and thus provides opportunities for participants to raise their views and arguments about a particular issue. This method is also important in giving participants opportunities to remind one another about issues under discussion. By putting them into debates concerning opinions that they do not commonly share, this gives the researcher the chance to discern the diversity of views within the group. In such cases, a guided interview with individual members of the group discussion helps to gain detailed views and explanations on their points of argument.

To gain diversified views, discussions were arranged among elders, youth and women. Six FGD sessions were held with peasants from three villages in Amaro district. Moreover, I conducted four FGDs with the agro-pastoral Guji community. Participants in the FGDs were selected based on their knowledge, experience and involvement in the contestations and negotiations over the Nech Sar National Park. The specific selection process was made deliberately through a snowball method. Local people knew one another and mentioned who could give detailed knowledge about the themes of discussion. To gain views from the youth, school youth from Darba Manana, Tifate and Yero of the Koore villages and Arda Guddina and Tore among the GujiGuji took part in three separate focused group arrangements. Moreover, discussions were held with out-of school youth to increase diversity.

An interesting part of this method, particularly among the youth, was that they became active in the debate after a few moments of being passive in the interaction, and began arguing against one another for some time. They were initially reluctant to pronounce their views due to a lack of experience and to being skeptical about the objectives of the research. The major themes covered by this method include the role of local communities in planning and management of the park, issues of eviction and resettlement, livelihood dependence and challenges of living close to the park, as well as the future prospects and questions of benefit to local communities. The discussions actually provoked additional issues such as inter-ethnic tension, hunting, cattle trespass and the role of local scouts in hiding perpetrators.

1.3.3 The Challenge of Encountering Competing Actors in the Field

In Ethiopia, social science research faces government scrutiny at different levels. Research topics might be labeled as 'developmental' or 'anti-development' and 'peaceful' or 'anti-peace', not based on critical investigation of the objectives and findings but rather haphazardly from the superficial impressions that the topics might convey. For example, according to Baye (2008: 57), Addis Ababa University has increasingly become a political center and lost its alleged claim "to be national authorship on sociopolitical developments of the country and on the collective consciousness of the people. Research in the Humanities and Social Sciences has always been prescribed and often under-funded and/or indirectly discouraged with stringent financial rules and inefficient financial administration."

Such denunciation of social science research as a threat to government political ideologies has created a tendency among local government officials, whom I consider as 'gate-keepers', to be skeptical about giving research permits. In the case of this particular research, the Nech Sar National Park has become one of the national parks located in volatile regions of the country due to the persistence of conflicts and claims between competing actors. This is why the government put it under the administration of the Ethiopian Wildlife Conservation Authority under the federal state in 2009 (Personal Communication with the Former Head, Culture and Tourism Bureau, SNNP, April 2012). Issues pertaining to the relocation of the Guji and Koore people and the subsequent criticism it generated from human rights organizations have compelled the government to inspect any research activity that involved the local communities closely. There were day-to-day encounters of hostility between the people and the park administration officials. Therefore, the park administration suspects that researchers would reveal to the public its actions against the local communities if the administration gives them access to the field. Below, I mention only a few of the complex scenarios on the ground, because space constraints do not allow me to describe all the complex circumstances regarding the politicization of the research and the control of access on the part of the 'gate-keepers'.

Based upon its relevance to the political ecology theoretical framework and actor-oriented perspectives that I employed as an analytical approach, my encounter with competing actors warrants further discussion, as provided below. The fieldwork challenges and experiences are presented from three different dimensions, namely bureaucratic challenges related to research permission, inter-group conflict and accessibility problems, as well as competing actors' attempts to manipulate the course of the research to their interests. Regarding the issue of getting research permission, the initial challenge was related to frequent turnovers of management of the national park, resulting from the oscillation of managerial responsibility between regional and federal administrations and a multinational company. Initially, the park administration was centrally managed, as was the case for other state institutions during the military regime (1974-1991). Following the downfall of the military regime, the park administration was given to the SNNP regional state as part of the decentralization of state institutions (1991-2004) until the government transferred it to the APF (2004-2007). Since then it has been reverted to the federal government, having been administered under the Ethiopian Wildlife Conservation Authority (EWCA). Such administrative changes created different sorts of discontent on the part of government institutions, who covertly claim the right of managing the territory, as in the case of SNNP culture and tourism bureau and the EWCA. I approached the regional Culture and Tourism Bureau in Hawassa, requesting research permission because it handles issues of tourism related activities within the region. After giving me two weeks appointment to decide on my request, the bureau later informed me that it had no mandate to give research permits, given that the park administration had been transferred to the federal government. Thus, it referred me to the EWCA in Addis Ababa, from where I finally got the permission within a few days. As it became clear throughout my research, there have been tensions between the Culture and Tourism Bureau (SNNPR) and the EWCA over the administration of the park. Typified by the saying "where elephants fight, only the grass suffers", the lengthy bureaucratic process of obtaining my research permission was part of the costs of conducting research.

The major obstacle in the process of getting a research permission, which continued throughout the research process, was posed by the park administration located in Arba Minch town. The park headquarter is located at the eastern outskirt of Arba Minch, in Gamo Gofa Zone. Unlike the federal authority, people in this local office knew more about the conflicts between the park and the local communities. They were also the central actors in displacing local communities, with the latest incident being the destruction of sixty-four Guji houses along the park border on the eastern side in March 2012 (see chapter eight for detail). As a result, they were not keen on allowing researchers to enter the communities via the park territory. This route is the only on-land route to the territory of the Guji Oromo from Arba Minch side, with the alternative involving a boat ride on Lake Abaya, which is often risky. As it has been described earlier in this chapter, the newly established Guji village is located between the park and Amaro district (see map 3 in chapter three).

There were only three possible routes to the Guji villages. While the first and perhaps 'safe' and easy way was crossing to the villages through the park from Arba Minch town from the west, the other option involved descending to the valley from Amaro district from the east. The third option, which was more risky than others, was crossing on Lake Abaya using a traditional boat called *Jaatara* or *Wogolo*. I was aware of the fact that without access to the Guji who are the major local actors in park-people relations, my project would not be complete. At the end, the two possibilities turned against my attempt. My first attempt involved requesting the park administration to cross the park territory to the people. However, the park administration gave me permission only to enter the park territory and observe the wildlife, but not to contact any group of people beyond its boundary. To ensure the terms of the permission were adhered to, park officials assigned scouts to patrol my activities. Nevertheless, given that my research is not on the wildlife per se, my entry permit to the park only enabled me to observe some cattle trespassing into the 'park territory', wildlife distribution and environmental degradation, as well as how scouts respond to cattle trespass from a distance.

Because of the authority of the 'gate-keepers' over information and territory, I had to suspend my effort to reach the Guji for some time. While it would have been an hour's walk from the park outposts to the villages, I had to travel for four days by public transport in my attempt to enter Guji villages through the Amaro district. Luckily, four Peasant Associations/villages (*kebeles*) in Amaro district (Gumure, Darba Manana, Tifate and Yero) were part of my research because many people from these villages had plots of land in the contested Sermale Valley called Tsalke/Golbo. Getting a research permission and a field assistant from Amaro went smoothly. Although the road from Amaro Kelle (administrative center) to the Koore villages was difficult because of the rugged Amaro Mountains, which took us eight hours to climb and descend, the hospitality of the Koore people in these localities was very good. The fieldwork in the Koore villages also went quite well. Once again, it would have been a matter of few hours to cross to the Guji community adjoining the park, but because the conflict that erupted between the two groups in late 2010, shortly before my arrival in the field, there was a sense of insecurity among the people crossing into each other's territory following. My Koore field assistant refused to guide me to the Guji villages and threatened to leave me alone in the field if I insisted on going to the Guji territory. Considering the repercussions this would bring during the course of my research, I had to accept his proposal not to travel to the Guji villages. Thus, my continuous efforts to get access to the Guji villages were entirely blocked by different actors on both sides. Meanwhile, I conducted several interviews with Guji individuals who came to Arba Minch town for their own business in 2011, during these failed efforts to reach the Guji. However, I felt that this was insufficient in gaining detailed knowledge about people's day-to-day encounters and lived-in experiences in terms of their interaction with the national park.

As a last resort, I had to return to Arba Minch, which took me another four days by public transportation. As part of this last option, I had to make one of the most difficult decisions in the fieldwork process, namely the decision to cross Lake Abaya with a traditional boat, which the Guji use when they travel to and from Arba Minch for market, healthcare and other purposes. The risk is attached to reports concerning the existence of dangerous crocodiles in the lake, often attacking people on the boats. Moreover, the risk is associated with the keeling over of the small, locally made boats. The most recent incident at the time was when a boat capsized in February 2012 on its way to Arba Minch town, killing nine passengers. These stories placed me in a dilemma for some time, before I finally decided to set aside my fears and to take the boat ride.

Figure 3: Photo Showing our Boat Trip on Lake Abaya (Photo: Asebe Regassa, May 2012)

My Guji field assistant, myself, two boat operators and sixteen Guji individuals were on board when a strong wave suddenly rose up to a height far above the boat in the middle of the lake, but luckily the two skilled operators were able to divert the direction of the boat and escaped the wave. When we were about to cross to the other side of the lake, a hippopotamus emerged from the water and was about to tip our boat over, had it not been for one of our Guji companions who slapped his machete into its head. While I did not see a crocodile, these two incidents were 'good' experiences in terms of showing the authority of 'gate-keepers' over the territory and the risk that the local people incur in their day-to-day interaction with Arba Minch town. This was mainly because of the blockage of the route in the mainland by the park administration.

Another challenge that I faced was after my field guide and I entered the Guji village, Arda Guddina, which was established in 2008 and is claimed by the park administration as its de facto territory. The incident was as follows: while they were patrolling around the villages, two park scouts asked me how I came to the village. They were not authorized for such inspection because the village was not formally under the park administration. However, I had to tactfully and diligently deal with the encounter in order not to obstruct my fieldwork. They asked if I had gained research permission from the park administration, which I did not get despite relentless efforts. In fact, I had a letter of permission from Galana district, to which the Guji villages belong. Finally, they reported to the park administration in Arba Minch about my entrance to the Guji villages and eventually the park administration ordered them to arrest me, having allegedly accusing me of entering the 'park territory' 'illegally'. The park administration asked the chairman of Arda Guddina village to extradite me for arrest, although the chairman and other local government officials angrily reacted against the intervention of the park into their administrative spaces. The chairman boldly reacted stating that it is neither a Guji custom (*aadaa* Guji) nor is it administratively sound to hand over an *imala* (guest) to a third party that might endanger the person's safety.

The Guji gave me the status of *imala* (guest), conferring on me a privilege of protection and respect. They defended my case, arguing that the park administration has no right to inspect who entered their village. The chairman warned them that any attempt to arrest their *imala* would lead to open conflict between the park administration and the people. One month later, when I came out of the fieldwork, I was informed that the park administration had appealed to Borana Zone administration accusing both the chairman and myself of flaunting official orders. They sought to politicize the case as it is often done in Ethiopia, stating that the village had hosted a 'stranger'. While the Guji received me as their *ima-*

la with respect, the park administration labeled me as stranger. Within this context, the term 'stranger' was used with the politically loaded meaning of referring to a person who is there with hidden motives. The case was finally settled after I presented to the officials the research permits and support letters I had received from different institutions. My field guide, who immediately returned to Arba Minch after introducing me to the Guji villagers and local government authorities, was later detained for three days in Arba Minch town, being accused of guiding me into the villages.

The second dimension of fieldwork challenge was related to the Guji-Koore conflict in late 2010. Fieldwork in a conflict zone poses significant methodological and practical challenges, particularly where research activities involve two or more conflicting groups (Robben 1995; Asebe 2010). There are tendencies to label a researcher either as a government agent surveying perpetrators or as a spy collecting information for the 'Other' group. Based upon such suspicion and a sense of insecurity, inter-group communication between the Guji and Koore was limited and subject to scrutiny. As I described earlier, I was unable to pass from Koore territory to the Guji villages because my Koore field assistant refused to guide me to the Guji border. Koore informants were also not comfortable with my plan to enter the Guji territory. During one of my fieldwork trips to the Koore villages, my field assistant warned us (my brother in-law and myself) not to speak the Oromo language as the Koore informants would associate us with their Guji Oromo rivals. In this regard, my own ethnic identity with one of the research groups, the Guji Oromo, had two different implications. Among the Koore, it would have given them an impression of mistrust had I disclosed my ethnic identity because they had open conflict with the Guji only a year prior to my fieldwork. Apart from suspicion regarding my impartiality, the Koore would have been skeptical about freely discussing their discontent concerning the Guji and Oromia regional state in general.

At this point, it is important to reflect on the implication of working in my own ethnic group from three different dimensions. First, when I entered the Guji, my ethnic membership at times enabled me to gain backstage information or, as Scott (1990) calls it, the 'hidden transcripts' and culturally embedded norms and practices that are difficult to obtain for outsiders. Accordingly, due to their experiences of marginalization and mistreatment in the hands of government and the park authorities whom they consider as outsiders, the Guji developed a sense of trust to share their grievances with me. Second, despite opening some opportunities of gaining access to their backstage information, my ethnic background did not automatically enable me to win the trust of the Guji, as they had had experiences with Oromo government officials who were allied with the park administration and who evicted them from their land. Therefore, it was only after long periods of rapport establishment that they became open in revealing their experiences, views and actions. Third, because they considered me as an insider who was supposed to know Guji values, norms and practices, they were not keen to explain some of their stories in detail. The Guji socialize their children into the customs, values and practices of their community. As a result, they found it surprising when I asked for explanation about some stories or practices, because adults of my age were not supposed to lack the knowledge of common practices such as herding culture.

In a different dimension, I encountered another fieldwork challenge, which I put here as a third category of challenges. Whenever I moved from one territory to another, I came across contrasting perspectives among the Guji and Koore communities. It was actually not my expectation to experience a uniform knowledge production from multiple actors. I was cognizant of the fact that multiple interpretations of realities characterize complex themes in cases such as my project. However, my encounter was more with the two groups' accusations of one another. For example, the Koore informants directed their blame towards their Guji rivals for transgressing park laws, whereas the Guji blame the Koore for hunting the wild animals in the national park. Far beyond this local context, government officials from both regional states blame one another for the human and environmental problems in the national park. Under such circumstances, members of different competing actor groups produce narratives that place the researcher into a methodological dilemma. This hints at how competing actors strategically craft different narratives to win the researcher's attention. As a social science researcher, I had to use different mechanisms of crosschecking information, sometimes by going beyond the specific locale to incorporate the views of outsiders and sometimes by probing into the backstage information of each actor group. In fact, my objective is not to find 'truth', but to investigate how actors construct narratives and discourses in their production of knowledge and counter narratives against their rivals. Even then, it was essential to interrogate how and why competing actors use a certain type of narrative and downplay the views of the other.

Interrelated to competing actors, particularly the two local communities, there were also continuous attempts by my informants to manipulate me to their ideas and opinions. In fact, working in one's own ethnic group compels a researcher to enter into what Robben (1995: 83) calls as 'ethnographic seduction'–the challenge of being seduced or manipulated by the views of the research subjects. In all such challenges, I employed techniques of impression management to control my own subjectivity and the pressure from my informants. As Robben (1995: 85) rightly describes, fieldwork on groups with memories of conflict puts a researcher in a condition of "ethnographic seduction", because both groups strive to recruit the ethnographer into their own camps by highlighting all possible 'justifications' for their actions, while also denouncing the actions of their rivals. Unless one receives such information cautiously, the data from the field would tend to favor the interest of one or two actors by silencing others. Despite persistent efforts by different competing actors to drive the course of the research to their advantage by accusing their rivals and concealing what they perceive as wrong on their side, I managed to balance the views while also reserving the right of my subjects for their views. This is why we need to interpret the impressions of our subjects cautiously by scrutinizing what has informed their reactions (Berreman 1962; Robben 1995). The analysis of such competing views becomes important to understand the complexity of the circumstance.

1.4 Actor-Oriented Approach as Analytical Concept

From the previous discussions of the involvement of different social groups and institutions in the struggle over the Nech Sar National Park, it has become clear that the study should encompass the characteristics, actions, interests, perspectives and networks of actors who could be place-based or non-place based with direct or indirect influences on the dynamics of the national park. In the study of human-environmental interactions, political ecologists have begun to adopt a comprehensive approach of researching the political and ecological conflicts and cooperation by analyzing how actors perceive, contest and negotiate environmental issues (Bryant and Bailey 1997). This approach is called an actor-oriented approach (Long 2001), touching upon competing environmental discourses and narratives espoused by different actors in their interactions over environmental resources. For instance, conflicts between pastoralists and state, peasant farmers and pastoralists, and farmers and state could be the subjects of investigation in political ecology to understand how these actors compete, cooperate and negotiate over environmental resources and what actions they take in the process of achieving or defending their interests. By recognizing the agency of social actors, particularly concerning grassroots actors, the actor-oriented approach scrutinizes how different actors respond to discourses of nature conservation. This includes how they define nature-culture relations and conceptualize development and community-based conservation because these ideas reflect the actors' power relations and ideological dispositions (Escobar 1996).

In this work, I use an actor-oriented perspective as an analytical approach to understand the complex dynamics related to the Nech Sar National Park and how different actors contest, negotiate or come to terms with conservation discourses and practices. In this context, I use the concept 'actors' to denote both social groups with similar interests and individual(s) who use their agency in articulating different narratives, discourses and practices to maneuver their interaction with other actors in the struggle over environmental resources. In the section where I problematized the notions of protected area conservation, I mentioned how the perspectives of nature-society dualism apparently brought the establishment of protected areas first in the United States and subsequently in different parts of the world under the 'fortress' conservation approaches. As it will be detailed in chapter two, nature conservation should be understood in terms of its economic and political dimensions, i.e. beyond the traditional apolitical approach in the study of environmental problems. In the case of the Nech Sar National Park, multiple social actors have been involved in the contestation over the territory for different purposes. For example, the state operates as a visible and invisible actor through its institutions, officials and policies. At the same time, non-state actors such as conservation organizations, human rights advocacy groups and tour operators have their respective interests and strategies of achieving their goals. Likewise, local communities, still internally heterogeneous, operate both as groups and as individuals in their contestation against other actors. This is why Few (2001) states that in defining degrees of access to formerly 'open' spaces, the process creates a struggle over place, with political, economic, social and cultural implications at different scales between different agents of varying interests and interpretations. Thus, as Long (2001) also suggests, an actor-oriented approach can be used as an analytical approach in the contexts of interlocking contestations to explore the interplay between social actors at local and external arenas in a series of intertwined battles over resources, meanings, institutional legitimacy and control. This approach is often employed to understand how meanings, interpretations and concepts about a particular 'reality' are socially constructed, contested and negotiated, thereby addressing issues of social interaction, and the constitution of power and co-production of knowledge (Few 2001).

Social actors do not operate as separate entities; rather, they create networks and develop social capital, in turn building on their legitimacy, trust and credibility, which increase their capability and knowledge of their cause. This means that legitimacy, trust and credibility generate the agency of social actors to maneuver a certain event or process to their advantage. Thus, in inter-actors or intra-actors relationship, the type of social capital built shapes the nature of the relationship. This relationship can be analyzed at different levels (domestic, regional, national, or international), as well as in terms of actors' interests in forming networks. An analysis of the actors' network enables one to understand the interconnections/disconnections between actors concerning the conception of nature-society (culture) relations. In this regard, the perspectives of the actors should encompass large-scale institutional frameworks, resource fields, networks of communication and support, collective ideologies, socio-political arenas of struggle and the beliefs and cosmologies that might shape actors' improvisations, coping behavior and planned social actions (Few 2001).

Actor-oriented approach cautions us from considering external intervention as smooth process of translating programs into practices unchallenged because as actors have the power and capability to define external interventions like national parks, it often faces counter knowledge production on the side of local communities. As Long (2001: 27) maintains, intervention into a certain social system entails a constant process of transformation that "is reshaped by its own internal organizational and political dynamics and by specific conditions it encounters or itself creates, including the responses and strategies of local and regional groups who struggle to define and defend their own social spaces, cultural boundaries and positions within the wider power field." As the empirical data from my fieldwork shows, the government faced challenges both from its own structure as well as from local communities in channeling down its conservation policies. As indicated above, actors' choices, strategies and networks are influenced by internal (local) and external (extra-local) phenomena. Thus, through actor-oriented approach, I attempt to understand broader economic and political structures and perspectives that have been shaping the government approaches of resource management and local responses to it. Understanding social relationships between individuals or groups requires a concept of human agency. In the process of contestation over a particular social space, individuals negotiate within differential power relations, including actors who are internal and external to a particular field of struggle (Gezon 2006). Therefore, I employed an actor-oriented approach to analyze the characteristics, interests, views, actions, power positions and responses of actors in the process of struggle over environmental resources as part of economic, ecological, cultural, political and spiritual spaces.

1.5 Ethical Considerations

Social science researchers deal with societies whose cultural, political, religious and economic practices are subject to different aspects of interpretations. Due to the agency of researchers and the participants of the research in interpreting a certain phenomenon based on their worldviews and experiences, the entire research process in social sciences is neither smooth nor uncontested. In developing countries such as Ethiopia, where researchers and their participants fall under government scrutiny for pronouncing their perspectives, it remains the researcher's responsibility to handle the entire process in order to not endanger the research participants. As part of the ethical considerations in this work, I explain issues of informed consent, confidentiality and impartiality as follows.

The primary task in the fieldwork process was to acquire the consent and permission from government institutions, 'gate-keepers' (local government officials), local elders and members of the local communities, who participated in the research. To this end, I presented my research proposal to the Ethiopian Wildlife Conservation Authority and signed an agreement, after which the Authority granted me a research permit. Likewise, I presented my project to all government institutions directly related to the study. I clearly presented the objectives of the study and its scope to the actors from whom I sought research permission or information relevant to the study. Local communities, particularly the Guji and Koore, had experience of outsiders' manipulation of 'research' findings for the sake of implementing the government's conservation programs. As a result, they were skeptical of welcoming researchers to their villages. Within such circumstances, it has been crucial to develop a rapport based upon trust, respect and transparency. Accordingly, this was why I had to invest my time, knowledge and skills in establishing rapport by ensuring the members that the study did not have any connection to government institutions or any non-governmental conservation organization. After we reached a common understanding on the objectives of the research at large, I had to seek permission from individual participants for audio or video recordings, photographs and for their consent on issues of anonymity.

In research work that focuses on political, cultural and religious sensitivities, the main role of researchers should be to ensure the security of their research participants. To this end, depending on their consent and interest, informants have the right to remain anonymous and their details to be kept confidentially. Due to the involvement of different competing actors in the struggle over the Nech Sar National Park and given that different claims and accusations had political implications at different levels, my informants are given pseudonyms, except on topics and issues where revealing the real identity of the informants was unavoidable yet does not threaten their security.

Another difficult aspect of social sciences research is the issue of impartiality, which is somewhat different from objectivity. Although the debates on objectivity and subjectivity are more related to the level of the researcher's interpretations vis-à-vis the information provided by his/her informants or from fields of observation and other sources of inquiry, impartiality rather has to do with personal inclinations to or the avoidance of a certain aspect of the study (issues, people, territories, time, discipline, approach, etc.). These sorts of bias might originate from the researcher's academic background, which favors a certain discipline, as well as his/her theoretical orientation, his educational background, ethnicity, religion, political affiliation and place of living. In this particular study, my academic background in interdisciplinary programs has given me the opportunity to mediate different methodological and theoretical frameworks from geography, anthropology, political sciences and environmental history. As a result, I have been able to avoid this challenge. Moreover, my previous fieldwork experiences among many ethnic groups in southern Ethiopia such as the Gedeo, Burji, Koore and Guji have also given me insights into research beyond my own cultural and geographical settings. Nevertheless, research that involves competing actors, particularly if one of them is associated with the researcher, falls under scrutiny. In this regard, it is neither the ethics nor the objectives of the research to favor any group at the expense of others through the research findings. Neither does it aim to criticize government policies and practices by putting the researcher on behalf of the local communities and vice versa. Rather, it seeks to bring different voices, interpretations and perspectives into dialogue through the researcher's own interpretive skills.

1.6 Structure of the Work

This research work is divided into three parts and nine chapters. In part one, the Research Settings, Methodology and Theoretical Frameworks have been discussed in three chapters. By articulating the research problems, questions and methodological frameworks in chapter one, the theoretical frameworks in chapter two and the study areas in chapter three, part one sets the basic building blocks in the research.

In part two, the political ecology of nature conservation is presented in chapters four and five. This part begins with the discussion on the exportation of national park ideals to Africa in general and its introduction to Ethiopia, and goes on to specific discussions on ethnic federalism and the creation of contested spaces in post-1991 Ethiopia in chapters four and five respectively. Chapter four specifically deals with the political ecology of national parks in colonial Africa by synchronizing multidimensional motives behind the establishment of national parks with the broader colonial policies of resource extraction and political control. The chapter further creates a nexus between colonial conservation discourses and practices in other African countries and the center-periphery relations in Ethiopia and how the notions of nature conservation were systematically exported to the periphery. Chapter five problematizes the political experiment of ethnic federalism by analyzing the rhetoric of local autonomy and participatory development against day-to-day experiences of local communities around the Nech Sar National Park.

In part three, I analyze the Nech Sar National Park as an arena of contestations and negotiations between competing actors with various views on human-environmental interactions. This part is divided into three chapters (chapters six, seven and eight). By employing actor-oriented approach in the analysis of human-environmental interactions and contestations, negotiations and conflicts that ensue through interplay between competing actor groups, the three chapters present and analyze actors' conceptualizations of the territory and their strategies in achieving their interests or deterring the interests of others. By taking the Ethiopian state as one of the major actors with pivotal influence on the contestations over the Nech Sar National Park and yet as an internally differentiated actor, chapter six presents and analyzes the state's conceptions of nature conservation and how it exercises power in achieving its economic, political and conservation motives. Chapters seven and eight take grassroots actors and their claims of interests and referents of contestation as important points of analysis. Finally, chapter nine gives some conclusions drawing on the theoretical and empirical discussions and analysis in the preceding chapters.

CHAPTER TWO: CONCEPTUAL AND THEORETICAL FRAMEWORKS

2.1 Conceptual Frameworks

The conceptual frameworks of this study have emerged from the prevailing competing conceptions of nature-culture relations, complex environmental, economic, cultural and political claims and motives behind nature conservation in the Nech Sar National Park, as well as the persistence of multilayer actors that shape the dynamics related to the park. More specifically, the territory now designated as the Nech Sar National Park represents different meanings and purposes for the competing actors involved. The territory signifies arenas of struggle over the symbolic, cultural, economic and political representations, as well as being a territory where competing actors deploy different forms of power in the process of achieving their respective interests or deterring the interests of others. Sharp et al. (2000: 26) indicated how social groups ascribe different meanings to a particular space, in that, "different social groups endow space, and of course place, with amalgams of different meanings, uses and values. Such differences can give rise to various tensions and conflicts within society over space for individual and group purposes, and over domination of space by the state and other forms of dominating social (and class) power." This hints at the fact that contestations between social actors not only occur due to competing interests (material, economic and political), but also due to disparities in their ontological representations of the space. In other words, competing views of nature and differences in meanings ascribed to a particular space or territory might reinforce tensions over environmental resources entailing deployment of different aspects of power, namely dominating power and resisting power (Sharp et al. 2000).

For instance, the area now designated as the national park represents economic, spiritual and historical meanings for the Guji Oromo agro-pastoral community. As a physical space, it provides the people with livelihood opportunities upon which their entire economic practices (cattle herding and farming) depend. The park was part of their grazing land, particularly during dry seasons. As a cultural and spiritual space, the Guji perceive some sacred sites in the park territory as places that once served as nodal points where they used to have their rituals to communicate with *Waaqa* (God) for the spiritual values attached to the spaces. In the past, elders performed earnest practices of rituals and prayers to establish harmony between the people and *Waaqa*, as well as between humans and non-humans (Baxter 1991). They also claim entitlement to the area, tracing their ancestral rights over the land since the sixteenth century (Getachew 2007).

On their part, the Koore peasants see the territory mainly from economic perspectives because the lowland Tsalke Valley (Golbo–as the Guji call it) on the eastern border of the park provides them with year-round crop production through irrigation and rain-fed agriculture. Besides the economic values of the area, the Koore seek to link their claim over the land to the political and administrative ownership that they exercised during the military regime. The same is true for the Gamo Gofa Zone in claiming for entitlement based upon the administrative authority it had on parts of the national parks under the past regimes.

On the other hand, the state considered the territory as 'uninhabited' land to be 'tamed' for tourism purposes. This mainstream representation of peripheral lands in Ethiopia is analogous to the Western notion of wilderness in dichotomizing the human and non-human (Turton 2011). The governments highlighted conservation and development discourses in the process of achieving its interests and representing the Nech Sar National Park as a space for economic and environmental purposes, while the process actually reinforced enforcement of hegemonic power from both the state and other actors involved in the contestation. Due to long history of center-periphery dichotomization in the country, the establishment of protected areas in southern Ethiopia should be conceptualized within the asymmetrical power relations between the highland Abyssinian state and the southern peripheries. Turton (2009: 2) argues that, "viewed from the center, with its mountain ranges, heavy rainfall, Christian tradition and plough cultivating peasantry, the southern periphery was seen as a virtual wasteland, inhabited by anarchic and violence-prone nomads." The state thus pursues conservation and 'development' agendas through 'ecotourism' programs in the protected areas of the south that it perceives as 'wasteland', 'uninhabited' and 'natural'. It has used the enclosure of the territories as instruments of control and mechanisms of gaining legitimacy in projecting its state power into the periphery. This perspective of the state, which emanates from the historical dichotomization of the north as source of 'civilization' and the south as 'backward' that awaited the hands of 'modernizers' to be tamed into useful resources, has been central in the conflicting views concerning nature-culture relations in the peripheral regions of the country.

The complexity in addressing conservation discourses of the country also lies in the fact that the state has become more heterogeneous particularly since 1991. The current Ethiopian state is internally differentiated along ethno-linguistic regional divisions that significantly affect how people in different local state structures respond to national policies and programs such as national parks and the relocation of people in their respective regions, from where they draw legitimacy in their political endeavors.

As has been discussed in the preceding chapter, the conditions of biodiversity degradation in the national park and the livelihood situations of the people who live adjacent to the park borders have deteriorated over time. While the government presents its 'modernist' conservation discourse–the establishment of protected areas free from human influence–and its ambition of generating income through tourism as viable pathways to environmental and developmental needs, local communities often challenge this from their own eco-cosmological perspectives. Synchronizing all these interconnected elements, this study fundamentally builds on two conceptual building blocks that enhance the understanding of the dynamics in the politics of conservation in the area. These conceptual frameworks are notions of nature-culture relations and the politics of nature conservation, which entails aspects of domination/resistance, negotiation, conflict and agency of various actors in mobilizing different resources at their disposal. In other words, the contested space is the center of struggle upon which social groups or institutions with competing perspectives on nature-culture relations espouse their conservation discourses, which in turn entail the deployment of different forms of power in the process of its enactment.

Thus, the study critically probes into the politics of conservation by investigating how the economic, political and cultural motives of different social groups shape their understandings of people-park relations. Therefore, in order to situate the discussions within broader academic debates, I have situated these conceptual frameworks in two interconnected theoretical approaches. First, debates on nature-culture relations will be addressed from the perspective of the social construction of nature. Second, the project moves beyond the 'traditional' approach of studying nature conservation and environmental problems from apolitical perspectives by interrogating the political, ecological, economic and ideological forces and processes behind such projects. Notions of setting aside territories in the form of protected areas are political and economic inasmuch as they are ecological and involve different institutions and actors from local to global arenas. Thus, I use political ecology as a theoretical framework to disentangle complex conceptualizations of conservation discourses, practices and motives. In both theoretical frameworks, the study investigates issues of power constellations among different actors, including how domination/resistance has been exercised. Accordingly, the sections below will focus on these theoretical frameworks.

2.2 Theoretical Frameworks

2.2.1 Notions of Nature-Culture Debates: Dualism vs. Mutualism

Debates over nature-culture relations have been sparked among scholars in the fields of anthropology, cultural history, environmental studies, geography, development studies and literary studies since the 1980s (Descola 1996). Apart from discipline-specific arguments and theorizations of the concepts nature, culture and their conjoint or disjoint relations, the nature-culture debates stem from ontological conceptualization of nature and its relations to humans. In other words, the debates oscillate between those who conceptualize nature as a given 'reality' independent and outside of the human knowledge on the one hand, and those for whom nature is inseparable from humans on the other. Broadly speaking, the perspectives concerning nature-culture relations can be categorized into dualism and mutualism (Pálsson 1996), which are discussed in the subsequent sections. However, prior to embarking on the debates on dualism and mutualism and how these notions have been appropriated by academics, human rights activists, environmentalists, politicians, indigenous communities and other local or international actors, it is vital to briefly highlight the nexus between these concepts and the politics of environmental conservation, particularly in developing countries.

With its notions of human-non-human separation, dualism entails setting aside territories in the form of protected areas–protected from human influences–for the sake of aesthetic, economic, tourism and biodiversity conservation purposes. The notion of alienating people from the environment under the guise of environmental conservation has historically resulted in poverty, marginalization and conflict among indigenous peoples in different parts of the world, where most people in developing countries depend on environmental resources for economic, spiritual, cultural and other socio-political purposes (Igoe 2004).

On the other hand, some scholars and indigenous peoples' activists promote the notion that indigenous peoples have developed splendid ecological wisdom, enabling them to live in harmony with nature (Berkes 2008). Given that it builds on the argument of mutual or conjoint constitution of the human and non-human nature, mutualism thus considers indigenous peoples or local communities in general as potential agents in enhancing environmental conservation. However, this argument tends to homogenize indigenous communities and simplify their livelihood practices to low-scale 'hunting-gathering', fishing, pastoralism and farming that were believed to be insignificant to harm the environment. Overall, both notions polarize our conceptualizations of human-environmental relations and do not comprehensively engage with how these understandings are culturally constructed and what value changes occur because of the interplay between competing notions of nature-culture relations. Nevertheless, these competing notions significantly influence practices of nature conservation. By discussing some ontological backgrounds on both paradigms, I will move to discussing how we should transcend the polarities in the following sections.

2.2.1.1 Nature-Society Dualism

At the center of the dualist paradigm lays the presumption that nature exists as an objective reality independent of human knowledge and outside the human realm. In traditional Western scientific thought, external nature was believed to be investigated, predicted, controlled and measured through scientific research that was taken for granted as the only ultimate source of knowledge (Descola and Pálsson 1996). According to the dualist perspective, the objective reality of nature can be revealed through scientific investigation rather than being constructed through human lived experiences, ritual practices and indigenous knowledge. This notion was deeply entrenched in Western philosophy of realism (Simmons 1993). According to the defendants of this paradigm, particularly the advocates of positivist perspective, humans' unique existence outside nature is a given reality due to distinct evolutionary processes and thus the most influential question is how to function within human uniqueness in order to fulfill humanitarian responsibilities to safeguard nature rather than attempting to abridge the human-non-human dichotomy (Willers 2001). This view is paternalistic in that it presumes humans as stewards or guardians of nature. The dichotomy places nature outside the human realm and presumes it as an object to be acted upon, yet subsumes any agency of the non-humans. This objectification of nature and more specifically the commodification of the environment through tourism can be identified as the ultimate foundation of power, repression and exploitation of local communities in many parts of the world by territorializing and enclosing their customary resource areas (Hornborg 1996). In the politics of environmental conservation, notions of dualism have been strongly linked to debates on 'wilderness', in an idea that presumes some territories as 'natural', 'uninhabited' by humans and devoid of any human influences (Cronon 1996; Callicott and Nelson 1998).

As a dominant perspective in the Western public and academic thought for long period, particularly since the era of enlightenment, dualism has constructed boundaries between 'wilderness' and culture, in the form of dichotomy between 'primitive' versus 'civilized'', 'savage' versus 'noble' and later 'sublime' versus 'polluted' (Cronon 1996). This dualistic notion considers wilderness as an imaginary space separated from and 'unpolluted' by human cultural practices. So far, there is no clear and conclusive definition of wilderness at present, except the most commonly cited definition from the American Wilderness Act 1964. According to this definition, "a wilderness, in contrast with those areas where man and his own works dominate the landscape, is hereby recognized as an area where the earth and its community of life are untrammeled by man, where man himself is a visitor who does not remain" (Callicott and Nelson 1998: 121). This definition states that, "earth and its community of life are clearly distinct from humans" (Ouderkirk 2008: 441). Under circumstances of 'wilderness'/nature conservation, human habitation was considered as an encroachment, whereas its socio-economic and cultural practices were portrayed as destructive and polluting the 'purity' of nature. In the Western popular imagination, wilderness represents the last remaining place where any form of human influence has not fully infected the Earth. Some presumed it as a place on earth that stands apart from humanity. This perspective advocates for the isolation of human beings from nature in order to keep nature pristine and in its natural settings. As Cronon (1996: 80) points out, "wilderness embodies a dualistic vision in which the human is entirely outside the natural."

Some scholars link the genesis of representing wilderness as opposite to culture in Western culture with the Judeo-Christian tradition that labels wilderness as spaces of temptations, desertion and abandonment (Cronon 1996; Merchant 1996, 2003). According to this perspective, people linked wilderness with the place where Adams and Eve were tempted by the Devil in receiving orders to eat from the Tree of Knowledge. They also likened it with the arduous areas where Moses traveled for forty years with his people and worshiped the golden idol (Merchant 1996). This perspective presents the Fall from Eden as the separation between human and non-human nature, which created boundaries of hostility between the two. From this dimension, it was likened with places where one goes without his/her choices–"to be wilderness was to be "deserted", "savage", "desolate" and "waste"" (Cronon 1996: 70). In fact, some scholars trace the modern conceptions of nature in European culture to the eighteenth century following the industrial revolution when people began understanding nature and its relations to humans from economic, romantic, aesthetic and moral perspectives (Bargatzky 1994). However, I argue that the post-enlightenment transformations in Europeans' understandings of nature-culture relations do not nullify the presence of dualistic perspectives in popular imaginations before that period. Rather, the transformations introduced different dimensions of interpretation of nature and humans' place in it.

From the seventeenth century onwards, people began to change some aspects of representations of wilderness as part of Enlightenment rationalizations, from fear and abandonment to sympathy, enjoyment, relief and care. At this junction, it is important to pose the question of what brought upon this change. With the transformations in the dominant European worldviews from religious dogmatism to scientific rationalizations following the eras of renaissance and enlightenment, people began to re-conceptualize their notions of wilderness, albeit within the dualistic paradigm. Wilderness got a new dimension of meaning as a lost garden to be 'reclaimed' and turned toward human ends and thus was considered as a place deserving protection from human encroachment (Merchant 2003). It represented parts of the earth that were 'unspoiled' by human influences, particularly through effects of industrialization. This shift in the Western perception of wilderness from images of lost space and threatening to untamed natural space brought sympathy in viewing such territories or spaces as the last areas on the earth, which human beings should protect before being lost into the archives of history. At the same time, the rationalization that human beings can control, tame, exploit and domesticate nature was also introduced into the discourse of Western enlightenment. Since then, "European colonists of the 'New World' have undertaken the massive effort to reinvent the whole earth in the conquered places in the image of the Garden of Eden but this time with the view of protecting, conserving and controlling the biodiversity as museums in natural settings" (Merchant 1996: 134).

However, as a conceptual representation of natural spaces that "ignores crucial differences among humans and complex cultural and historical reasons why different peoples may feel very differently about the meaning of wilderness", wilderness has attracted a series of criticisms (Cronon 1996: 85). The debate is mainly between modernist discourses and post-modernist deconstructions, rendering the boundary between the two a contested interface (Pálsson 1996). According to Cronon (1996: 69), wilderness:

> [I]s not a pristine sanctuary where the last remnant of an untouched, endangered, but still transcendent nature can for at least a little while longer be encountered without the contaminating taint of civilization. Instead, it is a product of that civilization, and could hardly be contaminated by the very stuff of which it is made. Wilderness hides its unnaturalness behind a mask that is all the beguiling because it seems so natural.

One important question to consider before delving into the details of wilderness and critics against the concept concerns whether wilderness is an idea or a place. If it is a place, what kind of place does it represent? Moreover, if it is an idea, what is the core idea embedded in the concept and whose idea is it? I argue that wilderness is both an idea and a representation of a physical place. It is an idea through which people conceptualize and give meaning to places. Wilderness represents an idea about an 'imagined' place on the earth where human influences, traces of activities and settlements are believed to be non-existent. This assumption had different historical roots beginning from Judeo-Christian tradition all through European Renaissance and Enlightenment eras to America's nation-building projects and the European colonial conquest of Africa, up until the rise of global environmentalist discourses.

According to Proctor (1998), the idea of wilderness is not an innocent description or representation of nature, environment, landscape or piece of land; rather, it has political, cultural and economic undertones and meanings attached to it. As with the Fall from Eden, it was associated with negative connotations where one goes without his/her will. As a picturesque site, it was labeled as an aesthetic and original remnant part of the earth where humans can gain relief from urban, industrial and polluted environment to feel 'back at home'. As a myth of frontier, it represented a boundary between the 'primitive' and the 'civilized', and particularly in the American myth of the frontier, it implanted memories of the birth of the great nation through conquest and mastery over 'wilderness'–despite of course being a home to indigenous Indians (Cronon 1996; Proctor 1998).

During the colonial era in Africa, colonial powers associated wilderness with 'wild' and 'natural' Africa, which also legitimated colonial intervention to 'save' the 'Edenic' garden (Neumann 1998). Proctor (1998: 356) further describes wilderness stating, "whether in a negative or positive sense, then, wilderness and civilization [nature and culture], have almost always been counterposed." According to this perspective, the issue at stake is not only about an objective reality of nature versus a socially constituted and culturally grounded nature but also whether our ideas "point fundamentally to wilderness "out there" or to the cultural predispositions that accompany our concept of wilderness" (Proctor 1998: 357). Delaney (2001: 490) further contends, "the nature that is constructed is a concept, an idea, a category, a set of conventionalized metaphors, and a trope for differentiation." This idea produced an understanding that wilderness not only existed apart from but also prior to human settlement in a particular territory.

On the other hand, wilderness represents places imagined as being devoid of human settlement, which were considered as 'unpolluted' by human cultural practices (with no traces of human socio-economic, cultural, and political exercises over the environment)–albeit the existence of such places is highly contested. In this context, the contestability of wilderness relates to the fact that areas such as national parks and game reserves, which are established under the assumption that the territories were free from traces of human settlement, for example, ignore human history by imposing the predisposition of human-nature dualism. In a similar dualistic position, Proctor (1998: 355) argues that, "wilderness is a real place, a compellingly beautiful place free of human imprint, an object of great worth precisely in having escaped human domination, where natural processes reign unimpeded." These assertions prompt us to raise some fundamental questions concerning whether or not humans were part of nature.

These notions of wilderness as an idea and imagined place were exported to other parts of the world partly through what Harmon (1998: 220) calls a "demonstration effect". Although historical dynamics and transformations shaped and changed meanings attributed to ideas representing wilderness, nature-culture dichotomy remained common across its history of transformations. The deep-rooted implication of wilderness dualism becomes a practical challenge to indigenous peoples, pastoralists and agrarian communities when other actors like the state and conservationists label their ancestral land as wilderness and put it aside in the form of protected areas. Because of its consequences in negating customary rights, as the case in the Nech Sar National Park illuminates, researchers and practitioners alike should see notions of wilderness under critical scrutiny.

2.2.1.2 Mutualism: Nature and Society as Conjointly Constituted

Both the academic milieu and those who have been dispossessed and dismantled from their homes, ancestral lands, sacred spaces and spiritual sites due to the protectionist acts emanating from the dualism discourse have challenged the notion in the past few decades. Contrasting the cosmological views of indigenous Indians against the Western popular imagination of wilderness, Chief Luther Bear (1933 cited in Callicott and Nelson 1998: 201) poetically expresses his views as follows:

We did not think of the great open plains, the beautiful rolling hills, and winding streams with tangled growth, as 'wild.' Only to the white man was nature a 'wilderness' and only to him was the land 'infested' with 'wild' animals and 'savage' people. To us it was tame. Earth was bountiful and we were surrounded with the blessings of the Great Mystery. Not until the hairy man from the east came and with brutal frenzy heaped injustices upon us and the families we loved was it 'wild' for us. When the very animals of the forest began fleeing from his approach, then it was that for us the 'Wild West' began.

For the Indians, as is common among indigenous peoples elsewhere, there was no distinctively demarcated boundary between the human and non-human, the social and the natural, and nature and culture; rather, all were conjointly constituted as common family. A striking contrast in the above quotation is that wilderness is differently interpreted by the newly arrived white men and Indians–as a place 'infested' by 'wild' animals and 'savage' people, and as a home for a family (human and non-humans), respectively. From the perspectives of mutualism, which is self-explanatory to denote a mutual coexistence between different subjects or objects, it has been argued that plants, animals and other entities of nature are considered as societies of nature (Descola 1992).

Descola and Pálsson (1996: 14) conceptualize mutualism as constituting humans and non-humans, in which they all "belong to a sociocosmic community, subjected to the same rules as humans; any account of the social life must perforce include these components of the environment which are perceived as forming part of the social domain." Some scholars move the argument further in depicting the conception of nature among indigenous peoples, pastoralists, hunter-gatherers and non-Western local peasant farmers as it is embedded and embodied in their cosmological schemes. According to this line of argument, the eco-cosmologies of most non-Western people holistically represent the human and non-human as inseparable (Escobar 1999). The perspective of mutualism traverses strict separation between the biophysical, human and supernatural worlds and considers the cultural practices in a dialogic relationship with nature (Braun 1997; Escobar 1999).

However, this perspective depicts 'non-Western' people as a homogeneous entity with common experiences, knowledge and ways of life. Besides blending diversity into an imaginary box of homogeneity, it mystifies the non-Western 'Other' as a 'noble savage' whose ways of life, modes of production, culture and knowledge are considered so 'primitive' that they cause limited damage on the natural environment (Berkes 2008). I depart from the above argument by Descola and Pálsson, because it blends humans and non-humans into the same rules despite the fact that the conjoint constitution of humans and non-humans is mediated through respect, reciprocity and trust, according to my empirical data from the Guji for example. I share the argument posed by Berkes (2008: 233) regarding the inadequacy of the dominant representation of indigenous peoples and their traditional knowledge as "noble savage", "close to nature", "communing with nature", and "harmonious co-existence between people and nature". This is because the understanding emanates from how Western culture has constructed dichotomies between the 'West' and the 'Rest' over time rather than reflecting the actual lived-in experiences of indigenous people. Despite superficially recognizing the nature-culture mutualism, such a representation is an artful cunning in homogenizing indigenous peoples and masking any internal transformation that these groups have undergone across history.

I would like to take the debate in mutualism one-step further and argue that most indigenous peoples, due to their iterative interaction with nature, understand their relationship with their surroundings in the form of respect, partnership, reciprocity and trust rather than domination and separation or even integration. This entails a different level of conceptualization of the relationship in the form of mutual recognition of partnership on the part of humans and non-humans, rather than the inseparability of the two. Taking the experience of Cree hunters and their relationship with the animals that they hunted, Ingold (2000) makes a strong argument that the hunters and animals interact based upon partnership, built on elements of trust and reciprocity. Ingold argues that the prey has the power to withhold its body not to offer it to the hunter if the latter tries to coerce it. Humans' knowledge and perception about their environment is also acquired through iterative interaction with the environment, as well as peoples' perception of environmental 'affordances' (Ingold 2005). Environmental affordance is a concept that refers to something that the environment offers/affords human beings–something that is either good or bad (Chen 1993). People interact with the environment or parts of it in anticipation of reciprocity or infliction based upon the nature of their actions.

People's understanding of environmental affordances shapes their behavior and builds their knowledge about the consequences of different aspects of engagement with nature. For example, drawing on the experience of Bhotiya people in India, Bosak (2008) argues that indigenous peoples' perception of nature is built on reciprocal relationship through their physical and spiritual connection to their environment. Bosak further articulates that the Bhotiya do something good for the environment in expectation that their god will reciprocate by looking after their crops, multiplying and regenerating their seeds, as well as sustaining life. Likewise, the Guji Oromo in the Nech Sar area had traditionally considered the 'Guji world' as constituting the people, livestock, other living and non-living beings and the divine being (*Waaqa*) and human and non-human intermediaries–the *ayyana* (spirits) (Baxter 1991; Van de Loo 1991).

In the case of the Guji Oromo in the study area, mutual interdependence, balanced or bilateral reciprocity and trust are the basic pillars of human-non-human relations. These dimensions of the relationship mark their self-representation as part of the space in which they situate their cultural, economic, political and social affairs. In this regard, it is not the physical landscape per se that is more meaningful, but rather the feelings, memories and meanings embedded therein. Likewise, these people represent nature as part of their culture. As a Guji elder once commented, "*Gujummaan keenya aada keenya. Aadaan teenna waa mara. Looni, jila, safuu, gaara, lageen fi mukkeen tunillee aadaa gabbisan*" (informant: Gagasa Adola, April 2012). This loosely translates as,

"Our Gujiness is confirmed by our culture. Our culture constitutes everything. The livestock, sacred journey [ritual], customs, hills, valleys and trees [sacred trees] all contribute to our culture". Such interdependence forms societies of nature, including the human, non-human and divine being (Descola 1996). Accordingly, this is why Ingold (2005: 502-503) argues that: "In reality of course, just as people have forever carried on their lives in the fields of their relations with others, so they always inhabited an environment including manifold non-human as well as human constituents. Social life has always been part and parcel of ecological life, if indeed the two can be sensibly distinguished at all." Ingold's argument is not more about conjoint constitution of the human within the non-human nature, but rather the interdependence between the two that manifests through important values of respect, trust, partnership and reciprocity.

Thus, it is sound to argue that both the dualist and mutualist perspectives depict nature-culture relations at two polar extremes by silencing multiple interpretations of 'reality'. Thus, any attempt to conceptualize nature-culture relations without contextual interpretation of multiple voices (the dominant Western culture of dualism and the marginalized views of mutualism among indigenous peoples and their advocates) persists as a debate between binary oppositions. In such binary approaches, the recognition of one perspective marginalizes the other and thus impedes comprehensive understanding and partnership between different forms of knowledge (Merchant 2003). In the next section, I will discuss another alternative approach to understanding nature-culture relation that recognizes multiple interpretations of the theme, in an effort not to deconstruct the previous perspectives but rather to contextualize both views within their socio-cultural contexts.

2.2.1.3 Beyond the Dualism-Mutualism Divide: Nature as a Socially Constructed 'Reality'

As I discussed earlier, since the 1980s, scholars in the fields of anthropology, environmental history and development studies have begun revisiting the old dualist perspective that dominated most of the Western academic and popular thought for centuries (Merchant 2003). As part of the global Indigenous Peoples' movement, scholars, human rights activists and environmentalists began a paradigm shift from the dominant notion of considering science as the ultimate basis of knowledge towards recognizing indigenous knowledge in the quest for sustainable environmental management. However, the seemingly 'politics of recognition' has also pushed the discourse to the other polar extreme, rather than creating a middle ground where the two perspectives could serve as complementary rather than competitive forms of knowledge. Descola (1996: 82) gives us some hints in this line of argument, "Many anthropologists and historians now agree that conceptions of nature are socially constructed, that they vary according to cultural and historical determinations, and that, therefore, our own dualistic view of the universe should not be projected as an ontological paradigm onto many cultures where it does not apply."

There is a growing understanding among scholars of embracing an epistemological shift towards recognition of multiple voices, interpretations and meanings about a particular social world though debates on constructionism continue to be unsettled (Philipmore and Goodson 2004). There are also two competing views within constructivist camp. The first is the view that nature is a mere social construction and thus is meaningless without interaction with humans who construct meaning to it. Second, some argue that the meaning we give to nature, objects, and living and non-living things are informed by our cultural milieu (views, norms, historical and political traditions, beliefs, etc.) (Kindner 2000). In this work, I will take the second view as an important point in understanding nature and the place of humans. I argue that it is not about the (non-)existence of nature as an objective reality per se, but rather that our ideas, meanings, knowledge and understandings of nature/wilderness are socially constituted and products of cultural, historical, political and religious dynamics in which we participate, all of which shape the meaning that people bequeath to nature. Along this line of argument, Cronon (1996: 25) notes that the perspective of social construction of nature "is not to say that the nonhuman world is somehow unreal or a mere figment of our imaginations–far from it. But the way we describe and understand that world is so entangled with our values and assumptions that the two can never be fully separated."

The second line of argument is highly relevant in this particular study because it is not the objective phenomenon, absence or presence of events and things such as environmental degradation, enclosure and physical landscapes that are the subjects of contestation as such. Rather, as the empirical data on representations of the Nech Sar National Park among competing actors reveal, interpretations and meanings about the territory have been built on political, historical, religious and cultural underpinnings that shape their understandings and narratives. While the dominant representation of the area by the Ethiopian state as 'wasteland' and 'uninhabited wild' was an extension of the historical north-south dichotomy, most of the local communities considered the land as their ancestral home in reference to history, culture and economic practices. Thus, the state builds its referents of representation on systematic historical and cultural dichotomy while cultural, historical and economic connectedness of the local communities to the land informs their understandings of their place in their surroundings.

The works of scholars in different fields in the late twentieth century have yielded abundant evidence that meanings ascribed to nature are profoundly human constructions. Some contend that, "our ways of thinking about the natural world are powerfully shaped by our time, our place, and our culture" (Cronon 1996: 35). Considering nature as naïve reality tempts us to presume that it had no cultural context, as if it were everywhere and always the same. Rather, one would argue that nature is culturally grounded, socially constituted, conceptually fluid and contested (Kindner 2000). Likewise, ideas about nature are "always constructed by our meaning-giving and discursive processes, so that what we perceive as natural is also cultural and social" (Escobar 1999: 2).

It is also essential to note that the existence of objective elements of landscape such as rivers, mountains, forests, valleys, rocks or plain fields are rarely contested. Rather, the representations, symbolic meanings and ascriptions attributed to these physical elements of a landscape are often the subjects of contestation. Slater (1996) describes such diverse ascriptions of meanings to a similar object by using the example of a forest, which stands for a timber in the eyes of a logging company, while the same object (forest) connotes a spiritual connection with God for indigenous peoples in Amazonia. While environmentalists might consider the same forest as wilderness that deserves protection, the indigenous people might see it as a home for people, animals, spirits and other non-human beings. In other cases, different actors interpret 'natural' disasters such as earthquakes, floods, droughts and epidemics differently in various contexts. While scientists might attribute the causes to geological factors or effects of climate change, indigenous peoples would see it as resulting from disruptions in relationship between humans, non-humans and supernatural power. Hence, all such instances hint at the socially constituted and culturally grounded notion of nature. In other words, nature embodies both natural and social characteristics as being "simultaneously real, collective and discursive–fact, power and discourse" (Escobar 1999: 2). In this sense, Escobar's conceptualization of nature as constituting fact, power and discourse is analogous to what Cronon (1996) describes wilderness as, namely a conceptual and spatial discourse that has been used as a latent instrument of subjugation of indigenous peoples. This is central in my work because, as political ecologists often argue, the perspective about wilderness imposes upon people an ideal form of nature that entails a construction of "nature without people" (Robbins 2004: 150).

At this junction, another profoundly important point to discuss within the politics of conservation is the discursive conceptual meaning attached to wilderness, which reflects the central theme of this research. During the European expedition to what they termed as the 'New World', as well as in the subsequent centuries of colonial establishments that followed, wilderness was used as strong pervasive instrument in legitimating European conquest and occupation by relegating indigenous rights of land and resource ownership (Merchant 2003). As I discussed in the preceding sections, European colonists portrayed the lands they encountered in the 'New World' as wilderness, although most of these territories were once homes for indigenous peoples.

Like the emerging epistemological and methodological recognition to research subjects in the social sciences, nature should also not be regarded as a passive object upon which humans could freely act. In the dominant Western notion of nature-society dichotomy, society's progressive stage of development was judged by its capability to maneuver, subdue, domesticate, dominate and control nature through science, technology and capitalism (Merchant 2003). Despite the perspectives of social construction of nature that open the door for multiple voices and interpretations on nature-culture relations, the politics of nature conservation still mingles the old paradigms of human supremacy over nature with emerging attempts of partnership. In the next section, I will discuss the paradigms governing human-non-human relationships hinting at how they affect conservation approaches. In doing so, I will discuss the relationship between wilderness conservation, development and customary rights accordingly.

2.2.1.4 Dilemmas in 'Wilderness' Conservation, Development and Customary Rights

Human interaction with nature is informed by people's conceptualization of the relationship from three different paradigms: orientalism, paternalism and communalism (Pálsson 1996). While the first two share common perspectives on human-nature dualism, communalism differs in that it rejects the radical separation of nature and society, and object and subject. It emphasizes the notion of mutual engagement between humans and non-humans in the form of dialogue. In environmental conservation context, orientalism promotes human dominion over the environment, which entails exploitation, control and manipulation according to humans' economic, aesthetic and political motives. Orientalism is fairly analogous to the post-Renaissance conceptions of 'wilderness domestication', which equate civilization to one's skill, capability and power to explore, conquer, control, exploit and dominate the perceived 'savage' wilderness. The transformations that occurred following the eras of Renaissance and Enlightenment regarding human-nature relations brought the view of equating and evaluating human civilization to the degree of its mastery and control over nature. Building on some narratives from Christian tradition that bestows human supremacy over nature yet placing it within scientific 'rationalities', European colonists of the 'New World' backed their expedition of subjugating wilderness based upon orientalist paradigm (Pálsson 1996; Merchant 2003). The conquest of nature had been practiced in the form of protected areas, although the rhetoric of nature preservation was often portrayed under the guise of paternalism for the diverse purposes of consumption, production, sport and emotional display.

Likewise, paternalism bestows a responsibility of guardianship upon humans to protect, safeguard and manage nature and its constituent entities (Pálsson 1996). Paternalism entails usurping the choice of a certain actor or subject by another actor and reflects the imposition of the latter's perspectives and interests upon the former (Archard 1990). While the practices of wilderness conquest, such as the case of European conquests of indigenous peoples' lands since the fifteenth century, were driven by the paradigm of orientalism, the practice and rhetoric of wilderness conservation and protection can be considered as paternalistic, given that it considered humans–of course not all humans–as rational guardians of nature. Both paradigms consider humans as masters over nature, whereas they see nature as a passive object to be acted upon and lacking any agency in the interaction with humans. The establishment of Yosemite and Yellowstone National Parks in the United States in the 1860s and 1870s marked the commencement of domesticating wilderness in the form of setting aside portions of territories for touristic consumption (Brockington et al. 2008). The Yellowstone National Park was the first national park in the world that was purposefully reconstructed into wilderness by removing indigenous Indians for the sake of American urban aristocrats. Subsequently, it was inculcated into American popular imagination as a symbol of the nation's conquest over the 'savage' (Cronon 1996).

However contested it might be, this notion of wilderness domestication and the representation of wilderness preservation as a sign of progress and humans' capacity of putting nature under control has led to the establishment of protected areas in most parts of the world (Brockington et al. 2008). Indeed, some associate the diffusion of national park ideals as driven by aesthetic, economic, religious and political motives. While the aesthetic aspects and their economic implications through tourism have been widely addressed, the religious and political dimensions are not commonly discussed in conservation literature. From religious dimension, some liken wilderness with the sublime–a rare place on Earth where people can meet God. Areas such as mountains, seashores, valleys and forests fall into the category of sublime nature (Cronon 1996).

The political driving force behind the establishment of protected areas beginning from the Yellowstone National Park all throughout colonial Africa, during which massive national parks were established intersected with powerful nation-building projects (Brockington et al. 2008). Moreover, national parks denote implicit and exclusionary sense of nationalism in which a certain territory is entangled with national identity (Haila 2012). Connotations of nationalism are embedded in the term "national" as used in national parks. For example, in some cases there are tendencies to use peculiar landscapes in national parks as symbols of national identity. As Haila (2012) mentions drawing on the Scandinavian case, fjords and mountains, and forests and lakes are symbolized as representing Norwegian and Finnish 'national natures', respectively. There are similar cases in the Nech Sar National Park, with the national park having been associated with identity of one ethnic group, the Gamo. As will be discussed later in the next chapters, the Gamo people proudly identify themselves with the national park and subscribe to it in giving names to hotels, university campus, colleges and streets in the Arba Minch town.

On the contrary, the notion of wilderness conservation contradicts the perspectives of non-Western societies who consider the land as their home. Cronon (1996: 79) clearly counters the notion of wilderness depiction as a reality stating, "the removal of Indians to create an "uninhabited" wilderness–uninhabited as never before the human history of the place–reminds us how invented, just how constructed, the American wilderness really is." The establishment of the Yellowstone National Park on the ruins of Indian homeland hints at the convergence of both orientalist and paternalist paradigms, both of which disregard the human-non-human coexistence.

Central to both orientalism and paternalism views is the tendency to perceive wilderness as an object that humans could tame and control using science and technology, before subsequently changing it into a commodity through capitalist development. In the eighteenth and nineteenth centuries, the triad of science, technology and capitalism accompanied European expedition to the 'New World' and enhanced the subjugation of territories that had been homes of indigenous peoples for centuries (Neumann 1998).

In explaining communalism, Pálsson (1996: 74) states that, "just as a child may expect the care of its parents, the environment provides unconditional support, irrespective of what happened in the past." It focuses on practice, reciprocity, trust and partnership. For example, the Guji Oromo in the research area believe that nature provides its bounty (fertility to people and livestock, rain, good harvest/yield, health and peace) in the form of reciprocity for the people's good care of wild animals, sacred trees, rivers and spirits. More importantly, they believe that such reciprocity is maintained through their spiritual harmony with *Waaqa* (God), which is expressed in the form of relentless prayers and rituals. It is in line with the Guji conceptualization of their surroundings based on trust and reciprocity that Pálsson (1996) and Ingold (2000) argue that representations of non-humans among indigenous communities in different parts of the world are not based on systematic recordings of ideas, events or phenomenon, unlike in the Western scientific tradition. Rather, they are expressed contextually through practical lived-in knowledge, interactions and rituals.

The Guji people do not create boundaries of otherness between themselves and non-humans. In their day-to-day conversations, greetings and interpersonal communications, they give priority to peace regarding their livestock, children, their surroundings and all forms of living and non-living beings, including spirits (*Ayyana*). The Guji use a common phrase in interpersonal exchange of greetings as "*alaa manni nagayaa*?" This literally equates to inquiring about the peace and wellbeing of the 'inside' and 'outside'. The Guji do not use the 'inside' and the 'outside' as classificatory spatial locations; rather, they have different representations. In such interpersonal conversations, *mana* (home, the inside/indoor) includes human beings, livestock and the *Ayyana* (spirits). The *alaa* (outside) refers to wild animals, forests, rivers and other living and non-living things with which the Guji have direct or indirect contact. On the other hand, on a more abstract level, the Guji label the inside as invisible things such as spirits and internal peace of human mind, whereas the outside are visible creatures including humans, livestock, wild animals, plants and rivers. Nonetheless, both the inside and outside are equally important for the Guji in maintaining peace and harmony with *Waaqa* and other non-human beings.

At this point, it is important to relate how these paradigms influence conservation policies. For example, during the colonial era in Africa, the colonialists systematically constructed the dualist notion of human-environment relationship to justify land alienation, dispossession and displacement of the native people. In this regard, "the environment is separated out from human agents and perceived as an exterior non-human habitat subject to appropriation, domination, attack, conquest and domestication" (Croll and Parkin 1992: 32). In such contexts, colonial agents used the systematic imposition of an 'externalized identity' upon nature to justify its subjugation by specialists who imposed outside distinctions and knowledge by relegating indigenous eco-cosmologies. By contrast, when the idea of the environment is recognized as "ontologically part of the people who give and draw sustenance from it", neither the person nor the environment is privileged or suppressed in the mutual relationship (Croll and Parkin 1992: 32). Such a holistic perception of human-non-human relations recognizes the agency of both actors and thus gives recognition to both in any development or conservation planning and implementation.

Bosak (2008: 215) explains the impacts of dichotomizing nature as externalized and internalized, stating that, "in conservation context, the protection of nature is necessarily a social question, in which nature can be seen as externalized or internalized." Externalizing or separating nature from humans entails seeking its protection from outside through scientific knowledge under 'experts' trained in modern science and technology. This often leads to 'top-down' conservation policies that privilege scientific knowledge over local expertise. Conversely, internalizing or considering nature as a holistic entity in which humans and non-humans are its integral part would enhance the recognition of local people and their knowledge as integral and essential components of the conservation endeavors. Indeed, this binary perception of nature significantly determines conservation policies (Bosak 2008).

It has been evident that the colonial system introduced the Western conceptions of nature to Africa, entrenching them into colonial policies of conservation by creating areas free from human influence in the name of protected areas. However, the notion of establishing protected areas through exclusion of human society should not be seen as a purely conservation scheme; rather, as will be detailed later in this chapter, it is highly intertwined with economic and political discourses on local, national and international scales. Political and economic schemes influence the relationship between humans and nature, particularly in the context of protected areas, embracing issues of rights and access to land and resources (Adams and Hutton 2007). Put briefly, the idea of pristine nature and un-peopled wilderness spread to Africa since the colonial conquest as an ideological framing of conservation, albeit implicitly as a mechanism of control of resources and the people. Accordingly, the conceptualization of environmental conservation in the form of protected areas as an instrument of control over the multidimensional spaces of local communities in Africa will be the subject of chapter four in this study.

Does Conservation Enclosure Enhance or Hinder Development and Customary Rights of Local Communities?

Protected areas in the form of wilderness preservation have been often in conflict with ideals of capitalist development on the one hand, and indigenous peoples' customary rights on the other. In some cases, protected areas have been established to deter development programs such as oil exploration, logging, mining, dam construction and extensive mechanized agriculture. In other contexts such as the experiences of the San people in Botswana, the Inuit in Alaska, the Maassai in Kenya and Tanzania, among many more indigenous communities, the struggle was between conservationists and indigenous peoples, with the latter struggling to maintain their land rights (Brockington et al. 2008).

These three schemes–conservation, development and customary rights–often operate on opposing ends, while at times working in a sporadic state of cooperation. In cases where development projects deem threatening the livelihood of indigenous peoples and the survival of the biodiversity, indigenous peoples and conservationists create an alliance against capitalist actors, namely the state and/or multinational companies. However, "when the alliance is between development and conservation–powerful as it is–[it] can overcome local objections and protests" (Brockington et al. 2008: 6). One aspect of alliance between conservation and development is in cases where conservation projects seem to generate economic development, as in the case of tourism. Despite its severe impacts on local communities, tourism and the economic values attached to this industry are among the important reasons for which protected area conservation schemes have been 'justified' and 'legitimated'.

The major justification for tourism promotion in protected areas relates to its convergence of development and conservation purposes, despite it often being at odds with indigenous values, practices and customary rights. It has brought together the neoliberal market oriented environmental management and local economic development. Neoliberal thought promotes economic diversification, which includes the tourism sector. Because of the North-South dichotomies and stereotypical representations of the global South as 'close to nature', "global financial institutions such as the World Bank and IMF portrayed developing countries of the South as having 'comparative advantages' in offering pristine nature" (Brockington et al. 2008: 133). However, the applauded benefit of ecotourism in promoting local empowerment and creating livelihood options for the locals and as environment friendly enterprise often obscures the cultural and social implications of tourism. While it promotes cultural commoditization among indigenous peoples, particularly in the form of cultural tourism, this enterprise overshadows internal heterogeneity within local communities. In such process, discrimination, domination and exclusion are also evident in ecotourism sectors, as is common in any interplay between actors with asymmetrical power relations. Moreover, it also encourages the appropriation of land and resources by few social elites while at the same time relegating other sections of a community to the margin.

With reference to conservation enclosures and the infringement of customary rights of local communities, the very normative perception of nature-culture dualism has entailed setting dogmatic rules of what is right and what is wrong in human-environmental interactions (Haila 2012). In this context, conservation promotes the idea of defending nature from destruction by humans and drawing strict boundaries between the two to preserve the pristine character of nature (Castree and Braun 1998). As a result, it has become evident that nation states, multinational corporations and conservationist organizations have seized indigenous peoples' lands under the guise of conservation, albeit mainly for economic and political purposes associated to it.

In a short-term 'alliance' in the 1990s, conservationists and indigenous peoples' movements used the rhetoric of conserving the 'world's first peoples' and the 'world's last places' in their respective struggle against encroachment from development corporations and state backed large-scale development projects. However, the alliances did not last long due to the incompatible value systems between the two. Indigenous peoples' ways of valuing and using natural resources, their ontological worldviews, cultural practices and beliefs happened to be in contrast to the expert-oriented scientific approaches advanced by conservationists. As a result, conservationists became skeptical of working with local communities and rather opted to what Adams and Hutton (2007: 165) called "back to the barriers" in the late-1990s. This term implies the resurrection of the strict protectionist approach of protected area conservation through the complete exclusion of people from 'wilderness'.

Indigenous peoples across many parts of the world have been victims of competing perspectives. These include ideals of developmentalism that encourage large-scale development projects such as mining, oil exploration, mechanized agriculture and logging on customary lands of local communities, as well as conservation projects (protected areas in the form of national parks, forest reserves, game reserves, and sanctuaries) that also restrict the customary rights of resource. In this regard, development and conservation projects have significantly restrained local communities' access to their customary rights and have been instrumental in dispossessing the people from their land and resources (Brockington et al. 2008). Modernist discourses of development denounce and denigrate livelihood practices, local knowledge and eco-cosmologies of indigenous communities. As Harawira (2005: 108) claims: "In the newly created states, indigenous peoples' traditional cognitive and social patterns, modes of governance and ontological world views were identified as 'obstacles to development' and at odds with those of an economically based society." Despite development institutions such as the World Bank and IMF defending conservation by suggesting that it contributes to nascent tourism, it has apparently become an instrument of dispossessing people from their land. Nevertheless, I argue that nature conservation through enclosure is one of the ways in which states exert control over local communities. For example, according to Brockington et al. (2008: 125), "in Botswana, Ethiopia and Thailand, ongoing evictions or attempts at eviction are continuing to disrupt the lives of indigenous peoples."

In the conservation-indigenous peoples' rights nexus, the fundamental problems of wilderness relates to its exclusion of people from their homes, both physically and conceptually, because it "negates long histories of association between people and places and thus excludes them historically" (Brockington et al. 2008: 49). During the colonial period in Africa, such representations were adamant among colonial elites, who constructed a separate category between long resident natives and nature, thus portraying the people as destructive. Fisher et al. (2008: 18) argue that this assumption underlies the "colonial origin of conservationist belief in 'wild Africa'." According to Fisher et al. (2008), the representation of nature particularly by the social elites in the 1960s and 1970s as spiritually charged wilderness strengthened the preservationist approach to nature through national parks. Such a preservationist approach reinforces resettlement of resident communities for establishing protected areas, which in turn exasperates the livelihood insecurity of the people by limiting their access to resources. Resettlement induces uncertainty amongst local communities, in some cases instigating conflicts between 'original' inhabitants and 'new' settlers. For example, a violent conflict erupted between the Guji and Koore in the study area in 2010 over a contested farmland from where the Koore were relocated in 2004. The dispute emerged when the Koore attempted to return to their former settlements, which the Guji partly possessed since their departure (informant: Gacha Atura, February 2012).

Put briefly, development and conservation persist as competing discourses, at odds with the needs and rights of indigenous communities unless both embrace issues of citizenship rights for the ownership of resources, as well as recognizing views and knowledge of local communities. For development, conservation and customary rights to go hand in hand, the rights of indigenous peoples to participate in decision making processes in matters that directly or indirectly affect them must be recognized. On the other hand, development and conservation schemes could become viable pathways towards ensuring sustainable environmental conservation and solving development questions of local communities or the general public if, and only if, development projects in particular recognize the intrinsic values of human-environmental relations. Unfortunately, the 21st century development path in developing countries is incompatible to the environment, as well as local demands, values and priorities. This is why Katz (1998: 45) states that peoples' engagement and interaction with the environment under "capitalism have operated as if nature were given, a free good or resource of wealth, an unlimited bounty awaiting only the "hand of man" to turn it into a bundle of resources."

To summarize the discussions under the previous subsequent sections, there are different notions of nature and humans' place in it. Apart from ontological debates concerning nature-culture relations, these competing perspectives are also embedded in political, economic and cultural views of social actors. Nonetheless, the way in which people understand human-environmental interactions fundamentally shapes approaches of environmental conservation, particularly in the form of protected areas. For example, debates on nature-culture dichotomy are expressions of asymmetrical power relations, knowledge and discourses, because they entail the nullification of the history of some groups, such as indigenous peoples, by inscribing the myth, narratives and knowledge of dominant actors such as states and business or conservation organizations. As a transition to the next major theoretical framework of this study, it is important to close this section by raising the following questions: Who represents a certain territory as 'aesthetic', 'uninhabited', 'wild' and 'natural'? What does it imply to set aside portions of land as protected areas and other parts for human settlement, utilization and management? The search for answers to these and other interrelated questions prompts us to understand the politics of conservation and politicized environment in developing countries, with reference to Africa. This transitory reflective note takes us to my second theoretical conceptualization concerning the political ecology of conservation.

2.2.2 Political Ecology of Conservation

2.2.2.1 Why Political Ecology?

In the politics of nature conservation or what Bryant and Bailey (1997: 27) referred to as the "politicized environment", ideas, perspectives and discourses surrounding environmental problems, approaches of conservation and the status ascribed to some parts of territories should be scrutinized against the backdrop of competing forms knowledge, power and worldviews. In the quest to understand conservation trajectories in Africa and particularly to comprehend how ideals of national park were introduced to the continent and sustained thereafter, research in the field needs to transcend traditional approaches of studying environmental problems from apolitical perspectives. The traditional approach focused on land-based resources and its relations to land-based societies or livelihood practices (Bryant and Bailey 1997; Robbins 2004). Early works on environmental problems and ecological changes in developing countries emphasized investigating the issues mainly from apolitical perspectives, placing the blame on population pressure, poor technology and the inadequacy of traditional knowledge concerning environmental management (Peet and Watts 2004; Robbins 2004). Apart from casting the blame on local people who are often victims of environmental problems that are mostly not their own making, these approaches fail to address the intricate socio-economic and political aspects of environmental issues. As a result, the approaches paid less emphasis to local and extra-local political and economic forces, as well as different discourses, ideas, knowledge and narratives related to environmental problems in the processes of planning and implementing conservation schemes.

In response to incomprehensiveness of the works to address environmental issues, scholars began to acknowledge the need for understanding environmental problems (land degradation, deforestation, soil erosion and air and water pollution) beyond apolitical explanations in the 1970s and 1980s. According to Peet and Watts (2004: 6, emphasis in original), "less a problem of poor management, inappropriate technology, or overpopulation, environmental problems were *social* in origin and definition." As early as 1980s, it became clear that local and extra-local forces have had significant contributions to environmental problems besides demographic, technical, skill and ecological factors. Thus, there was a call for a comprehensive and ample approach to understand the complex interplay between multiple forces, actors and processes. It has since become evident that political, economic and social dynamics, processes and institutions at micro and macro-levels shape power relations, questions of knowledge and contestation among different actors. These constellations of forces at various levels and in different forms not only influence environmental conditions, but also guide our understandings about the phenomena. Thus, the approach of investigating environmental problems and ecological changes from economic, social and political dimensions as an interplay between local and global forces is a political ecology perspective.

As Gossling (2004: 10) states, "political ecology thus seeks to investigate the interaction of international, national, regional and local actors at the interface of environmental change, economics and politics." It deals with how local resource use practices, environmental changes and ecological conditions are driven by and are products of political and economic motives (Robbins 2004). These explanations mainly draw on an actor-oriented approach of understanding competing perspectives, ideas and interests, framing the arguments around the centrality of political and economic forces at different scales in influencing how local resource ownership, utilization and management are defined, contested and negotiated.

Political ecology is a concept used in many disciplines, including geography, anthropology, sociology, development studies and environmental history. However, scholars differ in terms of their scope of usage and analysis of the concept. While some have focused on the dynamics of political power and resource distribution at local level, others emphasize analyzing macro level political and economic dynamics, paying little attention to localized case studies (Gezon 2006). Political ecology is thus as an approach of addressing environmental problems from political and ecological dimensions, bringing into the discussion the role different institutions play in promoting or curtailing environmental problems (Peet et al. 2011). Drawing on Blaikie and Brookfield (1987), Peet et al. (2011: 24) explain the concept that a "field of political ecology coalesced around investigations into: first the impact of capitalist development on the environment; second, the social and political implications of environmental protection, conservation and management; third, the political economy of the way new natures (species, landscapes and ecosystems) are produced."

As an interdisciplinary concept, some scholars use political ecology to explain soil degradation and water management (Blaikie and Brookfield 1987), protected area management/conservation (Neumann 1998; Gezon 2006), as well as global economic and political dynamics (Peet et al. 2011). Political ecology emerged as a field of study in its effort to address land degradation by integrating human and physical approaches to the problem. In its classical conceptualization, political ecology encompasses the interplay between society and land-based resources by recognizing internal heterogeneity within society itself on the perceptions of individuals or groups about environmental problems. However, the emphasis on local communities and physical elements in the analysis of environmental problems in the earlier works obscured the impacts of extra-local factors, political and economic systems and external actors in the understanding of environmental changes (Robbins 2004).

The major elements of analysis in the early studies on environmental problems drawing on the political ecology perspective were aspects of market integration, commercialization and dislocation of customary forms of resource management, all of which were treated as processes of transition to a capitalist mode of economy (Peet and Watts 2004). According to Blaikie and Brookfield (1987), political ecology builds on two essential assumptions: first, it presumes poverty and land degradation as mutually causal in such a way that poverty induces land degradation, which in turn deepens poverty. Second, it assumes that environmental problems are regionally chained and thus the analysis should consider these dimensions. Likewise, Peet and Watts (2004) emphasized the importance of a third dimension in the study of political ecology of environmental problems, which frames land management within the interplay between external and internal forces such as the state and world economy and local ways of interaction with the environment.

However, the assumption that attributes land degradation to poverty at local level and vice versa relegates the influence of powerful actors such as the state and business organizations, who significantly contribute to the environmental problems, yet bear minimal costs of their actions (Boyce 2002). Because it masks out the impact of dominant actors in exacerbating environmental problems, the notion of associating poverty and environmental problems as being mutually causal is close to blaming the victim by covering the villain. It is often due to the inadequate attention given to political and economic forces in the early works of political ecologists that brought the field under critique later in the 1990s. The other two assumptions in the early political ecology works became part of the major approaches of studying environmental changes under political ecology given that issues of human-environmental interactions constitute spatial and structural connections of power.

In response to such limitations of the works in the preceding decades, scholars began to acknowledge the role of multilayer actors, forces and processes that directly or indirectly influence environmental changes. More importantly, issues of power relations between states, inter-state, multilateral institutions and local communities over resource control and the recognition of multiple ways of knowledge and practices related to environmental management and utilization were integrated in political ecology in the 1990s (Peet and Watts 2004). In this regard, political ecology scrutinizes the politics of environmental knowledge and practices by opening its areas of spectrum to multiple interpretations of reality. To understand competing narratives about environmental knowledge and human-environment interactions, one might ask the following key questions: whose knowledge prevails in environmental management? How do different forms of environmental knowledge interact? While this section only outlines background discussions on the questions, the subsequent chapters will present empirically supported discussions concerning power relations and interactions among competing actors, based upon the case of the Nech Sar National Park.

A critical aspect of researching environmental problems within the political ecology perspective began to challenge the human/nature dichotomy that until recently dominated most academic discourses. This is because nature conservation in the form of protected areas is highly entangled with issues of power, knowledge and representation, which entail the imposition of some actors' narratives or perspectives by relegating the views of others. For instance, the eviction of over a thousand Guji and Koore households from the Nech Sar National Park (APF) during its transfer to the African Parks Foundation (APF) in 2004 was premised on the notion of delineating human-wildlife coexistence, although economic motives fundamentally triggered the Ethiopian government to take the actions (Dowie 2009). The APF aimed to ensure long-term economic and ecological sustainability in the park. However, "this framing of protected areas (PAs) in ecological and financial terms excludes any consideration of the social and political context of the establishment and management of PAs, despite the obvious importance of such issues" (Adams and Hutton 2007: 148). As I discussed earlier, the coalition between conservation and development often restrain local communities' customary rights. Political ecology thus addresses the politics of conservation from multidimensional approaches by incorporating the economic, ecological, political and social aspects.

Many scholars now recognize that treating the human and non-human nature as distinctive and mutually exclusive worlds would curtail complexities embedded in the interaction between the two (Descola 1996). Therefore, it is towards this accommodation of multiple interpretations of environmental problems that political ecologists moved beyond the structural approach of investigating the topic. On the one hand, recognizing the representation of local people's voice in environmental problems has become central to the works of political ecologists. On the other hand, political ecology tries to address the major challenge that is "to integrate the structural focus on state, society, and industry, and the poststructuralist attention to how interactions between such actors co-construct environmental discourses and narratives about the environmental change, and who should be represented as victims and villains" (Forsyth 2003: 8-9). It is essential to situate the study of conservation schemes within broader power relations in order to understand how different constellations of power inform people's narratives and practices of conservation schemes. In the Nech Sar National Park case, local people produced critical knowledge and challenged the state discourses of conservation and development. The interplay between such competing perspectives have produced different representations concerning environmental change or conservation practices in the national park, which are influenced by the nature of power (dominating power and resisting power) of the actors involved.

Despite different approaches to the study of environmental problems in developing countries, political ecologists agree on two major frameworks. The first framework attributes the problems to broader political and economic forces mainly manifested through capitalist intervention. The second is concerned with the impact of state intervention, which at times goes hand-in-hand with external forces, whereas in some other cases the state intervenes in local environmental practices as part of political control, national security and personal enrichment of government officials (Bryant and Bailey 1997). In other words, environmental problems in many developing countries are explained as functions of capitalist extraction in the form of logging, mining, fishing and cash-crop production, which was introduced during the colonial period and maintained thereafter. It has also been seen as a form of state exercise of power over its subjects through the control of their customary lands and resources. Moreover, due to their omnipresence and monopolization of coercive power, most states in developing countries promote environmentally destructive activities through different forms of natural resource extraction. For instance, large-scale agricultural investments, mining, logging and dam construction are among a few cases through which the state control of 'development' projects prevail; however, due to the limited recognition given to local voices, such practices often impede the customary rights of local communities over their resources.

By building on the second approach of understanding environmental problems, I will focus on the political ecology of protected areas conservation. Owing to its history as a field that emerged as a study of environmental problems mainly within the context of land degradation, political ecology has given less emphasis to discourses of conservation, given that it is produced and reproduced from different perspectives through interplay of competing social actors who build their narratives on different ontological discourses. With the exception of a few cases (for example, Neumann 1992, 1998; Gezon 2006; Igoe 2004), works on the politics of conservation in Africa have not comprehensively addressed the issue from broader historical, political, economic and social frameworks (Garland 2008). Even the existing ones do not address the complex historical trajectories of nature conservation from its genesis as part of the European imagination of Africa to the adoption of similar perspectives and practices by postcolonial Africa elites. Some works have focused on the socio-economic impacts of conservation on local communities, while others strived to position the case within European colonial conquest and its implications (Igoe 2004).

While it is not possible to dismiss these consequences of protected areas management approaches, a comprehensive understanding of the topic is only possible if we look beyond the notion of 'cause-effect' relations. This is because positioning our arguments within the presumption that environmental degradation perpetuates loss in biodiversity and exacerbates rural poverty would put us in a trap of blaming rural communities for the cause of the problem. Such a perspective also promotes the idea of excluding rural communities from resource management practices because once they are portrayed as causes of the problem, the system does not allow them to be part of the solution. The gap is particularly profound in the case of Ethiopia, because no single research has been conducted in this field to date, despite the existence of protected areas in the form of exclusionist and fortress conservation approaches.

Thus, I share the perspective that despite apparent evidence for changes in the physical forest coverage in the Amazon, dwindling biodiversity in Africa and an increase in climate variability elsewhere, ideas and narratives about these phenomena are socially constructed and reproduced by actors with dominant voices (Peet et al. 2011). Thus, a land-degradation perspective curtails actual environmental dynamics, for example, by relegating why powerful actors advocate for enclosure of resourceful territories. In response to this gap, some scholars have attempted to contextualize conservation enclosures as a focus of investigation, taking the impacts, logics and operations of conservation and environmental protection itself from multidimensional perspectives. In this regard, I use the political ecology of conservation as an appropriate theoretical framework because it probes into conservation politics by recognizing the issues of power, knowledge and discourses espoused by competing actors in the struggle over environmental resources.

2.2.2.2 Conservation Enclosure as a Form of Hegemonic Control

Moving the argument further beyond the classical political ecology perspective of studying environmental degradation, I argue that conservation is a form of hegemonic control over the social, economic, political and cultural spaces of resource owners, and thus, we should understand the politics of conservation as political inasmuch as it is social, economic and environmental (Adams and Hutton 2007). Within such a framework, the study focuses on the idea that, "conservation, purportedly an effort to create better conditions in the world [for the humans and non-humans] *can frequently be a mechanism for (or more cynically a "cover" for) powerful players to actually seize control of resources and landscapes, and the flow of values that issues from these*" (Peet et al 2011: 27, emphasis added). More specifically, this research applies the 'conservation and control thesis' (Robbins 2004: 150), drawing on the following four fundamental theoretical foundations: 1). conservation reflects a form of hegemonic governmentality 2). It depends on the growing understanding of traditional resource management strategies as institutional systems, where rules govern extraction without necessarily state intervention on individuated property rights 3). It draws upon the notion that wilderness–as imposed ideal and a produced material reality–is a social construct, specifically taking the form of nature without people; and 4). it reflects a prominent understanding of conservation territories as bounded, regular, polygons, as ecologically and socially problematic and inadequate to meet the goals of preservation of either wildlife or livelihoods.

In the preceding sections, I discussed the three theoretical premises of the conservation and control thesis but it is important to briefly discuss and contextually explain the issue of 'conservation as a form of hegemonic governmentality'. In the politics of environmental management and governance, which has become a focus of political ecology, environmental governmentality refers to a circumstance in which hegemonic environmental governance is normalized within the life worlds of resource users, commonly local communities (Robbins 2004). It refers to acts, tactics and strategies that governments and non-governmental institutions use to create behaviors of 'being governed' on the side of their subjects through persistent discourses and narratives of rules, ethics and principles. Despite some scholars treating environmental governmentality different from coercion and sovereignty (Li 2007), it becomes a form of hegemonic exercise of power, particularly in countries with history of state violence. As (Peluso 1993) notes, states use coercive power in their efforts to enact their claims over particular resources particularly when other actors such as resource users or non-state actors contest and challenge the states' control over a specific resource. For instance, the state coerces conservation "when the resources are extremely valuable, and/or when coercion is considered either the last resort or the easiest means of establishing control over people and resources" (Peluso 1993: 201). Moreover, by institutionalizing different technologies of control such as the establishment of a local environmental regulatory community, environmental education and the so-called community-based conservation approaches, nation states and other environmental organizations often attempt to create "environmental subjects" (Agrawal 2005: 15).

In Africa, colonial and postcolonial states resorted to use coercion in the process of controlling resources that have been extracted through tourism, mining, logging, fishing and large-scale agriculture. It is hegemonic in many aspects, marginalizing local resource users from the benefit and management of resources, imposing elitist approaches of environmental management at the expense of indigenous knowledge and aiming at the physical and psychological domination of the subjects (Peluso 1993; Singh and Houtum 2002). For example, enclosing common resources entails setting mechanisms that justify and enforce control, as well as a form of governance that inculcates environmental law among the community who previously enjoyed free access to the resources. By doing so, environmental governmentality attempts to promote a culture of internalizing certain forms of control and authority as normal and natural within resource users (Peet et al. 2011).

Similarly, colonialism has introduced to Africa a system that disabled, dismantled, and denigrated "old forms of life by systematically breaking down their conditions, and with constructing in their place new conditions to enable–indeed, *oblige*–new forms of life to come into being" (Scott 1995: 193, emphasis in original). By emulating colonial notions of delineating humans from non-humans, postcolonial states in Africa also replicated hegemonic forms of environmental governance that were intended to legitimize state and non-state interventions under the guise of conservation and development.

Building on the conservation and control thesis, this work probes into how the conservation and ecotourism project in the Nech Sar National Park has been articulated, defined and negotiated among different actors. In doing so, the study investigates how competing notions of nature reinforce contradicting perspectives of conservation. It also looks at strategies the government used in channeling down hegemonic environmental control over the local communities under the guise of conserving nature and improving the people's ways of life. Understanding the political and material meanings and interpretations ascribed to a territory is sine qua non to comprehending the motives of different actors, because "any effort to environmental conservation in its many forms, for better or worse, is basically *a form of environmental control*" (Peet et al. 2011: 28, emphasis added). Environmental control entails control over the material, cultural, economic and political spaces of the people who inhabit or claim to have inhabited a particular territory.

As discussed earlier, local Guji and Koore communities were restricted in terms of access to their farmland, grazing fields and ritual sites following the establishment of the Nech Sar National Park. Through territorialization of the park territory, the state has exercised its hegemonic authority and power over the people by controlling their economic, spiritual and social spaces. In other words, evictions and resettlement programs have not only displaced the people from the physical space but also dis-spaced them from the places that symbolize different aspects of their lives, including their modes of spirituality, emotions and sense of belonging.

Here, it suffice to note that political ecologists criticize the notion of wilderness that is behind the establishment of many protected areas in developing countries, albeit not based upon a lack of concern for global environmental problems or change in local biodiversity. Instead, it is because such 'conservation' efforts serve the interests of the few by sidelining the knowledge and rights of local people. Peet et al. (2011: 27) have discerned between genuine conservation concerns and those who have used the discourses for different purposes:

> While certainly the catastrophic decline in key species around the world is a matter of universal concern, it is notable that a generation of efforts to save animals (e.g. the African lions, Indian tigers, Chinese pandas etc.) have largely been built around the "fortress" model where urban elites call for the enclosure of lands long used and occupied by rural, indigenous, and local people, all in the name of protection.

In contrary to the accusations against indigenous people from conservationists, accounts of the local community rather emphasize that the traditional conservation practices of the local people have long provided a home for the animals that conservationists claim to protect, preceding the arrival of the conservationists themselves (Berkes 2008). The following statement from a Guji elder during my fieldwork illuminates this argument:

> If our ancestors or we did not care for the animals, would not it be the case that they would have been perished long time ago? Who cared for these animals before the coming of the government? Was it not the people? Who cared for them 50 years ago? Was it not that we lived together hundred years ago? We did. Our grandparents, our parents and we took care of them. However, these people [the park authorities] came yesterday and started to tell us what to do and what not to do. We rather know how to live with the animals. We care for the animals as we do for our cattle. We do that not because of the order from the government but because of the oath, we have with Waaqa [God] through our ancestors. We care for them so that our cattle would reproduce, we get adequate rain, our homes become peaceful and everything in our neighborhood stays safe and happy (interview with Gagasa Edema, April 2011).

Guji elders do not subscribe to modern conservation discourses and practices; rather, they inherently have values, beliefs, ethics (*safuu*) and ecological wisdom that govern their interaction with nature (*uumama*). In Guji worldviews, their surroundings constitute the entire physical world and divine beings that *Uuma/Waaqa*/God created as he did with humans. Although conservation practices have globally undergone some changes towards an inclusive and decentralized approach since the 1990s, political ecology still suggests great scrutiny of the practices, given that power differentials at local arena might also lead to inequalities in access to resources. Local social differentiations in terms of ethnicity, religion, language, gender, education and age are important elements of investigation because these variables affect the way in which people negotiate local-global interactions (Gezon 2006). The so-called community-based conservation or the politics of participation masks local power relations under the ambiguous concept of 'community' participation. For instance, in the case of the Guji and Koore local communities, the park administration and respective district administrations have established local regulatory committee out of the local communities under the name of participation. Although the idea was to engage the people in the management of the national park, it has created decentralized form of hegemony. In practice, only a few local elites who had a strong network with government officials were handpicked and assigned as members of the local committee, while ordinary men and women remained marginalized. This is why the applauded community-based resource management is often criticized for its representation of 'the community' as a homogenous entity with similar interests, claims and history.

Considering local people as a homogenous community always obscures internal or intra-ethnic differences within an ethnic group itself. As Gezon (2006) maintains, there is a danger in assuming ethnic divisions in the case of conservation and development interventions from a fixed ascription of economic activity as a marker of a group. The simplification and complexity emerge when conservationists or state officials ascribe a certain ethnic group with specific ways of life and accordingly blame it as a cause of environmental problems. This obscures internal diversities within ethnic groups as well as the engagement of members of other groups in the way of life with which the other group is 'identified'. It has predominantly been the case that the Guji were considered as pastoralists and any cattle trespass to the national park has been directly associated to this group, although other neighboring communities also herd livestock. In this regard, a former park warden of the Nech Sar National Park who later worked as head of Culture and Tourism bureau of the SNNP regional state described the threats to the park by associating livelihood styles and ethnicity with environmental degradation:

> You know, nomads are always nomads [referring to the Guji agro-pastoralist]. They move with their livestock. They are not worried about what will happen tomorrow. When they come with their livestock in a certain season, their cattle overgraze the land and move out to another location in search of another pastureland. This is their way of life but it is not a wise way of utilizing resources. This practice affects both the people and the park. On the other hand, agrarian societies [the Koore in this context] are much concerned about land use system. Their economic system favors resource management far better than nomadism. Of course, they also compete with the park over farmlands. That is why the government is concerned about both groups and the environment. We exerted maximum effort to improve the lives of the people, that of the wildlife and biodiversity in the park. The park is a national treasure. Both the local communities and the country at large would have benefited from tourism. Now the park is threatened because of overgrazing and encroachment from the people (Former Warden, April 2012).

Rather than ascribing ethnic identity to fixed markers such as economic activity and language, it is very important to look at how members of different ethnic groups tend to subjectively construct and contextually switch their ethnic identity, particularly within the contexts of contestations and claims over resources. In this respect, group members adjust their allegiance and membership flexibly based upon imagined or real material/objective advantages (Gezon 2006). For example, in the Guji-Koore relationship, the Koore often contextually frame their identity based on changing circumstances such as issues of access to resources. Both Koore and Guji informants agree on cases whereby the Koore switch their identity to the Guji through traditional rites of passage called *ilmofacha* (adoption). *Ilmofacha* literally means a process of being adopted as a son/daughter by a family other than one's own biological kin. As the account from my Koore informants shows, their members have been going for the *ilmofacha* tradition for two major purposes. For economic reasons, as it ensures the Koore with access to land and livestock from the Guji territories; and as socio-political mechanism, because it served both groups, and particularly the Koore, in seeking shelter among the Guji during times of social and political crisis at the highland. Once a Koore individual is adopted into a Guji clan, he/she is equally entitled to resources of the Guji family or clan, and thus ascribing pastoralism to Guji does not necessarily mean that the Koore were not engaged in herding activities.

As it has been discussed in the previous section, human perceptions of the environment, representations of different landscapes, ascriptions of meaning and the conceptualization of the place of humans in nature are culturally constructed and shaped by socio-political dynamics and historical trajectories. Thus, political ecology challenges the radical separation of humans from the biophysical environment by identifying human systems of meaning and socio-political dynamics as they shape the material contexts within which people act (Gezon 2006). As an analytical approach, examining the political and ecological aspects of environmental issues such as resource use, access right and claims of entitlement also helps us to analyze how, when and by whom a particular territory came under possession/control. It enables us to know the contested nature of a space and unequal power relationship in the process of interpretation and ascription of meanings to a territory. Gezon (2006: 15) further argues that "a concept of the political as embedded in everyday interactions reveals the process through which elements of the biophysical environment become resources, per se, as well as how they become distributed and contested." It is also possible to address contestation and debates between people of different interests and claims by assessing the dynamics of material and discursive struggle over power, entitlement and resources (Gossling 2004). Nonetheless, imposing a certain form of environmental knowledge upon another social group implies exercising a hegemonic form of knowledge that suppresses multiple voices and interpretations. For example, in the research area, the park authority's attempt to impose scientific approaches of conservation by separating humans from the wildlife, and monitoring the movements, reproduction and behavior of animals is part of the state's exercise of hegemonic form of knowledge that denigrates the indigenous eco-cosmologies of people-environmental interactions.

Ethiopian state officials, and particularly those in the SNNP regional state, used different rhetoric, accenting that the national park was designed for the benefit of the local people. The state presents itself as the only legitimate body responsible for safeguarding the people, the wildlife and the environment in general. For example, during the resettlement program of the Koore people in 2004, local government cadres were reported to have promised the people better living conditions (access to health centers, pure water, schools and roads) and the provision of production materials in the new settlement areas (informant: Shora Mamamcha, February 2012). However, the officials fulfilled none of these promises after relocating the people. This indicates the government's reliance on narratives in order to translate its policies into practice without challenge. Therefore, an understanding of the link between local resource use policies and the political, economic and social institutions at national and international levels helps us grasp conflicts and motives surrounding environmental change (Campbell 2005).

Economic and political forces and contestations over environmental resources at different scales cannot be understood without a thorough understanding of the life worlds of local communities and their environmental narratives. As Campbell (2005: 296) argues, in the study of environmental change, "place-based livelihoods need to be understood through accounts of life histories, social change and aspirations, rather than being slotted into the naturalistic narrative box of communities geared to 'survival' or simply replicating traditional values." Indigenous people's life worlds are built based upon the engagement, interaction, participation, experience, memories of past activities and the "perception of human and non-human life process in contexts of relational ways of knowing" (Campbell 2005: 299). These are important in understanding different discourses, contestations and claims of entitlement among social actors and institutions engaged in the struggle over environmental resources, such as the Nech Sar National Park case.

Political and economic forces that operate in local and extra-local arenas often influence the rules of the game in contested social spaces by creating an unequal power relationship in the process of interpreting and ascribing meanings to a territory (Gezon 2006). While the meaning people ascribe to biophysical environments largely depends on their economic, cultural and social interests and transformations, the process of distribution, ownership, control and management of resources reflects political forces and the power relations between actors who are involved in the contestation. As has been discussed in chapter one, actors of different power positions, agency and network define a particular territory based on their worldviews, economic motives and political aspirations. Under such a context, "the physicality of space should not be radically separate from the metaphorical space of human interactions" (Gezon 2006: 21), because it is through engagements and re-engagements that people make meaning and create their sense of identity, building on the past memories of landscape that are embedded in their culture.

In addition to cultural dimensions of a place, the very physical element of a place in the form of topography, geographical properties, the species of plants and animals and the distribution of its resources add to its significances. However, the question remains as to whose representations and interests of the space prevail in conservation schemes. Based on the meaning conferred to a particular physical space by individuals or groups, "elements of biophysical environment become resources as people appropriate them as means of subsistence and as commodities" (Gezon 2006: 85). To understand the process of conferring meaning to biophysical environments, material and discursive struggle over power, resource and entitlement reflect important elements of analysis to disentangle political and economic forces from environmental concerns in conservation politics. It is difficult to isolate local resource use practices, access rights, distribution patterns, customs and governance at the local level from national and international political, social and economic institutions, given that national institutions, which are influenced by global discourses and practices, significantly influence the mode of human-environmental interactions at local level (Gossling 2004).

The following three different narratives from a representative of the Ethiopian state and members of the Guji and Koore people hinge on how a particular phenomenon–in this case, the establishment of the national park–is presented differently by the actor groups. A former head of culture and tourism bureau of SNNP regional state claims that the territory was uninhabited during and before the park was established:

> When the territory was reserved for a national park in the 1960s, there was not any trace of human settlement except some nomadic people who used to come to the Nech Sar plains during rainy seasons. Apart from this, it was only a home for the wildlife. As you might have seen, the landscape is beautiful for tourism. According to our information, the government wanted to conserve the biodiversity and simultaneously generate foreign currency through tourism. The national park was established for the benefit of the local people, the biodiversity and the country at large (Almaz Beyero, April 2012, Hawassa).

Furthermore, the Ethiopian Wildlife Conservation Authority (EWCA) has posted a kind of business advertisement, depicting national parks in the country as wilderness. The main page of the EWCA's website states: "*Come and See the Real Wilderness in Ethiopia*" (http://www.ewca.gov.et/). Beneath the surface, this advert implies the official notion of the organization about national parks in the country, portraying the territories as wild and empty. We noted earlier that knowledge about the environment is produced through combinations of such narratives, discourses and frameworks that are invented by a particular social actor to legitimize a particular form of knowledge and suppress counter narratives.

By contrast, the Guji consider the territory as their ancestral home, having lived there for several generations. A Guji elder shares the opinion of many other members of the group in stating:

> Our ancestors lived here since time immemorial. This is our home. From generation to generation, we inherited the tradition of herding. We keep our livestock and make rituals of thanksgiving, prayers and blessings through which Waaqa [God] would reciprocate us with nagaa [peace]. Whenever there is nagaa in our homeland, we will get everything. Rain comes; cattle and people reproduce, and animals and plants get their food and water. However, we lost nagaa when those people of the government came and chased us from our home. They suddenly came and told us the land belongs only to the animals. They told us that we were not allowed to live with animals. However, how did we live together before the government came? Since the Haile Selassie period, and particularly during the Derg and the current government, we have been denied access to our homes. All ritual and burial places of our ancestors, malkaa [gate to a river for cattle] and pastureland were all included into the park. Before the park, the entire territory was our customary land. It included the present day Sinqalla [Arba Minch town]. They chased our people and now we are placed between the park, the Amara [Koore] and the two Lakes (interview with Ararsa Gumi, May 2012).

The Koore people, who are the rivals to the Guji over the contested farmland along the eastern border of the park share some of the aforementioned state official's positions, yet also produced their own version of the story. Both Koore state officials and local peasants share similar views in claiming the entire part of the park as empty except the lowland called Tsalke/Golbo during the park's establishment. According to a Koore informant:

> The entire Nech Sar was under Koore administration during Haile Selassie era. We lived in the highland and seasonally moved to Tsalke for irrigation. However, except few nomads in the desert [Nech Sar plains] there were no people. Only the Koore lied in Tsalke. These Guji nomads came to Tsalke after they learned agriculture from us. During that time, they used to pay government taxation to Amaro Awuraja [province]. Our land goes far up to Sanxaqa Dingay [Bonke Hills close to Arba Minch Forests]. During the last years of the King [Emperor Haile Selassie],

> the park issue came. Initially they did not come to our agricultural areas but the Derg destroyed our property and removed us from our home in Tsalke. By that time the Guji were also chased out (informant: Gudo Mamacha, February 2012).

From the three aforementioned versions of the story, it is clear to comprehend how the competing groups produced diverse narratives and the referents that the actors brought aboard to justify their claims. Apart from the construction of competing narratives over environmental resources in the Nech Sar National Park, there is also evident competition over forms of knowledge regarding conservation approaches and the place of humans in nature. Given that they were built upon the dominant Western discourses of knowledge that privilege science over indigenous or 'traditional' knowledge, conservation practices and policies in colonial and post-colonial Africa are noted for promoting the former at the expense of the latter (Peluso 1993; Gezon 2006). For a better understanding of why customary resource use approaches were criminalized following the introduction of colonial conservation practices, one should position the discussion within the broader debates about the representation of environmental problems and how we come to know these issues. Who defines the status of environmental conditions? Who holds the power to categorize societies as victims and villains?

It has been a common trend in capitalist economy that access to environmental resources or the bounty of nature is filtered through political and economic institutions in which, "those people who are relatively wealthy and powerful generally reap more of the benefits from uses of the environment, and bear fewer of the costs from its abuse, than do those who are relatively poor and powerless" (Boyce 2002: 1). The irony is that environmentally destructive economic activities produce powerful groups who reap the benefits of such destruction, while it produces losers who bear the costs perpetrated by the wealthy and dominant groups. In other words, the wealthy and powerful sections of the populace (multinational corporations, businessmen/women and states) control knowledge over environmental problems that are often reproduced through discourses, narratives and practices (Peet et al. 2011). By associating environmental degradation to poverty, much conservation literature in developing countries has continued to cast the blame on poor local communities. Although the reliance of the poor on environmental resources in their quest to survive is undeniable, particularly in developing countries, such an argument overshadows the strong link between wealth, power and environmental degradation (Boyce 2002).

By contrast, it is evident that knowledge about conservation practices in Africa has been co-produced through interactions between the Western forms of scientific knowledge and African responses to the external impositions building their critical knowledge on their culturally embedded belief, values and ecological wisdoms (Garland 2008). For instance, local conservation schemes, values, norms and knowledge that are embedded and embodied in people's beliefs, culture and worldviews (sacred indigenous cosmologies) informed the dominant conservation paradigm to adopt what has been called community-based environmental management approaches (Adams and Hutton 2007; Berkes 2008). Despite the continuation of asymmetrical power relations between the local communities and state or non-state actors, even within the so-called community-based conservation practices, it has at least brought a recognition that indigenous approaches of human-environmental relations would contribute to the global endeavors to ensure sustainable environmental conservation.

2.2.2.3 Conservation Enclosure-Domination/Resistance Interfaces

The establishment of national parks and other types of protected areas in Africa has criminalized the utilization of customary land and natural resources (Neumann 1998). Enclosures of common resources under conservation schemes have instituted new forms of compartmentalization and territorialization by reinforcing check posts, patrols and surveillance, ultimately enhancing categorizing territories as 'accessible' and 'inaccessible' and introducing regulations concerning "who is legitimate to enter the territorialized space and on what conditions" (Campbell 2005: 294). Enclosure has also entailed the imposition of scientific knowledge upon local communities through the externalization of nature from those who hold the knowledge–in this context the local people. As I discussed earlier, the dominant academic and popular imagination of considering scientific knowledge as a superior and the only legitimate way of knowing environmental problems has continued to marginalize other means of knowing. Thus, it is partial, reductionist and ultimately instrumental in achieving and maintaining control over nature and the people who traditionally inhabited a particular territory (Peet et al. 2011).

As Bosak (2008) articulates, environmental conservation is a social question in which nature is labeled as internalized or externalized from culture. Such categorization places the state or non-state environmental managers in opposing camps with local communities in the process of implementing conservation programs. This happens because the externalization of nature entails the employment of modern, scientific and expert-based knowledge rather than giving responsibility of conservation to the representatives of nature, namely the indigenous peoples (Bosak 2008). Evidently, historical trajectories in the conservation of protected areas in colonial and postcolonial Africa depict the domination of scientific knowledge over indigenous forms of environmental management.

As a result, most protected areas in the continent became arenas of struggle over perceptions of nature-culture relations, questions of customary rights and forms of knowledge. During the colonial period, the European appropriation of African territories took place partly for the purpose of aesthetic consumption and political control, which in turn enabled the colonial powers to strengthen their control over different resources. The postcolonial African states inherited most forms of the colonial system in the run for 'nation-building'. One such inheritance was the continuation of the protectionist approach of conservation, which presumes people as threats to the environment, especially to the wildlife. By contrast, local communities in many parts of the continent exerted relentless efforts to curtail external interventions as part of their struggle against the colonial system, which continued also in postcolonial eras. Accordingly, it is sound to argue that protected areas in Africa have become arenas of struggle over conservation discourses between traditional and scientific knowledge.

At this point, it is important to raise a fundamental issue pertaining to the nexus between enclosure and domination/resistance. Conservation enclosures are arenas in which different actors exercise various forms of power. The interest of state or non-state actors to dominate or subjugate the environment entails stifling or controlling resistance from the part of local communities who rise against the costs of environment-related development projects or conservation activities (Peet et al. 2011). While such practices of domination can be termed as dominating power, local communities also respond to the dominating power by exerting different forms of resisting power (Sharp et al. 2000). Against the general assumption of equating power with acts of the dominant characterized by domination and control, Sharp et al. (2000) urge us to pay attention to both entangled forms of power, namely dominating power and resisting power.

Although power in environmental management is crudely understood as a polity's capacity to control the actions of others, such as local communities, organizations and firms, it is also imperative to understand power as possessed by the subjects who are the targets of dominating power (Peet et al. 2011). Thus, the sources of power, resources actors use in the exercise of power and strategies they design in the process ultimately shape the outcome of the interplay between dominating power and resisting power.

Various actors with competing perspectives articulate their own narratives about a certain phenomenon to substantiate their claims of entitlement to knowledge, resource and territories. However, the power over influential narratives is not evenly distributed among social actors. Based on their capability and knowledgeability–the agency to articulate narratives–competing social groups accumulate power over influential narratives, myths and resources, which in turn changes the game to their advantage (Stott and Sullivan 2000). Such influences often stem from actors' economic and political power, because it has been noted, "political influence on environmental policies tends to stem from economic interests which are able to translate their needs into political outcomes" (Stott and Sullivan 2000: 59).

I have discussed how conservation enclosures enhance hegemonic control over resource users from the part of the state or non-state conservation or business organizations. This is why political ecologists call for great scrutiny of any form of conservation, particularly those enacted through establishment of protected areas, because it disempowers traditional resource users by dispossessing them from their socio-cultural and economic spaces. In this regard, conservation policies are often seen as political endeavors that produce winners and losers (Neumann 1992). Understanding questions of power and forms of rule/governance is important in terms of comprehending how certain sections of population become winners while others become victims of the politics of conservation (Robbins 2004; Peet et al. 2011).

Given that power is not uniformly or evenly distributed among competing actors, there is no doubt that hegemonic discourses of conservation and development are imposed on actors whose power constellation does not counterbalance that of the dominant actors. Simultaneously, the exercise of dominating power nurtures and endures resisting power on the part of the subject(s), upon whom the dominating power is exercised, prompting them to take different forms of resistance, in what Scott (1985) calls 'everyday forms of resistance'. Despite its suitability in explaining hidden forms of resistance from below, Scott's 'everyday forms of resistance' puts resistance within binaries of power: domination versus resistance. Moreover, it also considers covert resistance as a trivial practice by peasants who cannot change the course of action by the dominating actor(s). On the contrary, it has become evident from historical experiences in developing countries that grassroots actors have been successful in halting or undoing the implementation of hegemonic development and/or conservation practices in many instances (Bryant and Bailey 1997). As will be detailed in chapters seven and eight regarding the Nech Sar National Park, local resistance has been nurtured by acts of domination from the state and non-state actors, enabling the local communities to succeed in disabling hegemonic exercise of power by the state. Therefore, I share Long's (2001) argument that it is difficult to imagine a smooth translation of hegemonic discourses, ideas and practices into the life-worlds of local actors within the contexts where grassroots actors can mobilize different forms of resources (social capital, networks, knowledge and agency). This is because external intervention often encounters critical knowledge from local actors who define, articulate, contest and negotiate external intervention based on their worldviews. Although local communities might not opt to overt resistance due to repressive actions from the states, everyday forms of resistance at times compel the state and non-state actors to come to terms with the demands of local people.

CHAPTER THREE: THE STUDY AREA, THE PEOPLE AND THEIR RELATIONS TO THE LAND

3.1 The Study Area

As I outlined in the preceding chapters, the Nech Sar National Park has been contested from different dimensions. The physical settings, economic and aesthetic values ascribed to the space, as well as political, cultural and historical representations of the territory have been among the major elements triggering competitions over the territory. While all such characteristics of the territory will be discussed in this research, this chapter focuses on the geographical settings and some background knowledge on people-environmental interactions in the study area. The research project takes the national park as a core arena where multilayer constellations of social actors interact among one another and strategize different narratives, ideas, programs and practices related to the ownership, management, utilization and conservation of the territory and its resources. However, despite the park being represented as a central field of contestation among the actors, this study also pays due attention to the people who traditionally inhabited the territory or those who used to draw resources from the area for their livelihood, cultural and religious purposes. Therefore, the people and their land reflect the focus of this chapter.

The park is located in southern Ethiopia, at a distance of around 505km to the south of Addis Ababa (Briggs 2006). It is located within the administrative territory of the Southern Nations, Nationalities and Peoples (SNNP) regional state. However, the region's administrative authority over the park remains contentious because the park administration has been swinging between regional and federal states. Moreover, there is an ongoing claim between Oromia and SNNP regional states over its administration and ownership (Dessalegn 2004; Getachew 2007). In terms of geographical settings, the park is adjoined by Gamo Gofa Zone in the west, Borana Zone in the east and northeast[4], Amaro district in the southeast and the two lakes, Abaya and Chamo, in the north and south, respectively (see map below).

[4] Since 2008, the Galana district of Borana zone (Oromia regional state) adjoins the park entirely along northeastern and eastern directions by disconnecting the Amaro district's direct border contact with the park.

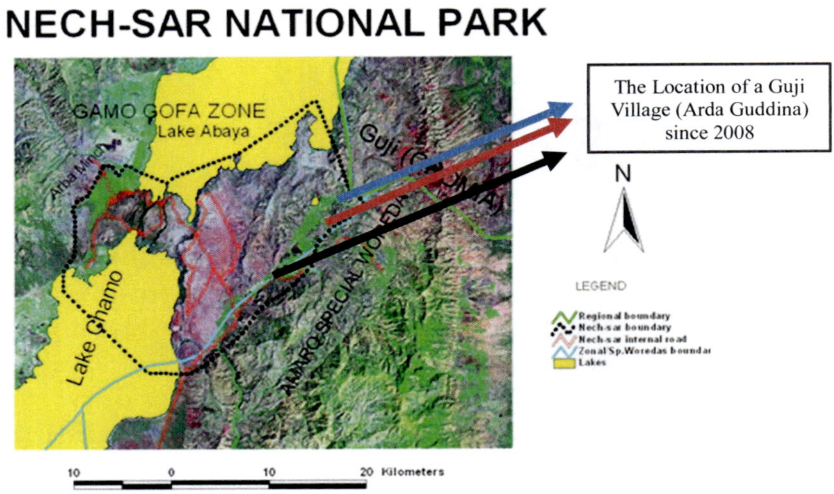

Figure 4: "Map of the Nech Sar National Park" as depicted by SNNP Regional State[5] (Source: http://www.southtourism.gov.et/Home/Nature/NationalParks/NNPBigMap.html)

With a distance of only three kilometers from Arba Minch town, located to the west, the western side of the park is easily visible if one gazes to the scenery from the gardens of hotels and lodges such as Bekele Molla hotel and Paradise Lodge in the town. According to the official figure, the park covers an area of 514km^2 (Dessalegn 2004) although other sources argue that it has been ever-expanding since then and thus the area is estimated to be over 700km^2 (Taddesse 2004).

The park includes different ecological niches, ranging from a highland forested zone located close to Arba Minch town at an altitude of over 1650m to the lowland Tsalke/Golbo Valley stretching from northeast to southeast of the park, adjoining the western slopes of Amaro Mountains. This lowland area is located at an altitude of 1108m (Dessalegn 2004). There is the vast grassland plain in between the two ecological niches, traditionally called the Nech Sar plains or Ergansa, as the Guji call it. The peculiar feature of Ergansa is the white grass

[5] Although this map shows some physical boundaries of the national park, it is highly controversial. A close insight into the map shows the contestability of the territory on the part of the actors involved. This map shows the 'official' depiction by the SNNP regional state and fully nullifies the territorial claim by the Guji Oromo by drawing the boundary of Oromia regional state outside the park territory. According to this map, Amaro district adjoins the park on the eastern and southeastern side, although it was actually the Guji village of Arda Guddina that adjoins the park in these directions except a narrow strip of land along southeastern side that connects to the Amaro district. This map does not show my position but I brought it here to show the complexity of the issues (Source: http://www.southtourism.gov.et/Home/Nature/NationalParks/NNPBigMap.html).

from which the park derived its name "Nech Sar", which literally means "white grass" in Amharic language.

Figure 5: "Nech Sar" (White Grass)

The Guji call the territory Ergansa, which means a land with abundant blessings (grass, water, fertility, etc.), although the new nomenclature is analogous to other areas in conquered regions where Abyssinians labeled Amharic names to places. For instance, the imperial regime changed the place names of Oromia towns such as Finfinne, Bishoftu, Adama and Ambo to Addis Ababa, Debre Zeit, Nazereth and Hagere Hiwot, respectively (Paulos 2011). Changes in place names or toponyms imply the denial of peoples' rights and their history related to the places. Place names have political orientations because changes are often undertaken in connection with nation-building processes (Horsman 2006). During the colonial periods too, colonial officials changed place names as a strategy of detaching the cultural, spiritual and historical meanings of the space from the people inhabiting the territory. This was why the Guji re-named their newly established village adjoining the park as Ergansa/Arda Guddina (a land of prosperity), as a mark of reclaiming their historical right to the park.

Returning to the discussion on the park, the local ecological zones constitute diverse fauna and flora specific to the topographies that have been used by local communities for different purposes. The dense forest covers the western side of the park and harbors the legendry "Forty Springs" from which the town derived its name in Amharic as 'Arba Minch'. These springs are the major sources of water supply to the town. Although the town is located close to Abaya and Chamo Lakes, it draws water from the springs due to lack of technology to purify (distil) water from the lakes. Its proximity to the town and absence of alternative fuel sources around the town made the forest to be the major source of charcoal and firewood for the town dwellers, organizations and service institutions such as hotels, colleges and University, hospitals (Freeman 2006). Low-income sections of the town population, mainly women and youth, depend on collecting firewood, charcoal and construction materials from the forest to sustain their livelihood (Bayisa 2011). Moreover, the forest is a home for arboreal animals and different bird species (Dessalegn 2004). The two lakes, parts of which were included within the park boundary in the 1990s, were fishing grounds for Arba Minch town dwellers and its surroundings. Thus, although quantitative figures are not available, these lakes remained the major sources of fish supply to southern Ethiopia and beyond.

According to the Guji oral tradition and other sources, the middle zone–the grassland plains–was a customary grazing land of the Guji Oromo. The Guji used this grassland for wet season grazing purposes by the time the Sermale Valley turned swampy following the commencement of rainy seasons. It also partially included the Bonke hills, where the Guji practiced rituals in the past. In terms of wildlife habitation, this zone was noted for its accommodation of diverse species of animals, including the endemic Swayne's hartebeest. The major tourist attraction site of the Nech Sar National Park belongs to this ecological zone. As the myth of 'wild' Africa mainly resonated around the exoticization of African wildlife, the Ethiopian state also capitalized on the global trends of nature commoditization and marketing in the national park. The third important ecological zone, the eastern escarpment of the park that adjoins the Amaro Mountains, remained contested among the three actors, namely the park administration, the Guji and Koore people. Besides its economic significance as irrigation land for the Guji and Koore communities, this narrow corridor has also been part of the park administration's claim for entitlement because the government considered the area as a dry season retreat for the wildlife.

3.2 The People and their Land

The park adjoins three different groups of people namely the Guji Oromo, Koore and Arba Minch town dwellers who traditionally depended on the resources from the park for different purposes. Although categorizing ethnic groups based upon their economic practices tends to overshadow internal heterogeneities, the first two groups traditionally had somewhat distinct livelihood engagements. The Guji in the study area have mainly been engaged in agro-pastoral lifestyle, where livestock herding constitutes their economic and cultural realms (Taddesse 2006). On the other hand, the Koore traditionally practiced peasant agriculture in the highland part of the Amaro district and some members of the group gradually moved down the slope to the lowland adjoining the park (Simeon 1995). The town dwellers in the Arba Minch town are multiethnic and their livelihood practices are diverse, as with elsewhere in towns in the country. Thus, this research only focuses on those people who depend on the national park for their livelihood sustenance.

For the sake of clarity, I discuss the socio-cultural history, livelihood practices, local conceptions of nature, and settlement history of the Guji and Koore communities within the subsequent sections. Moreover, changes in socio-political conditions in the area and their implications in altering people-environmental interactions will also be the focus of this chapter. It is evident that each group produces and reproduces different knowledge and claims drawing on different narratives. Although contestations among different actors for ownership, management and utilization of the territory and its resources will be the subject of a separate chapter later in this work, a brief outline of the settlement history helps us to reconstruct the roots of the claims.

3.2.1 The Guji: the People and their Economy, Settlement History and their Conception of Nature

3.2.1.1 The People and their Economy

The Guji people belong to the larger Oromo ethnic group and live in the southern part of Ethiopia. At present, the Guji live in Oromia regional state in Borana and Guji Zones with significant number of their members included in some zones and districts in the SNNP regional state (Dhadacha 2006; Taddesse 2006, 2009; Asebe 2010). The Guji in the Nech Sar area lived without any formal government administration from 1991 to 2008, until they were included into Borana Zone, Oromia regional state. During those years, the people were devoid of any social service from the government, apart from some sporadic vaccinations provided to their livestock by the African Parks Foundation during its administration of the national park (2004-2007). Their common encounter with the state was through the park administration that often threatened to relocate them from the vicinity of the national park borders.

Two mutually reinforcing institutions, namely the *gadaa* system and *qaalluu* institution govern every aspect of life among the Guji as a community and shape how their members interact with their surroundings. The *gadaa* system is a traditional generation set socio-political organization in which power is transferred to members of consecutive *gadaa* grades every eight years (see Hinnant 1977; Legesse 1973, 2000). Likewise, the *qaalluu* institution maintains Guji's spiritual wellbeing by serving as an intermediary medium between the people and *Waaqa* (God) through its leader, *abbaa qaalluu*. These two institutions shape Guji's internal relationship among themselves as a community, their interaction with their surroundings (humans and non-humans) and their spiritual connection to *Waaqa*. While the *gadaa* system ensures order through enforcement of customary laws and stable governance, the *qaalluu* institution maintains the spiritual connection between the people and *Waaqa* and also guides the people's moral and ethical values (*safuu*) in their interactions with other human and non-human neighbors (Asebe 2012a). For example, in human-environmental relations, Guji children are socialized into norms and values that regulate individuals' engagement with nature. Moreover, at every *gadaa* assembly, convened every eight years, the *Ya'aa* council (general assembly of the *gadaa* officials) promulgates laws including environmental regulations that govern people's ownership right, utilization and management of environmental resources (Dhadacha 2006). Due to the centrality of greenery in Oromo worldviews and the sacredness of different landscapes, both *qaalluu* institution and the *gadaa* system contribute to Guji's harmonious coexistence with their surroundings by providing the customary administration and spiritual basis of humans' connectedness to nature. However, as it will be discussed later, this does not mean that the Guji fall into the classical representation of indigenous people as 'close to nature'.

In terms of the economic dimension, the Guji traditionally practiced a pastoral way of life until the mid twentieth century, when they began combining farming with pastoralism (Hinnant 1977). Today, they are engaged in diverse livelihood practices (pastoralism, agro-pastoralism, farming, trade and other urban economies). However, regardless of such heterogeneity in Guji's economic engagements, herding has continued to signify social, economic and cultural values among the Guji until present, particularly in rural contexts (Taddesse 2006). Livestock embody Guji's social pride, self-esteem, worldviews, ecological wisdom and inter-societal relations, as well as shaping their relationship with their surroundings.

In Guji worldviews, everything is based upon symbolic and material representations of livestock. Cattle herding and the possession of large herd of cattle are associated with cultural pride, economic values (wealth), a sense of Guji identity and social privilege in marriage arrangements and inter-societal relationships. Taddesse (2006: 209) describes that although the Guji practice a mixed economy of animal husbandry and crop cultivation, "their real wealth consists of cattle, sheep, goats and horses. Emotions and pride are centered on stock. People who do not own cattle are not considered to be proper Guji." In Guji culture, beyond their economic values, cattle are also used for rituals, transition rites, gift, and bride price, compensation during reconciliations and as symbols of social prestige. Therefore, the Guji count cattle not in terms of heads of cattle but rather *moona* (kraal), which ranges from seventy to hundreds. This means that the possession of large herd of livestock represents social and economic privileges. However, owing to enclosures of pastureland and due to overall change in climatic conditions, the stock, which is the source of wealth and reflection of Guji identity, has been under serious threat putting the people into dilemma of socio-cultural transition(Taddesse 2009).

For the Guji, livestock herding has been rooted in their culture and self-pride. Livestock possession and engagement in livestock husbandry mark a crucial representation of Guji's social, economic and cultural capabilities/performances. The economic contribution of livestock among the Guji has both material and psychological dimensions. Livestock products are the major nutritional sources in the form of milk, yoghurt and meat. Moreover, the Guji exchange livestock and livestock products for cash with neighboring communities (McClellan 1988). The psychological empowerment that livestock husbandry provides to the people is related to how the Guji define poverty and wealth. Poverty is understood neither as a lack of food for daily consumption nor in monetary terms; rather, the Guji conceptualize it as an aggregate deprivation of a person from social, cultural, economic and spiritual involvements. In order to perform rituals or maintain inter-societal relations within the Guji and between them and their surroundings, possession of large number of livestock empower the owner with psychological capabilities. If a person does not possess over one *moona* (kraal) of cattle, he is not entitled to undertake rituals, prayers, blessings and sacrifices on behalf of their clan or village. The failure to undertake these culturally important practices impedes an individual's social relations in the marriage arrangements of his children, spiritual capabilities and other inter-societal relations. Restriction from rituals deprives the person from spiritual connection to *Waaqa* that in turn the Guji believe in affecting the person by blocking him from the possible blessings he would get from *Waaqa* in the form of the reproduction of the cattle, the fertility of women and peace to the people and the land. The Guji consider such deprivation as poverty. Thus, they understand wealth not merely from an economic dimension.

Moreover, cattle represent status symbols and serve as sources of social prestige. Therefore, the Guji believe that the possession of healthy, fertile and large number of cattle is a sign of blessing from *Waaqa*. It is from wealthy families that the Guji elect elders who undertake earnest rituals and prayers–which are paramount in maintaining a person's identity as being proper Guji (Baxter 1991; Van de Loo 1991). As Van de Loo (1991: 209) maintains, "numerous forms of animal sacrifices constitute the means of establishing and strengthening the complex vital interconnectedness between Waaqa-deity, and men, between the living and the dead, between marriage partners, or between a household and its livestock." It is in this context that a Guji elder, Danbala Badacha, described cattle as follows: "*Loon Gujiif waa mara. Loon jechuun jireenna. Loon kabaja, isaan waan jilaati, loon sagalee dha. Walumagalatti loon Gujummaadha.*" Loosely translated, he states, "Cattle are everything for the Guji. Cattle mean life. They are food, they are pride, and they constitute elements of rituals and part of sacred journey. In general, cattle represent Gujiness (Guji identity)" (informant: Danbala Badacha, May 2011).

The people broadly situate livestock herding and possession of large herds of stock in their creation myth as being born as herders with all rights and privileges linked to this way of life. Indeed, many pastoral communities in East Africa such as the Maassai consider themselves as the only people with the right to possess cattle (Igoe 2004). Likewise, the Guji also claim that *Waaqa* created them as livestock herders, but reserved the right to 'herd' other animals (i.e. these others consider as wildlife) for him. The people's positioning of pastoralism in their creation myth has developed a view of inalienable right to pasture lands, including the Nech Sar National Park, among their members. A Guji elder of over seventy years remarked the following about the intricate relationship between the Guji, their livestock, the wildlife and *Waaqa*:

> Waaqa [God] created us with cattle so that we look after them, care for them and use them for our needs. However, these other animals [wild animals] do not have shepherd except Waaqa himself. Waaqa gave us the responsibility to care for the animals on his behalf and he cares for our cattle, people. Waaqa gives us nagaa [peace of the Guji land]. If one kills those that Waaqa looks after, he would punish us by causing famine, drought, disease and instability that destroys our livestock and people. Nevertheless, when we care for the animals, Waaqa reciprocates us with fertility, abundance, rain, and peace. From our ancestors until today, we lived with these animals in peace. They are also peaceful to us. Our people know that taking away the life of these peaceful animals would invoke punishment. However, these people [the park authorities] came recently and began telling us not to live where we lived for centuries (informant Waaqoo Dhugoo, April 2011).

There are strong intertwinement between the customary conservation practices and their cosmological schemes, which are in turn embedded in their culture, beliefs and norms. Like other groups of the Oromo, the Guji have a strong cultural connection to their environments (Van de Loo 1991). The people's connectedness to their land is part of the complex eco-cosmologies and ecological wisdom in which the maintenance of peace and harmony between humans, non-humans and *Waaqa* is mediated and expressed through different rituals constituting their surroundings. Baxter (1991: 9) explains that, "Guji, like other Oromo, are keenly aware that the maintenance of their culture depends on the maintenance of *Nagea*: Peace, that is, amongst them as a community and between them and God. But, this Peace is not a free gift, its maintenance requires continuous, earnest application, and is never sure or certain." Thus, the duty of maintaining peace rests on the shoulders of elders, requiring them to provide continuous rituals, prayers, sacrifices, blessings and obeying the rules of *Waaqa*. Guji elders provide rituals and prayers to *Waaqa* on behalf of all people, cattle and their environment at large. The Guji believe that a failure to maintain harmony with *Waaqa* might result in affliction, principally expressed in the form of withholding rain, upon which all animals and humans absolutely depend. For the Guji, the fertility, growth and (re)productivity of cattle and humans are highly dependent on whether the Earth, the cattle and women are all moist (Baxter 1991). As Guji elders maintain, the nature of the relationship between humans and non-humans shapes their relationship with *Waaqa*. For instance, calamities/misfortunes such as drought, famine, flooding, disease and war are interpreted as being invoked by God for people's failure to maintain their ancestral covenants. Conversely, the people describe fortunes not as a random occurring phenomenon but rather as reciprocal provision from *Waaqa* for people's good deeds. A Guji elder by the name of Galgalo Jilo expressed the dynamics of Guji's relations with their surroundings and its implications as follows:

> In the past when our people obeyed ancestral safuu [norm/ethics], everything was fine. We did not experience problems like drought. Guji did not know hunger. Everything was plenty in Gujiland. Rain used to come in its proper time and cattle were fertile, milk was abundant, children and calves chanted in every village. By those days, elders respected their ancestors' customs, children respected the safuu of the Guji and women respected men. We practiced rituals according to specific customs prescribed to Guji tradition. However, now everything is changed upside down. As you observed along your way to our village [showing me the field], small rivers dried up, pasture degraded and now even big trees are dying. The rain did not come for two years. This is not natural; it is not normal occurrence. It is what Waaqa invokes as punishment of our people. I do not know how the future will be but if it continues this way, we are so worried. Something worse might

come (interview with Galgalo Jilo, April 2011).

In contrast to views from Guji elders who believe that maintaining ancestral covenants, customs and values is vital in seeking blessings and other forms of reciprocity from *Waaqa,* most Guji youth hold somewhat different views. The youth focus more on the utilitarian purposes of the environment rather than its cultural or spiritual values. While this hints at the internal heterogeneity within the Guji, such disparity also cautions us that issues of human-environmental relations are complex and cannot be understood through an 'all fits one' approach.

3.2.1.2 Settlement History, Livelihood and Conservation Practices among the Guji in the Nech Sar Area

Drawing on Guji oral tradition and other historical sources, some scholars trace back the group's settlement in the area to the sixteenth century (Dhadacha 2006; Getachew 2007). McClellan (1988) also mentioned the presence of the Guji Oromo around Abaya and Chamo Lakes prior to the period of the Ethiopian imperial conquest in the late nineteenth century. Guji elders trace their ancestors having settled in the places currently included into the park territory and far west up to Ganta hills to the west of Arba Minch town a long time before ten generations. Place names in the Arba Minch town such as Siqalla, Seecha and Bishaan Harre are all Oromo names and remain in use. The Guji used these places as ritual, sacred and burial sites, as well as for herding and farming purposes (Getachew 2007).

According to a European traveler who traveled from the coast of Somalia to Port Mombasa across southern Ethiopia in the late nineteenth century (1894-1895), the Jam Jams–as their neighbors called the Guji–were inhabitants around Galana Valley to the east of Lake Abaya (Smith 1897 [2008]; Briggs 2006). Although Smith did not mention Guji's presence in today's Nech Sar plains in his travel notes, their presence in the lowland part of the Galana Valley could be an indication that they used the pastureland around the Abaya and Chamo Lakes during their seasonal movement with their livestock. Lewis (1966) takes the history of Oromo settlement in the area further, arguing that the areas around Lakes Abaya and Chamo were part of the original homeland of the Oromo, from where they gradually expanded to other directions from the sixteenth century onwards. This argument seems convincing because most references to the Oromo history mention areas around Lake Abaya as part of the Oromo homeland.

The Guji were the only pastoral people in the area until they adopted a mixed economy of farming and animal husbandry in the mid twentieth century. Oral accounts from the Koore, Gamo and Burji people to some extent confirm Guji's presence in the Nech Sar area before the imperial conquest, but contradict with Guji's claim of settlement since the sixteenth century. Guji elders trace the original settlement area of their ancestors to Girja, a place located near Bore town in the present day Guji Zone (Dhadacha 2006). In a report submitted to the Ethiopian government in 1966, Blower, a British explorer, clearly indicated that the Guji pastoralist people lived in the area now designated as the Nech Sar National Park, stating:

> There were a number of small temporary encampments of cattle owning Guji people throughout the Nech Sar area. No permanent habitations were seen though it was understood that there were permanent villages in the hill to the east and signs of cultivation could be on slopes of the Amaro Mountains. There were several thousand cattle in the area belonging to the Guji, though for the most part these animals did not appear in very good condition and it was reported that mortality due to unknown disease was heavy (Blower 1967 cited in Getachew 2007: 48).

In the past, the Guji in the study area were a purely pastoral community and used the plain of the Nech Sar as wet season grazing land, whereas the eastern valley was used during the dry season. After the mid twentieth century, they began to cultivate crops and grow fruits in the Sermale Valley in the eastern side of the Nech Sar National Park. Areas such as Golbo/Tsalke, Gashe, Hiituu and Mi'o have been cultivated throughout the year by irrigation and rainfall. While the Koore moved down from their permanent settlement areas in the highland villages of the Amaro district, the Guji also moved to the Golbo/Tsalke valley during their seasonal encampment, practicing agriculture along the way. Apart from herding and farming, the Guji traditionally produced honey from the forest now included within the park boundary (Taddesse 2004; Abiyot 2009).

The geography of daily routines of individual peasants and pastoralists helps to illuminate how the people contest, negotiate, and challenge state intervention vis-à-vis their livelihood conditions (Neumann 1992). While the analysis of such negotiations and contestations forms the subject of chapters seven and eight, in this section I will outline some livelihood practices and pathways of the Guji between their seasonal encampment sites. Guji's movement between wet season and dry season encampment sites were fundamentally informed by two aspects of their ecological wisdom, namely Guji's knowledge of seasonal distribution of animal parasites and diseases, as well as their concerns of environmental conservation.

Regarding their knowledge of animal disease, Van de Loo (1991) acknowledged that herding has equipped the Guji with skills and knowledge of parasites, diseases, symptoms and prevention strategies. One of the major challenges that the Guji often faced was infestation by mosquito and tsetse fly, which cause malaria and sleeping sickness to the people and animals, respectively. In order to avoid such infestations, the Guji had to move from the Sermale Valley, which turns to swamps during rainy seasons (April-May and September-October). In the past, the people moved their livestock to the Nech Sar plains using mobility as a strategy of disease prevention.

In those early times before the enclosure of the territory by the government, the Guji on the western and eastern side of the two lakes converged at Bonke hill–traditionally called *Riqicha Waaqa* (God's Bridge), given that it creates a land outlet between the two lakes–to perform sacrifices and prayers for health and fertility of livestock and people. Rituals were made few weeks before the Guji moved back to their wet season campsites in April and September and were considered as rituals of purifying the land from possible infestation by different types of diseases and bad spirits by the time they were away to Sermale Valley in the previous season. They purify their surroundings in order that water, pasture and *hayaa* (mineral salt) suit the cattle, as well as ensuring the safety of their cattle from predators. They believe that predation on livestock is a sign of disruption in the human-nature relationship (humans, non-humans and supernatural power). For the Guji, any kind of cattle death is interpreted from cultural and spiritual perspectives, particularly in terms of their relationship with nature. This is why, as Baxter (1991) stated, Guji elders and religious leaders relentlessly invest their knowledge and wealth in rituals, prayers and blessings to maintain harmony with *Waaqa*-deity and ancestral spirits.

However, access to seasonal grazing lands and some sacred spaces was restricted after the enclosure. The people subsequently settled in temporary sites adjacent to the park boundary along the eastern side, from where their cattle would gain access to pasture in the de facto park territory. Children between the ages of 10-18 often tend cattle in the buffer zones between human settlement and park boundaries, from where they push their cattle to the Nech Sar plains by closely inspecting the whereabouts of park scouts. Their daily encounters with park scouts reinforce the children for further cattle trespass as signs of retaliating against the fierce measures they face from the state agents. A young herder by the name of Danbala remarked on the challenges that he and other herders encounter on a daily basis, as the park scouts often punish them both physically and in detention. He also mentioned that the reaction of the scouts to cattle trespass depends on their networks with the local people. The boy described his experiences with the scouts as follows:

> My grandfather told me that in the past they used to graze their cattle in all parts of Ergansa [the Nech Sar plains] including Bonke, Dhakaa Bule, Maddoo, Golbo, and Gode and as far as Gaattira and Mi'o. Now the government people, I mean these park people, confined us to this narrow land. Our cattle do not get enough pasture

> and access to water is limited. Sometimes they block the malkaa to the lakes. When they find us close to the park boundary, they chase our cattle, beat us, sometimes capture children, and take them to jail in Arba Minch. However, it is not all of the scouts who are cruel. Those from Amara [Koore] and Gamo do not tolerate us. We have few Guji scouts. We check whether the scout is a Guji or other and then decide whether we let the cattle into the park or not. Our Guji scouts do not chase our cattle. They do not capture the herders either (Danbala, May 2012).

In times of drought such as the 2010/2011 event that ravaged the Horn of Africa, Guji boys took the risk of herding cattle in the park at night (often from 9pm-5am). The boys never sleep because they have to watch over their cattle from predators. Another major tactic the youth skillfully design is how to herd their cattle without being detected by the park scouts in order to escape the risk of detention and punishment. While there is no fixed amount of payment for trespassing cattle into the park territory, Guji households are sometimes asked to pay 50 birr (about 2.5 Euro) per animal. This amount is huge for a household with over hundred livestock. Apart from formal payments in the form of fines, relatively equal amount goes to the hands of individuals who pretend to intervene in their case to minimize the level of fees imposed as a punishment. The worst aspect is that such charges of cattle trespass, hunting or farmland encroachment are made in district courts in Gamo Gofa zone, which the Guji always complain about lacking impartiality in the process of handling their cases. The people have to pay for translators in order to present their cases, because the court language is Amharic or Gamo, neither of which the Guji understand.

After three to four months of uneasy encounter with the park and its authorities, the Guji move fully or partially to the second encampment site during the dry season from November to March and again in June. Unlike those in the wet season encampments, huts in these sites are used only for another three or four months, after which they are fully vacated due to the danger of mosquito infestation. Some households move fully with all their family members and belongings vacating the wet season huts, whereas others maintain the one close to the park boundary as permanent site of settlement. In the latter case, only young girls and boys move with the livestock to the seasonal encampments. In this case, some members of the family bring milk, butter and cheese to the family who remain behind.

In order to ensure the health, reproduction and growth of their cattle, the Guji pay due attention to maintaining the normal flow of the life cycle of everything in their environment, which they believe highly depends on the myriad of relations between the people and their non-human surroundings, as well as the spiritual power. Besides the cultural and spiritual dimensions of maintaining the fertility and health of the stock, indigenous knowledge of resource management is integrated into their livelihood. Access to water and pasture is an essential aspect of pastoralism, demanding the proper management of the herd, pasture and water wells, which are all governed by local norms and customary laws. The Guji use three major strategies to ensure that a pastureland sustains the livestock throughout the year and continues without depletion for the future. The first is seasonal mobility during dry and wet seasons. At times when there is no drought, the mobility covers 15-20km between the Nech Sar plains and Sermale River valley that adjoins the Amaro Mountains in the eastern side. The people noted that remaining in a single location for more than a season affects the regeneration of pasture. Like fallowing and crop rotation techniques used by peasant farmers as mechanisms of maintaining soil fertility, pastoral communities also consider seasonal mobility as an alternative way of conversing pastureland and water wells (Igoe 2004; Berkes 2008). Guji elders were cognizant of the fact that fresh grass regenerates until they bring back their livestock from another seasonal encampment sites.

The second strategy the Guji use in maintaining the health and reproduction of their cattle is the use of local knowledge of conservation, specifically burning old and dry grass to ensure regeneration of fresh grass, as well as cleansing the pastureland from parasites such as mosquitoes and ticks. When asked why they burn the pasture, which the park administration considers as an illegal act, a Guji elder from Maddo village explained: "*Marra gubuun balleessuuf miti. Akka sonaan marguuf malee. Ennaa marra gubnu, isa misha ta'etu biqila. Sababni marra gubnu garabiraan immoo biinnii fi silmii akka hin horreef. Amma garuu warri paarkii kuni erga gubuu kana dhoggani marri bade, dhukkubni looni fi bineensota fixe*" (informant, Chari Gamasa, June 2012). Translated, the elder stated that, "Burning is not to destroy pasture. Rather, it is to enhance the regeneration of fresh grass. When we burn old grass, new and healthy ones grow. However, after the park people banned burning, pasture has been depleted, and disease infested cattle and other animals." In their limited pastureland adjacent to the park territory, the Guji still practice controlled burning (see the picture below). Guji informants explained that herbivores wild animals often move out of the park and compete with their livestock over the limited pastureland adjacent to the park boundary, because they get fresh grass once the old grass has been burned. On the contrary, the park administration considers burning as one of the causes of environmental degradation and punishes the local people in for setting fire to the park territory.

Figure 6: Guji's Pastureland Burning Practices (Photo: Asebe Regassa, February 2012)

The third strategy of ensuring the sustainability of their resources is through local resource management ethics and frameworks. As indicated above, institutions of resource governance and ethics pertaining to the utilization and access to resources among the Guji are entwined with their cosmological schemes. From religious dimensions in a traditional context, the Guji holistically understood their surroundings as part of their connection to *Waaqa*. The *Qaalluu* institution, the *gadaa* system and other social norms and values are important local frameworks that guide the nature of resource management among the people. Besides such normative prescriptions inculcating environmental ethics into beliefs, the Guji exercise resource utilization and access rights within locally defined frameworks. In principle, a Guji individual is entitled to access pasture and water for his livestock anywhere in Gujiland, although in practice access to 'common' resources is subject to local regulations. Local elders, particularly clan elders, oversee all aspects related to the utilization and management of pasture. In case disputes erupt between different clans, *gadaa* officials adjudicate the case and resolve before it turns into conflict. There is no private ownership of land; rather, households have small plots called *kaloo*, reserved for calves. Even these areas are not exclusive for private ownership. Against the Western notion of 'tragedy of the commons' that presents commons as being threatened by the irrational utilization of local community members who strive to maximize their utilization beyond the carrying capacity of the land (Hardin 1968), the Guji had traditional governance of the commons. Because pasture and water are life streams for the Guji, like any other pastoral communities in arid and semi-arid regions, the misuse, destruction and breach of locally defined rules are not tolerated, even among children (Tache 2013).

Any clan member who is accused of transgressing customary laws, for instance by killing wild animals, cutting trees from sacred spaces or cattle trespass into another clan's *kaloo* without permission, would be brought to the elders' *Ya'aa* (council). If they found the person guilty in violating local rules and norms, they would excommunicated him from social affairs in the community through local decree until reconciliation and dispute settlement practices will be undertaken. Once the *Ya'aa* issued the decree, even close relatives would restrain themselves from collaborating with the person. However, the people would not deny him access to common pasture and water, given that this would lead to the death of his livestock. In order to avoid the punishment, the transgressor pays fines in kind based on the degree of the crime. Sometimes, the person might be asked to pay the fines in terms of cattle (the number depends on the wealth of the person) to the *Ya'aa*, who conduct a feast of reconciliation, purification and blessing of the person in order that he will be morally and spiritually reintegrated to the people and reconciled with *Waaqa*.[6] This ritual is mainly performed not as a punishment but rather as a restoration of peace between the people and their surroundings, in cases of offenses such as killing wild animals and cutting sacred trees.

On the other hand, the Guji consider granting water wells and pasture to members of other clan or ethnic group(s) as an investment for times of uncertainty, resource scarcity, drought and war. Moreover, other social networks such as marriage and trade necessitate sharing resources. The Guji say that allowing livestock to die by blocking access to water and pasture is transgressing Guji's oath with *Waaqa*. Such an act is believed to invoke infliction by *Waaqa*, who would hold back rain or causes diseases.

By the time the horn of Africa was hit by drought in 2010 and 2011 (Tache 2013), the Guji around the Nech Sar National Park were prompted to move their livestock far south to a place called Gaattira, located between the Burji, Koore, Gumayde and Konso. Although the Guji claim that the land was traditionally under their ownership, it is now administered under the Amaro district. I inter-

[6] Punishments and rituals of reconciliation to ensure sustainable resource management systems are similar to the gondoro tradition that is common among the Guji in inter-ethnic conflict resolution (see Asebe 2010).

viewed a Guji elder, Danbala Badacha, who participated in negotiation processes with elders from the above ethnic groups to allow the Guji and their livestock to the area. He summarized the discussion with the elders as follows:

> *Caamni waan nutti hammaatteef loon keenna gara Gaattiraatti godaansisuuf marii deemne. Yeroo kana loon, keesumaayyu jabbilee fi bushaayeen nu jalaa dhumatte. Ijoolleen aannan dhaddee halkani guyyaa boochi. Amma utuma teennu dhumuu irra jennee fala barbaacha kaanee warra Amaaraa, Gumaayidee fi Koonso bira deemne. Warruma Amaaraa kanati nu didate malee Gumaayidee fi Koonsoon duruu fira keenna. Isaan kottaa nuun jedhan. Sana booda warra polotikaa sana bira deemne. Isaanis garuma jaarsoleetti nu deebisan. Mee ittuma deemnee ofirraa faccisaama lobbuu loon keennaa fi ilmoolee keennaa booru geessifanna malee maal goona? Bara kana mootummaanis Waaqnis waan nutti jabaate fakkaata. Kana keessa jirra ilma koo (Danbala Badacha, March 2011).*

The elder summarized the negotiation reads as follows:

> Because the drought has become severe, we went to negotiate on possibilities of moving our livestock to Gaattira. Currently our livestock, specially the calves and goats are perishing. Children scream/cry day and night because they do not get milk. Now, we decided to search for a solution in order not to die out. Thus we moved to the Amaara [Koore], Gumayide and Konso. These Amaara people always block us from using the resources. The Gumayide and Konso are always our friends. They are ready to host us. We also went to the politicians. However, they returned our case to the elders. Now we decided to go and defend ourselves from any possible attack and try to prolong the lives of our livestock and our children at least for a short time. What else can we do? These years both the government and Waaqa seem to have turned their backs against us. My son! We live under such terrible condition.

Unlike their hostile relationship with the Koore, the Guji built a strong social network with the Konso and Gumayde people through intermarriage and trade. Inter-group networks with other neighboring groups are used as strong social capital that both parties utilize during times of insecurity such as drought, war and state eviction. However, any aspect of inter-group relation with the Koore is only practiced during peaceful times (Awoke 1985).

3.2.1.3 Guji-Park Relations: An Overview

Following the enclosure of the Guji customary grazing land as the Nech Sar National Park in the 1960s, which entered into effect in 1974, the people's access to the entire Nech Sar plains was significantly restricted. As the next subsequent chapters will discuss, successive Ethiopian regimes enacted similar policies in alienating the people from their traditional land in the process of establishing a human-free environment for the wildlife, thereby enhancing tourism development. As a background for the subsequent analytical chapters, it is essential to raise two fundamental issues in relation to Guji's relation to the national park. First, the Guji strongly depended on the resources in the national park for their livelihood, because it was part of their seasonal pastureland and access to water. The people also cultivate food and commercial crops such as maize, sorghum, coffee, ensete (commonly called the Ethiopian false banana), banana, sweet potato and sugar cane along the eastern side of the park. Second, the Guji traditionally had a strong cultural connection to the land in terms of practicing different rituals at some particular sacred spaces.

The following piece of ethnographic note on a Guji man who was engaged in farming (subsistence and commercial crops), herding and other business activities illustrates Guji's livelihood practices and their interactions with the national park. It also hints at how the park administration discourages development practices on the part of the local people. In April 2012, two school teachers in Ergansa village and I wanted to buy some foodstuffs from a weekly market located in Golbo/Tsalke, between the Guji and Koore communities. Because it was the beginning of the rainy season, the Guji moved their cattle to the highland areas bordering the park (to areas such as Gode, Maddo, Arda Guddina, Hiituu, Mi'o and Malkaa), thus abandoning their houses in Golbo/Tsalke areas. Unlike in the past, Golbo has now become an important area for two major livelihood purposes, in addition to being a seasonal pastureland of the Guji. First, it is used for cultivation of crops and fruit trees throughout the year along the Sermale River. Both Guji and Koore cultivate maize and grow avocado, mango, banana, coffee and sugar cane in the area. The Sermale River flows from the Amaro Mountains to Segen, passing through Golbo/Tsalke, Gashe, Gode and Xabala. Second, it serves as a market center where the Guji, Koore and people from the Arba Minch town come together to trade different items. Indeed, it is due to the economic significances of the valley that the Koore, the Guji and the park administration struggle over the control of the territory.

Figure 7: Mango and Avocado Trees in Golbo/Tsalke (Photo: Asebe Regassa, May 2012)

Along our way to the market, we found a man by the name of Ayano Guyo, weeding his maize field. We gave him the common greeting, saying "Yooyya! Buliin nagayaa? Ooliin nagayaa? Loon nagayaa, ilmooleen nagayaa, jajjaboo?" The literal translation does not give what the Guji express in their greetings, but it roughly means, "Peace to you! How is life going? How are cattle and children? Are you strong?" Greetings have a strong place among the Guji, in which asking about the well-being of cattle and children is central. After giving us warm greetings, Ayano offered us to collect some mangoes and avocados from his field. He also gave us a stick of sugar cane to quench our thirst. Meanwhile, I had the opportunity to discuss with him about his livelihood conditions. It also gave me an insight into the trajectories of shifting economic practices among the Guji from pure pastoralism to agro-pastoralism and recently the introduction of a market-oriented economy. While we were seated under a Mango tree, Ayano joined us and we continued informal discussions on issues of living conditions, weather variability, relations with the park and inter-ethnic relations between the Guji and Koore people.

Ayano, who is a head of nine family members, lives in Arda Guddina/Ergansa village, which was established by the regional president of Oromia in 2008. The park authority still claims to include this village into the park territory. Ayano herds his livestock around the park, while supplementing his livelihood by cultivating maize and some root crops in Golbo where we found him. He also harvests sugar cane, mango, avocado, coffee and banana and sells to traders who come from Amaro or Arba Minch. These are the major sources of Ayano's income in terms of monetary form. However, due to a conflict between the Guji and Koore in 2010, traders rarely come from Amaro and Arba Minch because of security concerns in the last two years. As a result, it has become difficult for Ayano and other Guji individuals to sell their products. Since 2012, relative peace and stability has been restored and people resumed their business. Ayano plans to build a house in Golbo where he wants to begin a business. He has also prepared construction materials (wood and tin) to build a residential house in Arda Guddina, but he was skeptical due to frequent warnings from the park officials not to build permanent houses within the de facto park territory. Ayano was not the only one with this dilemma; indeed, I observed several people who had suspended building residential houses owing to tensions with the park.

Although the detail analysis of this issue will be discussed in chapters seven and eight, it suffices to mention here that the ethnographic note on the Guji individual demonstrates how the territory has been vital for the economic sustenance of the Guji as an agro-pastoral community. This is why the Guji consider the evictions in 1982 and 2004 as "common injustices by the governments because they [successive regimes] had given priority to wild animals than humans" (informant: Ayano Guyo, April 2012). Moreover, the Guji currently live under precarious circumstances because the park administration systematically blocked them off social services. In this regard, their efforts to tackle state imposed deprivation have been obstructed by the park administration's persistent threats to relocate them.

Because of Guji's diversification of their economic practices and sometimes due to sever food insecurity, trade relations between the Guji and other neighboring communities have become evident from time to time. During our visit to Golbo market–the only market where local products are sold and bought–we witnessed how the Guji and Koore people were interdependent, given that both were engaged in somewhat different economic activities. Guji women supply *areera* (cheese) and butter and buy in return *warqicha* (ensete) from Koore women. These are very important foodstuffs for both groups. Owing to ecological impacts and livelihood practices, neither of the groups produces what the other trade partner provides.

Figure 8: Guji and Koore women exchanging crop and animal products in Golbo market (Photo: Asebe Regassa, May 2012)

Another dimension of the people-park relations is related to hunting traditions, which has been one of the contentious issues during my fieldwork. The park administration accuses the Guji for killing the wildlife for food and business and thus considers as serious threats to the animals in the park. While the Guji mostly dismiss the state allegations, there have also been internal differences. While many Guji informants reject the state allegation stating that eating wild animals' meat has never been part of their culture, others acknowledge the involvement of some of their members, particularly the youth the in the act of hunting. As a point of departure, it is imperative to elaborate upon Guji's food culture in traditional contexts that contributed to the prohibition of killing wildlife for food in the past. In a traditional context and still in most cases, the Guji in the study area did not use meat as major foodstuff, despite owning large number of livestock. In most cases, their food constituted barley, maize, ensete (false banana), milk and milk products. The Guji eat meat mainly on special occasions such as festivals, the reception of a special guest, weddings and some specific rituals. In the forthcoming sections, I will discuss some aspects of restrictions, particularly regarding wild animals. While the reason for limited meat consumption culture in reference to livestock is related to the value they give to cattle, the Guji explain that traditionally they did not eat the meat of wild animals for many reasons. The prohibition was associated to belief, social implications and health factors. In fact, there are changes regarding the tradition of meat consumption, with most of the youth said to have been involved in the act of killing wild animals for food and business.

The prohibition related to belief can also be analyzed from a cosmological dimension. The people conceptualize the world in terms of three parts, which are interdependent, reciprocal and entrenched in one another. The first is the world of the Guji, which constitutes the people and their livestock. The second category includes all non-humans apart from livestock, encompassing the living and non-living creatures. According to the Guji cosmology, these two components were created, nurtured and safeguarded by the supernatural power called *Waaqa*. While the first two again join a common category called *Uumama* (creature), *Waaqa* falls in to the category of *Uuma* (creator). According to Guji cosmology, *Waaqa* gave the people ethical restrictions called *safuu*, which govern their interaction with other creatures, humans, non-humans and *Waaqa*.

The restriction in terms of the consumption of wildlife meat comes from the relationship between the three and particularly from the belief that their ancestors made a covenant/oath with *Waaqa* regarding their relations with non-humans. As part of the covenant, the people believe that their ancestors made covenants with *Waaqa* to safeguard animals that belong to him. Thus, they traditionally considered killing wild animals as transgression of ancestral oath and that would invoke infliction in the form of drought, disease, famine, war and the death or sickness of people and livestock. For the Guji, their relationship with wild animals is part of their connection to *Waaqa*. In this regard, Guji's worldview puts the biophysical, the human and the supernatural in one integral component of the environment, built upon interdependence, reciprocity, respect and trust.

While it is their deep conviction that they were born pastoralists to look after cattle, the Guji are conscious about the presence of other 'cattle' whose shepherd is *Waaqa* himself. These are what other people call wildlife. The Guji do not categorize "wild" and "domesticated" in the strict sense of the words; rather the dichotomy only prevails when it comes to the place of residence and ownership of the animals. Therefore, cattle husbandry is not only a livelihood style but also a marker of Guji identity as people, signifying their harmony with their God. At the same time, the Guji argue that they safeguard the life of the preys who seek refuge at their homesteads in fear of big predators. A Guji elderly woman said that, "We care for the animals by providing grass and water. For example, when we come across an animal that is giving birth or in case a predator attacks it, we rescue it. We do this because we want to save the life of the animals. *Waaqa* loves them as we love our cattle" (informant Galgale Aga, March 2011).

Similarly, Ingold (2000) explains such a relationship as trust and reciprocity in human-non-human relations. Drawing upon examples of interaction between a hunter and prey, Ingold argues that such an encounter is essentially non-violent. Rather, it is built upon trust in the sense that the animals have the power to withhold what they could provide if hunters attempt to coerce in the process of hunting (Ingold 2000). The Guji describe trust and reciprocity in a broader context, including the people, wild animals, ancestral oath (a belief in connection between the people and their ancestors) and the supernatural power. According to the Guji, the prey trust the people and approach them, seeking protection against predators. During my fieldwork among the Guji who live adjacent to the national park, I observed that zebra and deer stayed close to Guji kraals and homesteads at night. When they were asked why the animals come to the homesteads at night, local Guji informants were unanimous in stating that the animals trust the people rather than other animals such as lion, cheetah and hyena.

During the eviction of the people from the park in 2004, several wild animals migrated with the people. Although the park administration attributed the incident to a shortage of pasture and water in the park territory, the people described it from a different angle. They believe that the animals not only seek the protection of the people but also keep the natural harmony that pleases the supernatural. According to the Guji, perpetrating any harm on the animals, particularly when they have sought refuge at one's homestead, means transgressing ancestral oath and breaching the trust that has already been developed. The ultimate effects of such a transgression not only displease *Waaqa* but also create aggression of predator wild animals against humans, as well as their livestock. By contrast, maintaining the trust and covenant provides one with different forms of reciprocity.

Another restriction is related to the religious belief among the *Qaalluu* clan–a clan from which *Qaalluu* religious leaders are elected on hereditary basis. Among four major Guji clans–Uraga, Matti, Hokku and Alabdu–the latter is noted as the *qaalluu* clan, because it has traditionally been from this clan that *qaalluu* leaders assume power, while the other three are called *gadaa* clans (Dhadacha 2006). Traditionally, members of Alabdu clan subscribe to several food taboos. Due to their role as intermediaries between humans, non-humans and supernatural power, *qaalluu* leaders and clan members pay attention to 'purity', which includes the avoidance of poultry and all meat from wild animals, cabbage and some cereals, such as millet, *teff* and sorghum. Therefore, in the traditional context, they were prohibited from eating the flesh of wild animals. Social taboos also contribute to biodiversity conservation by imposing different levels of restrictions on members of a social group. Colding and Folke (2001) discussed that indigenous societies in different parts of the world subscribe to different types of social taboos. For instance, the avoidance of a specific animal or plant species is considered as specific-species taboo. Indeed, the Guji *Qaalluu* tradition categorizes the consumption of some specific-species as taboos. Moreover, the avoidance of specific food items, including wild animals, is meant to maintain their legitimacy as religious leaders. However, apart from religious restriction, Guji's prohibition of the killing of all wild animals, apart from those used for cultural pride, can be related to general social taboo, regardless of species specificity. Colding and Folke argue that such restrictions are mainly associated with beliefs that "in some traditional societies taboos are enforced through beliefs that spirits will sanction violators by invoking illness upon people" (2001: 589).

The restriction in terms of bush meat also has a social dimension. A person who kills wild animals for food is categorized among the poor because killing animals for food is perceived as being derived from poverty. Poverty implies low social prestige, which in turn has consequences on marriage arrangements and other interpersonal relations among the Guji. An elder from the Ergansa village recalled the tradition that "if a person is once labeled as killing animals for food, people would not give him their daughters for marriage. He would be given a different name as eater of bush meat. But now, many people abandoned the safuu [norms]" (informant Jilo Shota, May 2011). Moreover, health aspects were also part of the prohibition of killing wildlife for food, because they Guji believe that wild animals' meat spoils teeth and causes diseases (Getachew 2007). There is also a belief that if someone kills a wild animal, the infliction would be in the form of genealogical discontinuity. Among a patriarchal society such as the Guji, boys are considered as security for genealogical continuity in the kinship. They believe that killing wild animals induces punishment through the death of boys or the inability to create newborn boys. In short, whether because of the belief in ancestral oath, social taboo, health implications or a belief in its presumed consequences on genealogical discontinuity, the Guji had traditionally limited practices of killing wild animals. Nevertheless, as will be detailed later, there have been changes in the belief and practices regarding such restrictions.

3.2.1.4 Some Socio-Cultural and Political Changes among the Guji

The traditional livelihood styles and socio-cultural practices among the Guji that were characterized as fairly harmonious with nature do not necessarily place the Guji in the category ascribed to indigenous peoples by the dominant academic discourse as 'close to nature'. In other words, the Guji were not conservationists in strict sense of the term. As Berkes (2008) discusses, there are economic and cultural factors that work against the dominant representation of indigenous peoples as 'noble savage', 'close to nature' or 'communing with nature'. In many ways, indigenous peoples opt for utilitarian approaches in their interactions with the environment. Moreover, traditional ecological knowledge and cultural connectedness to nature do not necessarily guarantee that a certain group lives in harmony with the environment. Rather, the way in which the knowledge is translated into practice and the dynamism of traditional knowledge are important points of investigation, particularly in terms of understanding the political ecology of conservation. An important point to note here is rather how local communities respond to internal or external influences regarding their notions of nature-culture relations.

For example, there are apparent cases in which the Guji perceive the non-human constituents of nature as objects to be utilized for sustaining Guji identity, culture and economy. There were exceptions, changes and continuities in Guji's perceptions of their surroundings, including their relationship with the wild animals in the national park. The cultural, economic, social and religious dynamics that the Guji encountered across time enhanced changes in Guji's perceptions of their environment. Due to its centrality in analyzing changes and continuities in people-park relations, I focus my discussions on some exceptions in the restriction of killing wildlife, relating it to cultural diffusion, social implications of killing and the decline in Guji cultural and religious norms resulting from external influences. Traditionally, the Guji selectively killed the game for different reasons and motives. The first was killing wild animals for medicinal purposes, which targeted particular animals for specific diseases. In this category, they killed animals such as pig and porcupine for the purposes of curing malaria, cold, influenza and back-pain.

The second exception was when killing was practiced for *midda* (honor) tradition. Traditionally, the Guji celebrate killing big game animals such as lion, elephant, buffalo and rhino. The *Abba midda* (person of the honor) is respected within the community and thus gets higher social status and economic advantages (Taddesse 1994, 2006; Asebe 2010). The Guji interpret the act of killing game animals for honor from a cultural perspective as a marker of identity, pride and greatness as a community that is given to them from *Waaqa*. According to Guji oral tradition, they were given *midda* culture from *Waaqa* and it is through this tradition that they reveal their pride, self-esteem, bravery and identity. They also associate the ritual performed during the *midda* ceremony with the continuity of the life cycle of the animal because it enhances the fertility of the animals that remain behind. Although the custom is dwindling mainly due to the introduction of Christianity and sanctioning by state law, the Guji still believe that rituals of *midda* tradition enhance harmony between the people, nonhumans and supernatural power rather than distracting it. According to their perception, the *midda* tradition would please *Waaqa* because it is through rituals associated to the tradition that the Guji display their cultural and spiritual connectedness to their surroundings. From modern conservation perspectives, the *midda* tradition might be blamed for the extinction of the big game animals (buffalo, elephant and rhino) from the Nech Sar area.

The third aspect of killing wildlife has been related to economic values associated with such practices, namely involving food and business. According to most Guji informants, cultural diffusion from neighboring groups has introduced a culture of eating wild animals' meat, as well as the idea of killing animals for money. My informants explained this in two contexts. First, due to their closer interaction with neighboring communities such as the Gamo, Koore, Burji, Konso and Gumayde, for whom the meat of many wild animals was not culturally prohibited, Guji young generation began to question the traditional restriction. Therefore, interaction between the Guji and other groups has shaped some of the people's values, norms and customs related to man-animal interaction. Guji youth who herd cattle with children from the neighboring communities have been 'socialized' to the meat eating culture of the 'Others'. Moreover, there are also rumors within members of the Guji and Koore communities that the Golbo/Tsalke market is used as a clandestine business place for 'bush meat', in which women are at the forefront in the trade, although the killing is actually undertaken by men.

Secondly, the introduction of Christianity has denigrated and demoted the beliefs about infliction to be invoked because of killing wild animals. Some of my informants mentioned the verse from the Christian Bible stating, "Be fruitful and increase in number; fill the earth and subdue it. Rule over the fish in the sea and the birds in the sky and over every living creature that moves on the ground" (Genesis 1:28). In people-animal relationship, Guji Protestants who are mostly the young generation emphasize the paternalistic and utilitarian notion in the Christian tradition. Informants in this category argue that God has given man the authority over all creatures on the earth, as well as those in oceans/lakes.

This perception of the youth often brings them into conflict with elders, who consider themselves as guardians of Guji identity (culture, religion and ways of life). For instance, non-Christian elders complained that bad omens and misfortunes such as the flooding in 2009, drought in 2010 and 2011, and violent predation of predators over their livestock and fertility problems of cattle are partly attributed to the disrespect of the youth to traditional customs, norms and taboos. Elders mentioned cases in which the young people killed wild animals, roasted the meat and ate in the bush where they were herding the cattle. Besides the anger generated on the side of *Waaqa*, according to the elders, such practice also spoils the pastureland and water sources, which might in turn cause disease to their cattle. As Van de Loo (1991) states, the Guji are cautious of the natural order of things in their environment to the extent of considering even laud voices such as shouting at night would disturb the stability of spirits, wild animals and livestock.

Apart from specific exceptions in Guji's restriction of killing wild animals, the change in their conceptions of human-environmental interactions should also be broadly positioned within some socio-political changes they encountered following the advent of the Ethiopian state to their land. The major political and administrative change was the incorporation of the territory into the Ethiopian empire and the subsequent imposition of Amhara/Ethiopian culture (McClellan 1988). Gujiland was conquered and incorporated into the Ethiopian empire in 1895 by the forces of emperor Menelik II (1889-1913) (Taddesse 2009). The conquest introduced Amhara/Ethiopian culture, religion (Orthodox Christianity), a feudal system of economy and oppressive political system to the subjected regions (Vaughan 2003). In his study of "*The Gada System among the Guji of Southern Ethiopia*", Hinnant argues that the assimilative policy of the imperial regimes strongly worked towards eroding indigenous cultural practices and denouncing it as illegal and backward. They appointed their own patrons in place of *abba gadaa* and Orthodox Christianity and the priest instead of the great *Qaalluu* (Hinnant 1977). This dynamics had introduced new values and notions of people-environment relations among the Guji by denigrating indigenous ecological wisdom among the people.

Moreover, indigenous cultural practices, beliefs, governance, and livelihood conditions were denigrated as backward and obstacle to development during the military regime (1974-1991) (Dereje 2011). Referring to the 'modernist' discourses of postcolonial states in Africa, Stewart-Harawira (2005) argues that these nation states adopted Western scientific knowledge and popular culture as reliable sources of knowledge, while indigenous knowledge, modes of governance and worldviews were identified as obstacles to development and at odds with views of the mainstream society. Although the Guji Oromo culture and traditional religion continued without significant change in terms of the Guji worldviews despite internal and external influences, the impacts of these administrative and religious influences should not be underestimated. People were discouraged to take their cases to customary courts and the *Qaalluu* institution. Rather, local state courts and the Orthodox Church adopted the role of these traditional institutions. Later on, Protestantism was also introduced and played no less a role in denouncing Guji traditional beliefs as 'backward' and devilish. For instance, traditional ritual practices were considered as worship to Satan/devil (Van de Loo 1991).

To summarize, the Guji Oromo around the Nech Sar National Park considered the territory as their customary land, to which they had strong cultural, religious, economic and historical claims of entitlement. Despite changes in their views towards wildlife partly due to external cultural impacts and partly because of internal transformations, the Guji maintain strong connection to the land, mainly because it is at the center of their pastoral ways of life. Pastoralism is much more than herding livestock and utilizing their products; rather, it has cultural, social, historical and economic purposes and more importantly has been considered as a reflection of Gujiness (Guji identity).

The livelihood of the people, which is often characterized by seasonal mobility, informs their land use system, traditional conservation practices and knowledge of the environment. As Van de Loo (1991) indicates, the Guji possess deep knowledge of the anatomy, disease and remedies that they acquired through religious practices, experiences and largely because of their strong attachment to the stock. For the Guji, the concepts of conservation and development are understood from cultural perspectives. For instance, while caring for the environment is part of their cosmological schemes of local knowledge and belief, what they consider an appropriate development scheme is something that is compatible to local values, customs and livelihood traditions. Despite having expectations of securing schools for their children, road connecting to the nearest markets, a health center for the people and livestock, flourmills and access to pure water, any 'development' program that disrupts their pastoralism is not acceptable to the ordinary Guji men and women. As stated earlier in this section, livestock signifies beyond mere economic purpose among the Guji. Thus, most Guji consider the state's development conception, which places emphasis upon settled agriculture and ecotourism project in the area as a challenge to their livelihood and a restriction to their customary rights of resource ownership and utilization.

3.2.2 The Koore and their Land

3.2.2.1 The Koore Settlement History

The Koore ethnic group belongs to speakers of the Omotic language family, who predominantly live in southern and southwestern Ethiopia along the Omo River. The Koore people speak Koorete language (Hayward 1982; Binyam 2010). The people prefer to identify themselves by the name of Koore, although different groups have given them various names across history. For instance, the Guji and Borana Oromo called them Baditu (highlanders or people of the mountain) and Amara in the past (Hayward 1982). Likewise, other neighboring ethnic groups such as the Burji, Gamo, Konso and Gumayde identify the people as Koyra, which is the name labeled to the Koore by Menelik's soldiers after the conquest (Bediru 1986). Some still use the place name Amaro and the old name Koyra interchangeably while referring to both the people and the territory. During the imperial and military regimes, the Koore were administered under Sidamo Province, although they secured 'autonomous' self-government as Amaro Special District under the SNNP regional state in the post-1991 political and administrative reconfiguration.

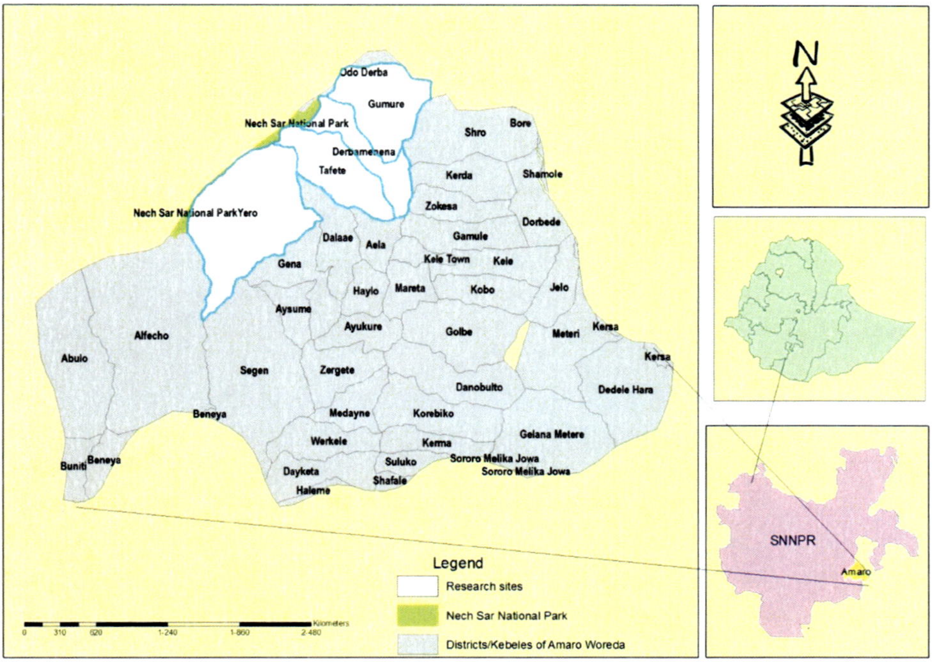

Figure 9: Amaro District Map (Source: Abiyot Legesse, February 2014)

According to oral tradition from the people, the administrative jurisdiction of

Amaro district during the imperial and military regimes stretched up to Bonke hills and included parts of the Chamo and Abaya Lakes. Accordingly, almost the entire Nech Sar National Park was under the Amaro district in terms of geographical administrative locations until 1991. This is why they maintain the view that Amaro district is entitled for administering the Nech Sar National Park because the territories up to Bonke hills were under its administrative jurisdiction in the past though the people did not settle beyond Tsalke lowland. The district adjoins the Borana zone in the north and east and the Burji and Konso in the south. Until 2008, it also shared borders with the Nech Sar National Park in the west, although the Koore were disconnected from direct border contact with the park boundary following the establishment of Ergansa village under Borana zone (Galana district) in the same year, with the exception of a narrow strip of land along the southwestern direction of the district.

Oral traditions and scanty literature available do not avail a concise account about the settlement history of the Koore people in the area. Despite such limitations, it is important to trace how the Koore moved to the present day contested territory because the group's claim over the land and the conflict with the Guji over the same territory remains blurred, without a clear understanding of their historical process of settlement in the area. The origin and settlement history of the Koore people is contentious among both scholars and elders from the group. However, the dominant speculation refers their original homeland to north Shewa around Menz, from where they were believed to have migrated to the south sometime before the war of Ahmed Gragn (1529-1543) (Bediru 1986; Simeon 1995; Binyam 2010). According to this line of argument, political instabilities in northern Ethiopia during the medieval period might have pushed the people out of their original homeland. Moreover, some speculated the southward movement of the Koore might be part of missionary activities by Orthodox Christianity (Bediru 1986). The people were believed to have initially moved to Kaffa and Gamo Gofa, from where they gradually moved to the present day Amaro district. Some mentioned that the people were followers of Orthodox Christianity when they moved to southern Ethiopia. They came with 'tabot' (arc of the Covenant of God) and established three churches in Gamo Gofa and Amaro (Bediru 1986).

Although such long distance migration was difficult during those early times, given the inadequate transportation system and the fact that the highland Christian state did not have administrative influence over Gamo Gofa and Kaffa areas at the time, Levine (2000: 44-45) describes the persistent inter-societal relationship between different ethnic groups in northern Ethiopia and those in the south as follows:

> More intimate relationships among different ethnic groups in Ethiopia have been formed through processes of migration and intermarriage. Settled or nomadic, the Ethiopians are great travelers. To find new land, go on hunting or raiding expeditions, seek fortune at royal court, escape enemies, study at religious center, carry on trade, or make a pilgrimage, Ethiopians have long accustomed to moving from one part of the country to another, in many cases settling there. The Amaro, Janjero, Konta, and other peoples have plausible traditions concerning the immigration of Amhara settlers who came from the north several centuries ago.

In terms of language and religion, the earlier immigrants were believed to have been speakers of Amharic language and followers of Orthodox Christianity, until they adopted traditional religion juxtaposed with Christianity following the destruction of churches and persecution of Christians by the forces of Ahmed Gragn in 1530s (Awoke 1985). The Koore people re-embraced Orthodox Christianity following the incorporation of the region into the Ethiopian Empire in the late nineteenth century. For those who support the story of immigration, similarities between the Amhara and Koore in terms of some religious practices, rituals, language and names given to crops, animals and other objects are cited as plausible evidence of a connection between the immigrants and their Amhara tradition in northern Ethiopia. In this regard, the type of animal sacrifice and rituals practiced among the Koore, in which they offered sheep or goat (Simeon 1995), resembles what Levine describes about sacrifices and rituals among the Amhara in the north (Levine 2000). This includes animal sacrifice in connection to Christian tradition (e.g. sacrifice of a particular type of chicken or sheep on first day of a new year) being a typical cultural element among the Amhara of northern Ethiopia. However, according to the author, this practice is shared among other groups in the country including the Koore. Although Levine's construction of homogeneity among diverse peoples in Ethiopia is a gross simplification of multidimensional meanings attached to such practices, some sorts of similarity can be drawn between the Amhara and the Koore in terms of the belief behind animal sacrifices.

In contrast to the above narrative, Awoke (1985) and some local informants associate Koore's original homeland to Tsinki Korso in Gamo Gofa zone. This seems an emerging argument among the young generation and a few writers from the group. My field assistant, who was a native from Amaro, challenged the story of the group's migration from northern Ethiopia. He studied Linguistics at a bachelor level and attempted to posit his argument from linguistic dimensions. Linguistically, Koorete language belongs to the Omotic (Ometo) language group, which is only spoken in the southern and southwestern part of the country (Hayward 1982; Binyam 2010). My field assistant once asked me:

"If the assertion that the Koore migrated from north Shewa and were speakers of Amharic was true, how it could be that the people completely changed their language to non-Semitic language family?" Methodologically, I was aware that his position about the story of migration would influence the course of my inquiry by directing the informants to his views unless systematically resolved. Therefore, I promised him to consult professionals in the area and communicate back to him, but to suspend the question until then. Despite such 'agreement', he continued preempting the informants with some guiding questions that would lead my informants to the 'Tsinki Korso' story.

Shado Udado, a Koore elder in Darba Mamana village, strived to trace the original homeland of the Koore to Tsinki Korso in Gamo Gofa as follows:

> Our grandfathers told us that our ancestors lived in Tsinki. You know Tsinki Korso in Gamo. They had churches in Tsinki. Birbir Mariam was our church. It is still in Gamo land. According to our elders, few people came to Amaro for hunting and gradually liked the place. It was good for human settlement. It was fertile, it had animals for hunting and the weather was good. Gradually, they settled on the highland of our Amaro land. Our grandfathers then began cultivating maize, coffee, ensete and other crops. Actually, the people introduced coffee much later. After the highland was overused, we moved to the desert like Tsalke, Yero and Abulo (Informant: Shado Udado, March 2011).

Drawing on linguistic similarities between the Koore and its neighbors, Awoke (1985: 17) also argues that: "Besides the oral traditions of the migration of the Koore from Gamo Gofa or descent of Kore from Gamo, the linguistic and cultural similarities with other Omotic languages indicate that the Kore most probably migrated from Gamo Gofa not from Menz. However, the coming of few immigrants or group of families cannot be ruled out and these might have come to Korso as Orthodox missionaries, hunters or conquerors." In fact, some Koore elders only recall the migration of their ancestors from Gamo Gofa, and thus attribute their origin to a place called Tsinki Korso.

It is neither my objective nor the scope of my research to deal with the controversy about the origin of the Koore. However, one important point to be noted is that, particularly within the current political articulation in Ethiopia, the people's views of their original homeland is framed along regional or ethnic identities, which in turn imply issues of political and administrative entitlements. As Taddesse (2009) argues, the post-1991 political articulation along ethnic lines has redefined boundaries of alliances, with former enemies having become friends while ancient allies became foes in most cases. My major focus in this section is rather on two important aspects. First, however contested it might be, Orthodox Christianity had a long history among the Koore, thus influencing their worldviews and perceptions of nature and shaping their interactions with their environment. Secondly, their settlement on the highland part of the Amaro district gives us an insight into how the current contestation over the Tsalke/Golbo valley has gradually evolved. Many Koore elders agree that the people inhabited the Amaro highlands until some families expanded to the lowland in the second quarter of the twentieth century.

3.2.2.2 Koore's Livelihood

In terms of economic practices, the Koore traditionally practiced agriculture (crop cultivation) and were engaged in trade with their neighbors such as the Burji, Konso, Borana, Gamo and Wolaita (Smith 1897 [2008]; Bediru 1986). While the trade with the people to their south was easily conducted by caravan route, the trade with the Gamo and Wolaita across the semi-arid areas of the Nech Sar by crossing the Abaya and Chamo Lakes was neither secure nor easy. Their hostile relationship with the Guji, presence of dangerous reptiles and poor boat transportation over the lakes were major challenges limiting their interaction with the people to the west of the lakes. The only land transportation was through a narrow land strip between the two lakes called "God's Bridge", which is now part of the Nech Sar National Park. According to a British traveler, Davidson Smith, who traveled in the region from Somaliland to Mombasa and visited Amaro and Lake 'Abaya'[7] between 1894 and 1895, the Koore only inhabited the highland part of the district (Smith 1897 [2008]). Smith (1897 [2008]: 214) reported his witness about the Koore (he referred as Amara in his text), their livelihood and settlement as follows:

> The Amara begged me to stay as long as I could among them, and promised to trade with me. The chief [Amaro chief] returned on April 25[1895], and with nice boys I accompanied him up the high mountain. I did not expect to be gone long, and took no provisions of any kind with me, or extra clothing; but the scenery became so interesting that I continued climbing higher than I anticipated. We could now see the Galana far below us, winding through a very wide valley, and then cutting its way through a great mountain range. […] I was now three hours away from camp, but the Amara told me their village was very near, and I was tempted to go still farther. We pushed on over the undulating plateau, passing one cornfield after another, in which hundreds of natives were working away with their three-pronged hoes.

According to oral information from Koore elders, the people did not settle in the lowland areas for three major factors. The first was related to mosquito and tset-

[7] The British explorer, Davidson Smith mentioned the Lake he visited as 'Abaya', although the lake he described was rather Lake Chamo (Briggs 2006).

se fly infestation, which rendered human habitation impossible. Secondly, due to the low population density in the highland areas during those times, there was no demographic push factor. Thirdly, because of conflicts with the Guji, it was not safe for the Koore to settle in the lowland. During the period of Smith's visit to the Koore land, the highland part of the district was sparsely inhabited. Smith witnessed the engagement of the people in the cultivation of crops such as Indian corn, durrha, beans, pumpkins, coffee, cabbage, cotton, tobacco and banana by terracing the mountaintops and its slopes. The traveler described that the Koore were confined to the highland areas (Smith 1897[2008]). Smith camped at a place he described to the south of Lake 'Abaya', which he mentioned as being uninhabited. Due to the timing of Smith's travel in the region (i.e. during the rainy season in April and May), pastoral communities might have been at their wet season encampment far to the west or east. In any case, Smith's account and some oral traditions from the group hint at the argument that the Koore did not inhabit Tsalke/Golbo lowland during the late nineteenth century. However, the people had a long history of contact with the Gamo across the two lakes through trade. Many interrelated factors reinforced the down-slope expansion of the group during the course of the twentieth century. The next section explores these trajectories as the background information for how the eastern part of the park became a contested space among the Guji, the Koore and the park administration.

3.2.2.3 The Downslope Expansion of the Koore and their Relation with the Park

Many interrelated socio-political encounters among the Koore shaped their relations with the environment, laid ground for their claims over the Tsalke valley and brought them in direct contact with the agro-pastoralist Guji Oromo, who inhabited the territory that later became the Nech Sar National Park. The incorporation of Amaro into the Ethiopian empire, the 'reintroduction' of Orthodox Christianity, the Italian invasion (1935-1941) and restoration of autocratic rule in the post-liberation period, as well as Guji's shift to agro-pastoralism, had significant implications in influencing the Koore's settlement history, environmental views and claims over the land. In this section, I will briefly discuss these issues as background historical antecedents for the current contestation over parts of the national park.

Historical antecedents of the late nineteenth century in Ethiopia that coincided with the European scramble for Africa drastically shaped the history of the peoples in the southern part of the country (Vaughan 2003). It was the period when southern Ethiopia faced what some historians call it as 'internal colonialism' under the oppressive and exploitative feudal system of the Amhara/Ethiopian Empire (Markakis 1974). Through alliance with Western colonial powers of the time, emperor Menelik II (1889-1913) managed to conquer the hitherto autonomous states in southern half of the country, thereby introducing a kind of 'settler colonialism' and in some cases indirect rule system by appointing local chiefs. The conquest and the subsequent introduction of hegemonic state power (culture, language, religion and economic and administrative systems) transformed the life worlds of the peoples in the subjected regions to dominant-subordinate relations. Land alienation in the form of transferring ownership right to the northern landlords or the locally appointed chiefs dispossessed the people, not only from their material landed resources but also detached them from their spiritual, cultural and cosmological attachments to their land.

Amaro, as the land is commonly called, was subjugated and incorporated into the Ethiopian empire in 1895 (Awoke 1985). As was common in areas where local chiefs submitted to the conquerors without fierce resistance, the Koore chief by the name of *Kate* Amole retained his traditional authority under the supervision of Amhara regional lords who oversaw administrative and economic matters in the district (Bediru 1986). Nevertheless, retention of traditional authority in the hands of the local chief does not corroborate with local autonomy. In the conquered regions of the south, local chiefs became autocratic over their subjects to convey loyalty to their Amhara masters (Markakis 1974). As a result, the Koore ordinary people faced suppressive and exploitative political and economic systems under the imperial regime. The people lost their land rights, with land becoming solely the property of the crown and its feudal landlords who were appointed to a different state structure. In the context of investigating Koore's relations to their land and the subsequent claim over the Tsalke/Golbo valley, which will be detailed in chapters seven and eight, some important consequences of the conquest should be singled out. The conquest dispossessed the Koore from their land and thus impoverished them through a tenancy system. Unlike Smith describing Amaro as a land of plenty few years before the full-fledged institutionalization of the feudal system, it turned to be a land of misery for the majority of ordinary Koore, as one informant noted:

> Our land was known for its bounty of honey, ensete and all types of crops. Our grandfathers told us that the Koore never knew hunger before the Amhara came. After Haile Sellasie took power, we lost our land. The balabats [feudal lords] exploited our labor and harassed our women. We lost the right over our produce let alone over our land. Many people migrated to Kenya and others escaped to the desert to shelter among the nomads. It was the time when we produced and shared to the balabats who presided over us as cirri [a predatory bird that lives by sucking

blood from cattle] (informant Gashe Bufa, February 2012).

It is common among many agrarian societies with landed economies that people interact and communicate with their environment in the process of cultivating the land for their sustenance (Escobar 1999). In the process of iterative interaction and communication between nature and society, people produce, articulate and transmit knowledge from generation to generation. For example, in the Koore case, they had developed knowledge of resource conservation through long engagement, particularly in mountain terracing (Smith 1897 [2008]; Bediru 1986). The Amaro district is mostly mountainous, steeply descending and rugged. Despite such topographical difficulties, the people were noted for their traditional knowledge of irrigation, and water and soil conservation practices. They grow coffee, ensete, banana, sugar cane, mango and avocado in their limited plots of land by irrigation (Smith 1897 [2008]; Bediru 1986). Nevertheless, once they were dispossessed from their material resource, which in turn dis-empowered them in all aspects of life, the people were disconnected from the process of knowledge production and transfer. The privatization of land rights through its allocation to local chiefs and settler landlords was a process in which the state enacted its power over the society.

In Ethiopia, the privatization and commercialization of local resources in the south during the imperial regime eliminated the subjected people's customary right to their land and resources, as well as undermining indigenous practices of resource management. Due to the erosion of their customary rights, peasants in the south neither were encouraged nor had the right to manage their resources. As Markakis (1974) argues, most of the fertile agricultural land in southern Ethiopia was under the control of absentee landlords and local *balabats* (feudal lords), whose primary objectives were increasing their personal wealth and satisfying their masters by extracting maximum labor and taxation from the subjects. As is common among most southern Ethiopian peoples, the Koore were exploited in the form of heavy taxation, providing labor for landlords, payments in kind (honey, coffee, butter, sheep, etc.). This leads us to the second consequence of the conquest, namely the administrative and economic burden that it perpetuated and the subsequent down-slope expansion of the Koore people to escape the heavy hands of the feudal lords.

In pre-colonial and colonial Africa, people in the peripheries of the state used strategies of escaping or exit as mechanisms of avoiding repressive regimes (Herbst 2000). This case was common in Ethiopia, where the people in the subjugated regions migrated to areas where the authority of the central state was limited, such as the arid pastoralist areas (Donham 1986; Markakis 2011). There were two major reasons for people's choice of lowland or peripheral regions as exit zones. First, there were strong level of state presence in the highland, which made the people in these regions highly susceptible to heavy taxation and other forms of administrative oppressions. Second, due to dense population settlement in the highland areas, land remained scarce in these regions, although it was largely abundant in the lowlands. Because of land dispossession and the imposition of oppressive feudal system, the Koore were also prompted to move down slope to Tsalke since the 1930s. The downslope expansion of the Koore was a gradual process undertaken as a mechanism of escaping from the exploitative and oppressive system installed following the conquest. Interrelated with this, individuals along the adjacent villages such as Darba Manana, Tifate, Yero and Gumure expanded to Tsalke valley owing to land degradation on the highland parts of the district. According to local informants, the expansion was particularly accelerated following the restoration of imperial regime after the end of the Italian occupation in 1941.

> Ayele Machamo, a Koore resident in Darba Manana described how his parents moved to Tsalke in the aftermath of the restoration of the imperial regime in 1941 as follows: Life became very difficult in the highland. The government took almost all of what our parents produced. They also asked our people to work on the plots of the balabats [feudal lords]. People suffered a lot even to survive. That was why some households moved to Tsalke. Tsalke was fertile but life was again difficult there. Mosquito infestation was the serious challenge in the lowland. Conflict with the Guji was another big problem. This is still a problem. My parents and others moved down to Tsalke long before the park issue came. However, they actually went to work there during the seasons of cultivation by keeping part of their family on the highland (informant Ayele Machamo, March 2012).

However, Tsalke/Golbo was not an empty space waiting for the Koore. As both Koore and Guji informants agree, the latter used it as dry season grazing land for several decades before the expansion of the Koore to the lowland. Gradually, the Guji and Koore established inter-group networks through mutual interdependence, facilitated by trade, intermarriage and adoption (*ilmofacha*) by the Guji. The two communities exchanged livestock, livestock products and agricultural products at local markets located in Darba Manana (Amaro) and Tsalke/Golbo (now in Guji territory).

Figure 10: Vegetable Market in Golbo/Tsalke (Photo: Asebe Regassa, May 2012)

The role of local trade in creating and enhancing inter-group interdependence among the Guji and their neighbors has been noted in other studies. For example, McClellan (1988) having once described that local trade interdependence was one of the major factors for harmonious relations between the Guji and Gedeo particularly because of their engagement in different economic activities. Likewise, the engagement of both Guji and Koore in different livelihood styles increased mutual interdependence between the two groups, although competition over land resources gradually triggered conflicts over access to resources, as in the case of conflict that erupted between the two over Tsalke farmland in 2010. Although intermarriage was limited from both sides, it is practiced specially during peaceful relations between the groups. The Guji complain about Koore's expansion to their land that the latter came as guests, in-laws and adopted 'children' (*Ilmoffacha*) before eventually settling in Gujiland. *Ilmoffacha* involves a process through which Koore individuals are adopted into a Guji family or clan and are thereby entitled to get land, cattle and considered as part of the Guji clan. Oral narratives from Guji and Koore recall that members of the latter sought asylum in the hands of the Guji pastoralists in the Nech Sar area who were marginal to state administration. As a result, the tradition of *ilmooffachaa* was developed.

It has been clear from some scanty sources and oral information that the Koore moved to the lowland adjoining the park during and after the Italian occupation. Administrative pressure from the central government, demographic factors and land degradation were the major factors for peasants in nearby villages (Yero, Tifate, Gumure and Darba Manana) moving to Tsalke for temporary encampment for agriculture, while keeping their plots and homes at the highlands.

Another historical incident that shaped the course of Koore's down-slope expansion to Tsalke/Golbo was the five years of Italian occupation. However, the role of the Koore people during the resistance movement against Italians remains contentious. As some local informants and a study by Bediru (1986) under the then Ministry of Culture and Tourism describe, the Koore were determined to fight along with patriotic fighters, despite the history of oppressive system that they suffered under the imperial regime. In fact, local elders have memories about patriotic leaders such as Abebe Mengesha who used the mountainous Amaro topography as defensive spaces in their war against the Italian colonial invasion. When the rebels clashed in the Amaro highland areas, they escaped to the inaccessible lowland areas such as Tsalke, where they gradually established settlements in the post-liberation period.

By contrast, like most peoples in the subjugated regions, the Koore used the opportunity of foreign intervention to fight against the oppressive Amhara domination in their region and thus collaborated with the Italians (Awoke 1985). In any case, the Italian invasion and the subsequent restoration of imperial administration in the aftermath of the liberation in 1941 enhanced Koore's expansion to the lowland. As it became common in the majority of the conquered south, where people mostly allied with Italians against internal Amhara domination, the imperial regime enacted cruel and exploitative economic and administrative policies in the post-liberation period in Amaro. As a result, the people were forced to move to the lowland to escape from the state or at least subsidize their production to pay the heavy taxation.

Another important point of discussion related to Koore's relation to their environment is that of religion. The Koore people were followers of Christianity before they moved to southern Ethiopia. Like their Orthodox counterparts in northern Ethiopia, the Koore understand human-environmental interactions from 'modern' religious perspectives. According to this tradition, all living and non-living things were created by God, but human beings were given supremacy over other creatures to subdue, use, control and manage them. In a purely religious context, they have dichotomies of what are allowed for humans to eat and what are taboos. For instance, animals categorized as Equine or those without cloven hoofs and those that do not chew the cud were classified as unclean among the Koore. This goes in line with some prescriptions in the Bible that orders Christians to eat only from animals with cloven hoofs that chew the cud.

The Koore people conceptualize their environment from an economic perspective, except in some cases where some members of the group turned to traditional religion after their arrival in the new place. Thus, the Koore dichotomously understand the human versus non-human relation in a similar way to the mainstream Abyssinian societies. Immediately following the incorporation into the Ethiopian empire, Orthodox Christianity was re-introduced as part of the state mission of conquering the people in the south through religion (Bediru 1986). It was also reported that the Koore people embraced Protestant Christianity in the early twentieth century (Simeon 1995). Therefore, both sects of Christianity have shaped the Koore people's perception of their environment to a utilitarian approach more than the Guji, who traditionally understood their surroundings as conjointly constituted with the human nature.

3.3 The Establishment of State Farm and Arba Minch Town: State expansion to the Periphery

Two important developments that took place in the 1960s should be discussed in the context of people-environment relation on the western side of the Nech Sar National Park and in reference to the current contestations over the territory, namely the establishment of Arba Minch State Farm and Arba Minch town in 1958 and 1962, respectively. I discuss these developments and the subsequent establishment of the national park according to the central state's continuous expansion to the periphery by seizing land from the local people in the form of development projects. Moreover, the state used the establishment of state institutions as a strategy of broadcasting its state authority over areas that it considered as nodal points of resistance against its presence. In such cases, the imperial regime established some towns to facilitate the administrative and economic purposes. For instance, towns in the south were established mostly in the form of garrison to enhance the easy pacification of resistance in the early twentieth century (Markakis 1974). Like the colonial conquests in other African countries, southern Ethiopia in the post-1890s encountered a massive influx of predatory northern settlers who moved to the newly established towns in the incorporated regions as soldiers, administrators, feudal lords and businessmen. These settlers served as the core catalysts of administration on behalf of the central government. Roads were built, linking provincial towns to the capital city to facilitate administrative and economic purposes of the state (Markakis 1974; Donham 1986).

The second strategy of the state's extension of its authority to the periphery has been through the so-called development projects. Since the mid nineteenth century, successive regimes in Ethiopia have been caught between emulating

Western ideals of modernization on the one hand and maintaining local conservative traditions on the other (Clapham 2006a). Beginning from Emperor Tewodros (1855-1868), different governments have attempted to adopt different models of 'modernization' in the form of state structure, urban development, social infrastructure, economic system and, more recently, nature conservation. Analogous to the nineteenth century European colonialists who camouflaged colonialism by the so-called mission of 'civilizing' the 'uncivilized', the Ethiopian empire also launched empire-building projects and subsequently introduced new forms of administration, economy, culture and ways of life to the newly subjugated regions of the south under the guise of 'modernization' (Markakis 1974). Following the conquest of the south, the Ethiopian empire introduced extractive state machinery in different forms, with state institutions, structures and socio-political traditions of the Amhara/Ethiopian Empire systematically used to extract material and human resources in the subjected regions.

Agricultural mechanization under state-owned farms was sought as one of the ways to 'modernize' the country's economy in the post-World War II period. Despite its feudal characteristics, the Haile Selassie regime began to amalgamate capitalist notions of development (such as granting land to multinational corporations for agricultural investment) with feudal ethos, neither of which ultimately benefited the society as whole. Rather, the emulated development model served the interest of the aristocrats, landowners and the nobilities, while exacerbating the destitute living conditions of the poor. For instance, the feudal regime introduced one of the first large-scale commercial farms in Awash Valley project and the Chilalo Agricultural Development Unit (CADU) in the 1960s. While the Awash Valley project led to the displacement of pastoralists, blocking their seasonal grazing land, CADU brought massive resettlement and displacement of peasants among the Arsi Oromo (Clapham 2006a). Such a 'modernist' approach of development, which valued sedentary life style over pastoralism and state farms over small-scale peasant agriculture, undermined the local livelihood conditions of the people and neglected their local perspectives of development.

Similarly, the government established Arba Minch State Farm in 1958 by confiscating land from Guji agro-pastoralists and some Gamo peasants (Getachew 2007). The farm, which covered over 97,000 acres, was initially established for cotton production by diverting Kulfo and Harre rivers, the two major rivers traditionally used by the Guji for their livestock. While it restricted the Guji from traveling between their dry season and wet season grazing areas, it attracted the Gamo people from the highland areas who moved down as daily laborers and gradually established permanent settlements in the area (Dessalegn 2004; Getachew 2007). Guji elders recalled the incident as a tragic denial of their customary land right and expressed that successive regimes in Ethiopia had followed similar approaches in displacing them from their homeland. A Guji elder from Abulo, a village close to Arba Minch University, exclaimed that:

> All the land from Ergansa [Nech Sar plain] through Bonke to Siqalla, Secha, Shara, Abulo and Harre were our ancestral land. Our ancestors lived here. They died here. The Amhara government [Ethiopian state] came and first chased our people from the farm and then from the town. They took our land and established town. But, the life styles in towns do not suit the Guji way of life. The smoke and noise in towns were not good for our cattle. As a result, our people moved to Ergansa. This happened during the period of the king [Emperor Haile Selassie]. The Derg [military regime] came and introduced the park. Our Guji brothers in Ergansa were evicted twice. Then came Meles [EPRDF government] and he also continued what his predecessors did. The government pushed our people to a narrow territory between the park and the Amara [Koore]. Can you imagine? None of these governments thought of Guji's right to the land. The prioritized the rights of animals than of the people. I do not know if they are governments of the people or of the animals (Informant Edema Aga, April 2011).

Like the state farm, the establishment of Arba Minch town also brought an influx of people from the Gamo highland and many other places. It created a new social setting and high demand for resource utilization, which was quite different from the past. However, I would argue political and economic interests were behind the establishment of most towns in the subjugated south. Such understanding helps us to single out territorial claims and environmental problems such as the Nech Sar National Park. Menelik's administration of the South relied on a network of garrison towns, manned by northern soldiers (Freeman 2002). Towns including Dilla, Chencha, Yirgalem, Hagere Selam, Kibre-Mengist, Yirga Chaffe, as well as highways connecting these towns to other provincial towns or directly to Addis Ababa, were built for economic and administrative purposes. In Gamo Gofa province, the provincial town of Chencha was established in a similar manner shortly after the conquest by appropriating land from local communities. In Ethiopian history, Ethiopian Orthodox Church and garrison towns were the two major institutions enhancing the conquest and the administration of peoples in the periphery of the Abyssinian state and were used as extractive tools. The location of Orthodox Churches and garrison towns was often on hilltops for strategic convenience and their symbolic representation of higher authority of the state. While garrison towns were residential areas for the northern settlers and local chiefs, the church was a basic institution that backed the conquest and administration through spiritual conquest. It taught the legend of Abyssinian "great tradition", representing the old empire as the cradle of civilization and Christianity. By doing so, the church inculcated a culture of submissiveness, obedience and loyalty to the church and state. Through the progress of time, the town of Chencha lost its significance and thus the provincial town moved down to Arba Minch in the early-1960s for different reasons.

According to oral tradition from the area and some historical records, Arba Minch town replaced the former provincial town in 1962 due to three fundamental factors. First, the location of the new town was chosen for effective control of sporadic conflicts between the Guji pastoral communities and the Gamo peasants (Getachew 2007). As previously mentioned, the Guji inhabited the lowland parts to the areas east and west of Lakes Abaya and Chamo. During these early periods, their relationship with the neighboring communities such as the Gamo, Wolaita, Koore and Burji was often marked as the history of conflict, although there were also some aspects of harmonious coexistence in the form of trade, intermarriage and participation in rituals. The people from Gamo highland used to move down to the semi-arid lowland areas adjoining the Guji for hunting, fishing, trade, as well as collecting forest resources. Elders from Guji and Gamo ethnic groups recall memories of insecurity when they moved to the present day Arba Minch town due to the killings from both sides. Therefore, the state moved the administration much closer to the contested territory to contain inter-ethnic conflicts.

Second, the shift of the administrative center from the highland Chencha to Arba Minch was due to abundant natural resources near Arba Minch, in contrast to the overpopulated and over-exploited Gamo highland. The dense and impenetrable forest and the legendary 'Forty Springs' were the area's major attractions, given that they provided fresh water and wood supply for the town dwellers. Moreover, Arba Minch is located at a distance of only 3-5km away from the two lakes (Briggs 2006). After the establishment of the national park, the scenery of the park also contributed to the development of the town following the building of hotels, lodges and other government institutions. As Briggs (2006) describes, Arba Minch town would have not attracted residents to the extent of its present status had it was not been for its location close to the Nech Sar National Park. Owing to the scenery of the national park, the forest and the two lakes, which are easily accessible from Arba Minch, Hotels, lodges and resorts, were built to attract tourists who stop overnight along their way to the most popular tourist destination in Ethiopia, the Omo valley. The third important factor for the establishment of Arba Minch town was its location along the lucrative trade route, connecting the peripheral southwestern regions such as Jinka in South Omo to the central administrative and economic systems of the country (Arthur 2002).

In short, the establishment of Arba Minch State Farm and Arba Minch town had remarkable consequences in terms of altering the socio-economic and demographic settings in the area. Firstly, the two state projects brought the Ethiopian state close to the periphery, particularly to the Guji agro-pastoralists. Secondly, by enhancing a massive influx of people from Gamo highland and other parts of the country, these two state projects brought to the scene some demands for resources, which later happened to be included into the national park.

To sum up, I have discussed some background data regarding the settlement history of the people and their conceptions of human-environmental interactions. The chapter has brought to the forefront discussion that the Nech Sar National Park has become a field of contestation between the Guji, Koore and the park administration. Different historical references and claims of entitlement on all sides depict how social groups politicize environmental resources and employ different narratives and forms of power to achieve their interests or deter those of other rival actors. Moreover, the chapter has briefly positioned the discussion within the center-periphery relations, whereby the center encroached into the periphery through 'development' and 'conservation' discourses and practices.

The discussions on Guji and Koore local communities challenge the notion of homogenizing local communities because these two groups not only had different socio-economic engagements with their environment, but also were quite different in terms of their understandings of human-environmental interactions. While Guji's understandings of their surroundings was traditionally based upon values of respect, reciprocity and trust between humans and non-humans, the Koore largely shared similarities with the followers of Orthodox Christianity in their understandings of the place of humans in nature, as a creature with uncontested power over nature. The state on its part used historical notions of north-south dichotomy in representing the space as 'natural' in justifying the establishment of the national park. Therefore, it is these kinds of competing narratives and claims of entitlement that made the contestations over the Nech Sar National Park more complex than a mere struggle over environmental resources for livelihood purposes.

PART TWO: POLITICAL ECOLOGY OF NATURE CONSERVATION

Introduction

It has become evident that discourses and practices of environmental conservation constitute political, economic, ecological and ideological interests on the part of different actors at different levels. Since the 1980s, studies on environmental conservation entered a new phase of scholarship with political ecologists calling for the study of multidimensional aspects of the political, economic and ecological aspects in the study of conservation projects and a shift from the traditional criminalization of grassroots societies for environmental problems to a holistic and comprehensive study of actor-oriented approaches. In chapter two, I have discussed that the politics of environmental conservation entails the production of different narratives, knowledge and discourses by actors in the process of achieving their interests or in a pursuit of deterring other actors from the same environmental resources.

In this part, I will discuss the exportation of the national park ideals to Africa as part of the colonial project. Moreover, despite Ethiopia's unique history as a country without external colonial experience, the importation of the national park ideals to the country makes it to share much in common with other African states and deserves critical interrogation. More importantly, the birth of the modern Ethiopian state through internal conquest and annexation in the late nineteenth century has brought the construction of center-periphery relations between the northern conquerors and the subjugated south. This historical incident introduced a north-south dichotomy where the north was represented as the source of history, civilization and modernization whereas the south was portrayed as 'natural' and awaiting the hands of 'civilizers'. Based upon the clear analogy between the internal center-periphery relations in Ethiopia and the broader dominant-subordinate relations that were established following the colonial conquest in other parts of Africa, these subsequent chapters illuminate similarities and differences with particular emphasis to discourses of nature conservation in both cases. While chapter four deals with the exportation of national park ideals to Africa and how Ethiopia imported similar notions in the 1960s, chapter five exclusively situates the Nech Sar National Park within the post-1991 political reordering under ethnic federalism in Ethiopia.

CHAPTER FOUR: THE EXPORTATION OF NATIONAL PARK IDEALS TO AFRICA

Introduction

In chapter two, I have discussed the debates on nature-culture relations where I presented that the perceptions of wilderness conservation are politically and ontologically loaded inasmuch as they seem driven by environmental concerns. Drawing on the theoretical conception of political ecology of nature conservation, I have argued that any attempt to set aside land in the form of protected areas should be analyzed within underlying political, ecological, economic, cultural and power positions taking the state as one of the major actors. It also goes that the way people define, conceptualize and articulate environmental problems largely depends on their ways of understanding nature and the place of humans in it. However, the chapter did not address some fundamental questions regarding why and how the Western notions of wilderness conservation were exported to Africa during the colonial period and why despite the long awaited visions of independence, most post-colonial African states maintained and even strengthened the colonial legacies of nature conservation.

This chapter addresses these questions on two levels. First, it seeks to establish a nexus between Western views of nature and its exportation to Africa. Second, by discussing how nature conservation in the form of protected areas was used during the colonial period as a form of hegemonic environmental control, the chapter creates a link with the continuation of the colonial conservation model to postcolonial periods, thereby hinting on the emulation of colonial models by postcolonial states. In this regard, it seeks to interrogate changes and continuities that prevailed in the representation of nature in Africa during colonial and post-colonial periods. In addition to investigating the process of exportation of ideals of national park to Africa, discussions will also focus on the features of nature conservation in the continent during the colonial period to probe into the motives behind colonial powers in championing for the project. Such understanding also provides some reasons on why postcolonial states maintained and built upon power, discourses and practices. The chapter builds upon the theoretical arguments of political ecology of nature conservation, specifically contending that apolitical positions to nature conservation are naïve as long as conservation functions within contexts of asymmetrical power relations. In addition, it builds on Robbins' (2004) argument that modern environmental conservation discourses and practices particularly through protected areas are forms of hegemonic control of people and their territories.

In doing so, the chapter specifically posits the notions behind the establishment of the Nech Sar National Park within the wider discourses of representation of Africans and their territories as wild. However, readers who are cognizant of the history of Ethiopia as a country never colonized by Europeans might wonder how the colonial experience can be applied to Ethiopia's context. I would like to make two important points clear from the onset. First, Ethiopia had a history of emulating Western models of 'modernization' since the mid nineteenth century until present. As a result, its exposure to different models brought the country close to the colonial powers in shaping state-society relations (Clapham 2006a). In addition, independence does not mean isolation from global dynamics, given that different developments in neighboring African states and in the general world at large had their own influences on the country's political, economic and environmental policies. Dominant colonial ideologies, perspectives and ideas traveled beyond national borders notwithstanding actual colonial practices.

Second, and perhaps the hidden side of Ethiopian history, which this chapter will attempt to unveil, is that the formation of the modern Ethiopian empire shares much in common with colonial establishments in other parts of Africa (Donham 1986; Markakis 2011). This second point illuminates center-periphery relations that were commenced following the birth of the modern Ethiopian empire in the late nineteenth century and hints at how the state imported the notions of nature-society dichotomy and then internally 'exported' it to its peripheries. Therefore, the Ethiopian experience is important in illuminating how the national park ideals were appropriated by the state, irrespective of the absence of presence of colonial rule in place. The fact that no research has been conducted on Ethiopia within these broader understandings of conservation discourses to date makes this particular study specifically important in bringing a different scenario to the discussions. It also depicts some features of similarities and differences between Ethiopia and colonial Africa in general in terms of how notions of national parks were articulated and used. As a point of departure, the chapter interrogates the question whether the myths of 'natural' Africa and 'natural' south (in Ethiopia) were the major driving forces behind establishing national parks or whether these narratives were used to legitimize economic and political interests of the state in these two different but interrelated contexts. In order to draw some lines of similarities and differences between colonial Africa and imperial Ethiopia, the chapter pays specific attention to how both represented their respective subjects and the territories under their control.

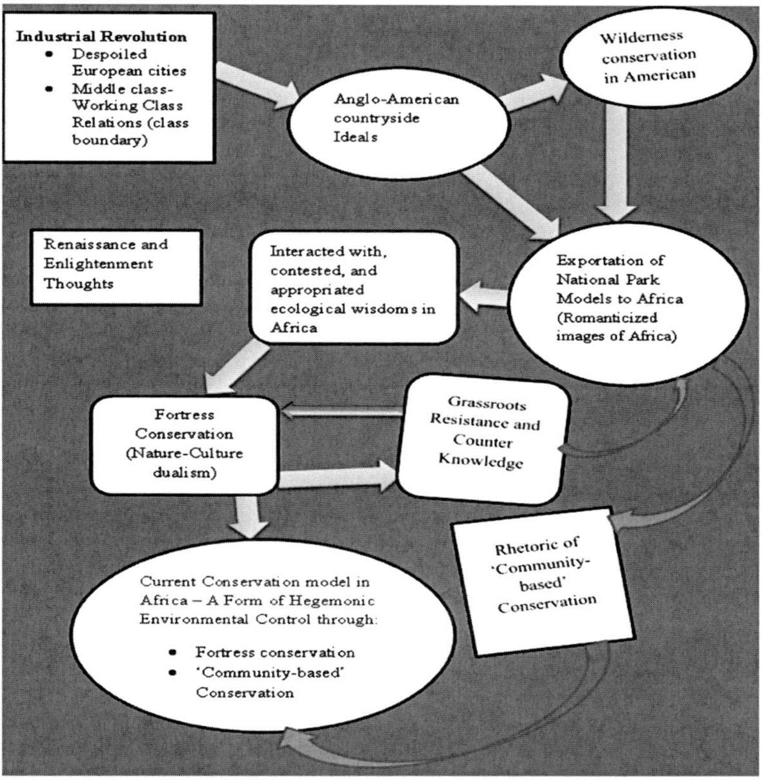

Chart 1: From Countryside Ideal to National Park Model in Africa (Source: my own graphic)

4.1 The Countryside Ideal and 'Wilderness' Conservation

In an attempt to create a nexus between the establishment of protected areas in colonial Africa in the form of 'wilderness' conservation and its origins in Europe, I will discuss the dominant conceptions of nature in European academic and popular thought. The discussion concerning nature-culture relations has been given due emphasis in chapter two. However, European conception of nature was partly nurtured by dichotomous views concerning city-countryside relations, whereby the city was symbolized with culture, while the countryside was considered as nature. Following industrial revolution in Western Europe in the eighteenth century, Europe experienced massive urbanization, urban pollution and despoiled industrial cities that continued throughout the nineteenth and twentieth centuries. Thus, the commencement in advanced industrialism and accelerated urbanization had enhanced city-countryside dichotomy not only in terms of spatial differentiations but also in areas of social, emotional, spiritual and class representations (Bunce 1994).

In his seminal work titled "*The Countryside Ideal: Anglo-American Images of Landscape*", Bunce (1994) illuminates the British and American images of countryside as ideal representations of human's harmony with nature that was partly developed as escapism from despoiled urban life and partly as a discursive marker of class boundary between the middle class and working class. In fact, there were differences between the European and American imaginations of countryside. In Europe, and particularly in England, the ideal was more of preserving the aesthetic values that the countryside would provide and it was used as an identity marker between the middle class and working class following the era of industrial revolution. The transition from rural agrarian society into urban life disrupted old social structures, instead redefining new social orders built upon class relations. For instance, industrialism introduced hierarchical social relations between the bourgeoisie and the working class in which work ethics had become defining regulatory factor in the relationship between different social classes (Bunce 1994). On the other hand, countryside ideal in America developed more as a marker of national identity because conquering and 'domesticating' 'wilderness' were expressions of nationalism and the triumph of science and technology over 'primitivism'.

In the context of emerging new social relations within the hierarchical structure, three broader factors enhanced the development of countryside ideal in Europe, which was later exported to North America and Africa, albeit in different contexts and for various missions. The first was associated with some sorts of nostalgia among the people who moved from rural life but missed the kind of life they used to enjoy in rural areas at least in terms of social relations, family ties, rituals, festivals and so forth. Second, the process of industrialization had created a middle class who gained social prestige by virtue of education, wealth and political power. This class sought to establish boundaries between themselves and the working class in terms of both social representation and spatial areas of settlement. Their economic and educational status and later the transformation in transportation schemes enabled the middle class to enjoy leisure time in the countryside, which the working class people were unable to afford. Third, the development of countryside ideal in Europe was linked to Renaissance and Enlightenment thinking about nature and the place of humans in it. Enlightenment transformed how people related themselves to nature because the old taken-for-granted conceptions of nature were being questioned by science and technology. Thus, Europeans began to equate human civilization with mastery over nature and the ability and capability of human beings in controlling, domesticating and changing nature into a consumable objects (Merchant 2003). In this regard, people began setting aside parts of territories in the countryside for the purposes of sporting, hunting, nature tour and vacation. Bargatzky (1994) also argues that the modern concept of nature in Western culture was built upon the eighteenth century transformations that shaped philosophical orientations of Europeans from religious dogmatism to scientific rationalization. Nevertheless, this last argument does not trace the genesis of nature-culture dichotomy to the Judeo-Christian traditions.

In the next few paragraphs, I will discuss how the trajectories in the development of countryside ideals specifically shaped the popular imagination among Europeans in conceiving country-city relations. Before delving into this discussion, however, it is important to make clear what the countryside ideal meant at least in the context of European and American imaginations. Bunce (1994) states that in the European context, the countryside ideal was a perceived or real representation of countryside as harmony with nature, strong social cohesion and quiet and spiritually intact social life where the connection between human and supernatural power was enhanced due to its sublime nature. While agreeing with Bunce in the above characterizations of the countryside ideal, I would further argue that the countryside ideal had both emotional and identity undertones. It represented people's sense of emotional attachment to nature and their longing for 'authentic' nature, which the European countryside could not provide. As a form of identity construction and boundary maintenance mechanism, the middle class people used the countryside ideal in their pursuit of self-representation as a higher-class vis-à-vis the working class.

Bunce describes the first factor as escapism from despoiled industrial cities where social relations, discipline, spirituality and work ethics were quite in contrast to the countryside. Although it becomes too simplistic to assume eighteenth and nineteenth centuries' social structure in Europe as egalitarian, social cohesion in the countryside was stronger than the type of social relationship in industrial cities (Bunce 1994). The clear contrasting features of the emerging urban life and its rural counterpart in Europe during that period were that the machine replaced the human aspect of interpersonal and inter-societal relations. People had to conform to a new work discipline, time frameworks, hierarchical relations between the proletariat and the bourgeoisie, limited cultural and social spaces of engagement and detachment from 'nature' that constrained their spiritual as well as emotional attachments to the land. 'Back to nature' took a form of social and cultural movements in Europe and North America later in the nineteenth and twentieth centuries as a form of environmental movement. Nonetheless, nostalgia concerning the countryside was partly a reaction against raising urban miseries, particularly in the case of working class, whereas it was partly a sign of leisure time escapism among the middle class. While the countryside ideal was initially exported from England to North America, it was adopted in accordance with the motives of the immigrants. Bunce (1994: 23) observes that:

> Most migrants to the New World in the seventeenth and eighteenth centuries were seeking new beginnings; escapists from religious intolerance and political repression, idealists and adventurers looking for new ways of living and opportunities to make it rich, above all people seeking freedom from the social hierarchies of Europe. These values had to be adapted to a quite different geographical setting and accommodated within the broader mercantile and political objectives of the colonizing powers. The pattern of rural settlement and land use which emerged from this emphasized the virtues of individual freedom.

The settlers in the Americas encountered new ecological understandings and new perceptions of its inspirational powers, completing them to view the natural world as salvation for alienating conditions of industrialism. Partly inspired by the new liberating condition of nature in the newly conquered regions of the world and partly nurtured by enlightenment philosophy, the settlers emphasized countryside-city dichotomies but opted for the domestication, control and commoditization of nature for the maximization of human leisure. In both the British and American cases, countryside ideals have been socially constructed as physical and social spaces with meanings ascribed to them. As I described earlier, it has been ascribed with representations such as harmony with nature, people in strong social cohesion, sublime and pastoral landscapes, as well as health and quietness.

The second aspect of the countryside ideal that I will later link to its exportation to Africa was that the middle class established lines of boundaries between themselves and the working class in the form of gentrification of residence patterns and by using landscape aesthetic as a form of class identity markers. In areas of spatial boundary formation, residence areas were sharply divided between the two classes, with all markers of social status and opportunities assigned to each settlement area in industrial cities. Access to education and other social services were differentially distributed among the middle class and working class settlement areas. Likewise, the rising middle class began to create another marker of class boundary by inculcating landscape aesthetic, countryside tours, summer vacation, mountain hiking and later nature tourism into its class culture. Backed by their class status in making use of wealth, social services and power, the middle class people were able to sharply construct a social-cultural and class boundary against the working class by subscribing to such practices, from which the 'other' was deprived.

Social groups establish boundaries of identification between their members and members of other group(s) by ascribing specific subjective and objective markers of categorization, which often constrain or enhance inter-group relationship (Barth 1969). Differences that are defining elements of group relations do not always constrain relationships but rather shape how members of the group perceive themselves and categorize the other. In this regard, the middle class built markers that put them at a higher-class status in relation to the working class and thereby defined the nature of class representation. Later during the colonial period, colonial officials and white settlers in Africa utilized this strategy of boundary formation, which postcolonial elites also adopted and used it as a mechanism of asserting control over the socio-political and economic spaces of the people. I will return to this discussion in the subsequent sections.

The third aspect in the development of countryside ideal mainly in the Anglo-American popular imagination was the notion of wilderness/nature domestication that was nurtured by enlightenment thinking that assumes nature as an object to be domesticated, controlled and consumed through science and technology. In chapter two, I have discussed the genesis of popular imagination of nature-culture dualism in Western tradition. The transformations in the representation of nature-culture dichotomy passed through different stages. It evolved from the Judeo-Christian tradition, within which nature, mainly represented as wilderness, was viewed as a desolate area where humans would enter without their will. Later, the enlightenment thinking transformed the perception to viewing wilderness as sublime, untamed and unspoiled territory that could be domesticated, controlled and consumed through science, technology, human rationality and capital (Merchant 2003).

Despite Bunce (1994), having treated the discussion of notions of nature in the British and American traditions within a common trend of development, these notions and the discursive representations of these notions were at times different. In the European context, particularly after the seventeenth and eighteenth centuries, positive attributes were assigned to the concepts as areas of sublime characters, harmony between humans and God (spirituality), unspoiled by influences of industrialism and so forth. Once they settled in the conquered regions which the Europeans labeled the 'New World', European settlers forged and redefined their relationship with nature in a way that placed them at a dominant position over what they labeled as 'wilderness'. This dualist and asymmetrical relationship between humans and non-humans had bestowed power upon humans to conquer, subdue and tame the non-human nature to use it for human needs (Merchant 2003). In all three different dimensions of countryside ideals discussed in this section, countryside-city dichotomies were constructed in similar ways as nature-culture dualism in Western imagination. While the countryside ideal was first exported to North America through European escapist settlers, it was later adopted by colonial administration and white settlers in Africa as a form of legitimating territorial appropriation, asserting control over the people, as well as maintaining Europe's 'modernity' by capitalizing on African 'traditionalism'. The following section will discuss how the model was exported to Africa, as well as how it was systematically adopted to fit the colonial contexts.

4.1.2 From Countryside Ideal to National Park Model in Africa

Since the era of colonial conquest, different development and conservation models, ideas and discourses have been experimented in Africa. In discussions concerning the success or failure of such experiments, points commonly missing are how discourses, models and ideals travel between different contexts and how such interplay between different ideas informs global and local discourses. Because the global is in the local and local developments and perspectives also inform global dynamics, it is essential not to treat the two as disentangled spatiality (Gezon 2006). In this section, I will briefly elaborate on how the national park models that were rooted in Western imagination of 'wilderness' conservation were exported to Africa during the colonial period and what underlying motives compelled colonial administrators and white settlers to introduce the model. The section also addresses how the interaction between different perspectives informed changes in value systems on both sides (Western and African actors).

During the early periods of contact between European explorers and Africans, the former were confronted with different ambivalence about the people and territories in the continent. Among intriguing competing encounters were their images of Africa and their experiences that rarely conformed to the popular imagination that most Europeans held about Africa at home. More specifically, from the onset of early contact to the heydays of colonial rule, Europeans' perception of Africa was a mixture of two different but not exclusive myths. These were the myth depicting Africa as "a land of pestilence and death and as a glorious garden of "Eden" in the interior" (Adams and McShane 1996: 6). Despite being seemingly contradictory, these perceptions were mutually complementary in the process of the conquest and later in shaping conservation policies in Africa. Whether it is used in policy articulation or narrated as a public discourse, the notion of wilderness is powerful because it implicitly constitutes the idea that "Man should subdue nature and that the world could be a man-made paradise" (Adams and McShane 1996: 6). Thus, both mythical representations of Africa were part of the driving forces for the conquest and later in shaping conservation policies in the continent.

According to Adams and McShane (1996), the establishment of protected areas and crisis related to nature conservation in Africa were rooted in how the West perceived the continent–its people and territories. In their book titled "*The Myth of Wild Africa*", Adams and McShane (1996) illuminate a Western imagination of Africa that was built upon a mythical invention of the land as the 'Dark Continent'. Based on scant information they got from early travelers and drawing on their own representation of Europe as a center of world civilization, Europeans invented a mythical Africa as wilderness. This soon claimed a place of privilege in the Western imagination of Africa. Such imagination prompted early explorers, missionaries, travelers and later colonial administrators to perceive and believe that there was a virgin Africa, although the continent had never been free from human influences. Later in the twentieth century, the role of media in portraying Africa as among few spots on the earth as the 'Edenic' nature has been paramount. For instance, channels such as "Nature", "Nova" and "National Geography" repeatedly feed their Western audience with the images of lions, zebras, giraffes, rhinos and elephants in Serengeti, Maassai Mara and Kruger national parks–the process that reduces diverse ecological and biodiversity in the continent only to the 'empty' and vast plain fields of Serengeti style (Adams and McShane 1996). On the other hand, Africa is indeed the continent in which the big predators have survived and it was sparsely populated at the time of the colonial conquest. This self-contradiction of Africa–a continent that is considered as the origin of human species, yet is also linked with the slow mastery of humans on nature in contrast to Europe and the United States–has contributed to the representation of the continent as 'wild' and 'natural'.

Although man has been integral part of African landscape, neither these media nor Western travelers recognized the presence of people in or adjacent to the areas later designated as national parks. For instance, to justify the myth of wilderness, European settlers, travelers, missionaries and explorers used the notion of 'the noble savage' thereby lamenting both the people and wildlife into similar categories. In other cases, humans were made invisible whereas 'nature', especially the wildlife has been magnified as a dominant image of the continent. In other cases, people were represented as cruel poachers, destroyers and threats to nature. For example, in the documentary titled "*Serengeti Darf Nicht Ster-*

ben"[8] or Serengeti Shall Not Die, the Maassai people who lived in and around Serengeti national park were brought to the scene only as poachers, while the spectacular wild beast occupied the full attention of the documentary. Thus European imagination of Africa as 'Edenic' garden and romanticized wilderness in opposition to the despoiled industrial metropolis in Europe underpins the historical development of national park ideal in colonial Africa (Anderson and Grove 1987; Neumann 1998).

Likewise, in his two volume books about historical reconstruction of encounters between British and Western African states from 1780-1850, Philip D. Curtin (1973) gives us a thorough discussion on how British image of Africa was constructed. According to the author, British image of Africa (its people, territory and wildlife) was informed partly by its contact with West Africa during the slave trade, 'legitimate' trade and missionary activities between the sixteenth and mid nineteenth centuries and partly by the mythical image of the continent based on 'common knowledge'. Despite increased degree of contact, the popular British image of Africa during the nineteenth century was a general portrayal of the continent as 'Dark Continent' devoid of 'civilization'.

According to Curtin (1973), drawing on the assumptions that civilization moved from Mesopotamia northward through cultural contact or diffusion, the British attributed African 'backwardness' to its isolation from centers of civilization. Although this assumption is wrong as ancient African civilizations were almost contemporary to Mesopotamian civilization and by far older than the European ones, the potency of the narrative in influencing British colonial policies in Africa should not be ignored. Thus, beginning from the era of 'legitimate' trade all throughout the colonial period, the British followed formal and informal policies of 'civilizing' the 'uncivilized' Africa mainly through cultural conversion in the form of Western education, Christianity and inculcating Western values in African socio-cultural life worlds (Curtin 1973). As it happened in North America during the eighteenth and nineteenth centuries, territorial conquest in Africa was also presented as a victory of 'civilization' (science, technology and capital) over 'barbarism' and 'backwardness'. It also signaled the triumph of nationalism in the sense that empire-building was partly informed by nationalist orientations inasmuch as it was driven by economic motives of the imperial regimes.

As part of colonial control and expropriation, Europeans imposed the notions of wilderness conservation upon Africa that they earlier exercised on the indigenous peoples in North America. In the African context, Europeans had a presumed image of how Africa should look like. However, who defines how a

[8] http://www.youtube.com/watch?v=8PDX_Ix6-rQ

certain place, territory and its natural environment should resemble? What does landscape vision mean for Africans and for European colonial settlers? The search for answers to these questions would take us to areas of power relations and ontological notions on nature-culture relations, and how these meanings and ideas were prescribed upon Africans from the Europeans' perspectives.

In Neumann's wordings, Europeans perceived that Africa should look like 'natural'–albeit this might again call for further discussion on what the 'natural' constitutes. In this particular case, however, the image of natural Africa was portrayed as wilderness where people were absent or if there existed, they were assumed to be close to nature, the landscape was aesthetic, sublime, and nature was presumed to be unspoiled by human influences (Neumann 1998). In this regard, the "Edenic" representation of African landscape had been an integral part of European imagination of the continent.

Nevertheless, one fundamental question worth addressing is the underlying motives of colonialists in introducing the national park ideals to Africa. There is no single and comprehensive answer as to why Europeans introduced the national park ideals to Africa, largely accompanying colonial administration. However, a set of possible arguments focus on the notions of aesthetic African landscapes ("Edenic" garden), conservation as a means of asserting control, protected areas as markers of class and racial identity and environmental concerns (Anderson and Grove 1987; Adams and McShane 1996). In the following sub-headings, I will attempt to discuss these prepositions.

4.1.2.1 The Myth of 'Edenic' Garden

The 'Edenic' discourse and the aesthetic dimension of landscape consumption have dominated discussions on national park ideals in Africa during the colonial period. According to this line of argument, Africa was among rare the places on earth where nature was believed to be unchanged by human influences. In other words, Africa could provide the 'authentic' Edenic features of nature that Europe lost due to advance in industrialism since the eighteenth century (Adams and McShane 1996; Neumann 1998). As I discussed in the previous section, landscape appreciation or landscape consumption (countryside ideal) began in Western Europe and the United States during the eighteenth and nineteenth centuries as a reaction against the despoiled life in industrial cities and as part of the class boundary. However contested the motives for the establishment of national parks in Western part of the United States might be, some would like us believe that the aesthetic aspects of nature and people's desire for emotional attachment to wilderness–back to nature–profoundly contributed to the exportation of the national parks ideal in the 'New World' (Neumann 1998).

For example, despite his thorough work on contestations between local people and the colonial and post-colonial states in Tanzania over questions of customary resource rights, Neumann (1998:17) contends that, "Whether we argue for the pastoral or for the sublime as the primary inspiration for national parks, aesthetics remain central to our understanding of what constitutes nature and its preservation." In fact, this perspective reduces the motives for the establishment of national parks as simplistic tourism attractions, which significantly blurs the inherent contradictions within the discourse of conservation. It is understandable that under capitalism, nature is packaged in the form of national parks and marketed through nature tourism or what some call 'eco-tourism'. In the process of commodification of nature and its marketing, human history and agency has been curiously denied as a means to magnify the wilderness or pristine dimensions of nature. In this regard, the historical narrative in European imperialism in relation to wilderness conquest and conservation denies the role of human hand in shaping landscape.

Under the myth of 'Edenic garden', there were two broader dimensions of motives for the establishment of national parks in areas annexed by Europeans since the eighteenth century. These include emotional and religious motives. As Bunce (1994) explicates, romantic views of nature resulted in the desire to tame, domesticate and control it. Moreover, the religious perspective that depicts natural environment as a salvation for the alienating conditions of industrialism, had strong potency in shaping the European perception of nature, particularly in the worlds they annexed. In this regard, Africa was thought to offer wild and natural environment. Nonetheless, the emotional dimension was a response and reaction to despoiled environment Europe has been facing because of industrialism since the eighteenth century. Particularly among the middle class elites, nostalgia to the countryside style encouraged them to re-discover a lost harmony with nature and natural environment (Anderson and Grove 1987). From this angle of argument, some scholars link the establishment of national parks in Africa to the desire for nature tourism as a sign of civilization on the part of colonial officials and white settlers. There is a continued push for nature marketing and commodification in Africa as some argue that the old world drew their strength and legitimacy from history and the 'New World' can likely invest in nature (Neumann 1998). Therefore, within the context of aesthetic dimension of nature as a driving force behind national park ideal in Africa, it is fair to argue that colonialists combined the economic and aesthetic values of nature to satisfy their economic interests through ecotourism and their emotional desire of 'back-to-nature'.

4.1.2.2 National Parks as Instruments of Control

Another possible explanation for the exportation of national park ideals to Africa was for its instrumental purposes enabling colonial authorities and settlers to exert control over the social, cultural and economic spaces of Africans. Through protectionist approaches of protected areas management, colonial powers controlled resourceful territories from which local communities used to draw different kinds of resources for their livelihood needs as well as cultural and spiritual purposes. As I discussed earlier in this chapter, "colonialism and the global expansion of European economic and cultural hegemony involved the transformation of vacant wilderness into productive fields, factories, and cities" (Neumann 1998: 29). Conventional conservation discourse is a form of hegemonic environmental control that puts resources, territories and the people who inhabit the territories under the control of powerful actors–the state and non-state actors (Robbins 2004). For the sake of understanding, I analyze the "conservation and control thesis" (Robbins 2004: 150) from three mutually interrelated areas of arguments. Broadly speaking, the notion of national parks was a form of asserting hegemonic political, economic, cultural and knowledge control and imposition over indigenous peoples in creating a system whereby the colonial powers and later post-colonial states have asserted their authority over the society.

In the context of the colonial system in Africa, the very idea of a national park is anchored in different dimensions of notions of power relations. The first dimension, as elsewhere in other parts of the world, is that notions of national parks have placed humans above other species to legitimize resource expropriation (Ramutsindela 2004). Through national parks, humans assert control and authority over other species by restricting their movements and putting them under human surveillance and exploitation. National parks put animals under human gaze, constrain their movement/migration, change their behavior and adjust them into objects of consumption. The second dimension concerns the idea that national parks create asymmetrical power relations between humans–those who have right of access into the parks and those who are restricted from access. As colonialism created racial and class boundaries, national parks also contributed to ongoing boundary construction processes because park legislations and policies privileged colonial officials and white settlers whereas native Africans were excluded from their traditional rights.

Indeed, it is very important to note the differences between the objectives and approaches in the establishment of national parks in the colonies of the two major colonial powers–the British and French. In British colonies, national parks were largely extended from the models of the Yellowstone national park as expressions of national identity, markers of class and racial boundaries between colonialists and the subjects and as forms of exclusive rights for colonial hunters. Tourism was also part of the objectives of setting aside land as protected areas in British colonies. On the other hand, the French loosely adopted the Yellowstone national park model and rather focused on forest reserves for preservation and scientific purposes (Ford 2012). In both cases however, eviction, relocation and restriction of indigenous communities from their ancestral lands were common practices.

At this point, I would argue that although the colonial powers did not always and not everywhere succeeded in imposing territorial rule, they attempted to implement territorial control as a mechanism of control over the economic, social and cultural spaces of Africans. This colonial project had multidimensional consequences. It disrupted human-environmental relations by dispossessing Africans, criminalizing customary practices of resource utilizations and detaching the people from their cultural and spiritual connectedness to their land. For example, among many pre-colonial societies in Africa hunting was practiced not only for subsistence but also for its symbolic representation in signaling local elites' dominance over the environment and as means of asserting territorial control (Adams and McShane 1996). Thus, the destruction of African hunting rights following the imposition of colonial hunting laws had disrupted economic and social relations as well as broken up the bonds between people and the natural environment. As Ingold (2000) states, individuals' engagements with the natural environment enable them to socialize with environmental elements because through hunting, people would interact, communicate and develop sense of trust and reciprocity with the animals and the environment at large. By disrupting African customary rights, the colonial system had transferred the right of resource ownership and utilization to a new emerging class–the white colonial ruling elites and white settlers–who happened to become at a privileged position to exercise power over the people and their land.

In contrast to the general portrayal of the struggle between natives and colonists during the colonial encounters and all throughout the colonial era as if it was a mere competition over control of resources, there are evident cases that most conflicts were mainly results of ontological contradiction on human-environmental relations (Mitchell 2012). However, simplifying the struggle to resource competition blurs the complex ideological and cultural contradictions between colonial powers and the Africans. Rather, the struggle revolves around control over space, territories and meanings inscribed in elements of nature (Brownwell and Falola 2012). It was a struggle between forces of domination and resistance. By setting aside large tracts of land such as the Kruger national park in South Africa, Serengeti and Ngorongoro national parks in Tanzania and Maassai Mara in Kenya, the British exerted domination under the guise of nature conservation. Restricting the natives from access to their traditional land was meant to make the people dependent on colonial administration and prompted them to engage in labor economy or cash crop production for their livelihood sustenance. In other words, control over economic schemes, cultural spaces, sacred sites and socio-political affairs of the people are tantamount to getting the monopoly over their entire life. The third dimension of national parks in Africa was related to its representation of nationalism and control over nature. Like the conquest of the American West, which was described as part of nation-building, national parks in Africa were also strongly related to sense of nationalism on the part of the colonial powers (Brockington et al. 2008). I would reiterate that despite differences between settler colonies and extraction colonies in terms of their the objectives and approaches in the establishment of national parks, such territorialization, compartmentalization and enclosure of land entailed the restriction of indigenous peoples from their customary resources. Whether they were established for the purposes of tourism, game hunting or scientific research, preservation of fauna and flora, national parks in Africa enabled the colonial powers to exert their hegemonic forms of colonial rule over the subjects.

According to Mitchell (2012: 43), the establishment of farming settlements by the Dutch speaking settlers in southern Africa in the seventeenth and eighteenth centuries had disconnected the Khoisan, not only in terms of access to material resources (water, grazing land and hunting grounds), but also detached them from their sacred ritual spaces. For the indigenous communities, cutting off access to their ritual spaces was "comparable to capturing a church, synagogue, or mosque". Mitchell (2012: 43) argues that, "the fundamental tensions between Khoisan and settlers were rooted in different ideologies of nature and conflicting notions about the utility of natural resources." Despite direct competition for land, the intensity of the violence was fueled more due to competing meanings ascribed to nature by the two societies as it was to the struggle over the material aspects of the land. Mitchell (2012: 46) further argues that the "Settlers' ideas about nature, people's place in the landscape and the appropriate use of resources stem ultimately from a Christian logic."

In terms of their relationship to the land, the settlers viewed it from a material perspective as an object to be owned, controlled, exchanged and commoditized. In contrast, the Khoisan inhabitants of the region understand nature as inextricably linked to humans. Africans' dependence on their natural environment for food, rituals and other aspects of their life entrenches their connectedness to nature. Therefore, the difference between colonists and Africans in their relation to the natural world was a fundamental point of contention during the colonial encounters. As will be discussed later in this chapter, the inheritance of similar dualist notion about nature-society relation by post-colonial states had also contributed to the continuation of the conflict.

4.1.2.3 National Parks as Markers of Identity

The third argument behind the motives for the establishment of protected areas, specifically national parks in British colonial Africa was related to its role in serving as maintaining class and racial boundaries of difference between colonists and native Africans. This dimension of national parks was mainly true in settler colonies. We have seen earlier that the countryside ideal (nature tourism, mountain trekking, hiking, vacation retreats to countryside and in general landscape consumption) was used among the middle class people in the eighteenth and nineteenth centuries Europe as markers of class boundary (Bunce 1994). Likewise, designating land in the form of protected areas created lines of categorization between colonialists and Africans but in this context drawn based upon racial and class membership and assigned differential rights and obligations on access, utilization and ownership of resources. Boundaries between Europeans and native Africans were subjectively constructed partly drawn from the images of Africa among European popular imagination and partly as a result of early encounters which provided the European audience scanty information about the people and the land at large. For instance, British colonial officials and missionaries constructed a sharp racial boundary between themselves and native Africans drawing the lines based upon civilization. In some cases, the people, territories and wildlife were commonly lumped together as "nature" in contrast to the civilized "self", which was represented as culture (Curtin 1973).

To elaborate further on how sense of nationalism and identity construction informed control over spaces in colonies, colonial nationalists have been intent to annex the homelands of others in their identity myths as mechanisms of reconstructing imperial national identities (Daniels 1993). Images, landscape iconographies, myths and narratives of landscape transmit sense of human history and thus contribute to production of national identity. At the same time, landscape iconography depicts boundaries between human and nature, as well as between different human groups (Ramutsindela 2004). For example, in the Yellowstone national park case, the landscape painting of the park partly depicts American national identity but it masks the history of indigenous peoples in the region. By annexing the homelands of others, colonial settlers worked hard to deconstruct the attachment of the natives to the landscape and sought to inscribe their own identity myth. The example of the conquest of the American West and the establishment of the Yellowstone national park is always illustrative because it has become a founding model for most national parks across the world. The conquest and the establishment of the park was symbolized in

American national history as a process of victory of 'civilization' over 'backwardness', which implicates indigenous people and their home as adversaries of civilization against which the conquerors waged relentless war (Brockington et al. 2008). In fact, national parks in Europe were more expressions of the aesthetic and pristine dimensions of nature than those in the United States, which were considered as symbols of national identity. Nevertheless, once they exported the national parks ideals to Africa, European colonial powers, and particularly the British, adopted the models from the United States in depicting them more as expressions of national identity and mastery over nature.

As I noted in this work, apart from being symbols of national history, national parks were also symbolized with victory of civilization over wilderness (Reichart-Burikukiye 2012). Backed by science and technology and inspired by capitalism and enlightenment philosophies of human dominion over nature, Europeans annexed the lands of indigenous peoples since the eighteenth century. For instance, in the late nineteenth century when a British company was constructing a railway in East Africa, one of the challenges railway workers faced was from lions that almost compelled the company to stop the construction. The British company and its Indian workers were in fierce struggle with man-eating lions that contested against the new technology that would inevitably threaten its territorial privacy and autonomy. In 1898, an Irish engineer, John Henry Patterson, killed two of the lions and was celebrated as a hero marking the victory of European technology over African wilderness (Reichart-Burikukiye 2012). Similarly, some of the national parks in British colonies were established as markers of British colonial victory in Africa, marking important landscapes in British colonial history (Ramutsindela 2004).

In this regard, park legislation and polices have been serving as instruments for administering the boundaries between parks and people as well as between different categories of people. Landscape enclosure in the form of protected areas were used as strategic impulses to draw lines or boundaries between areas accessible to Africans and those exclusively reserved to colonialists. It was a marker of power relations in which aspects of domination and subordination manifested which in turn implies asymmetrical class and racial relations (Neumann 1998). Hunting rights were best cases in point where clear boundaries were made between Africans and whites regarding access to wildlife. Although most national parks in Africa, and indeed elsewhere, were presumably established for conservation reasons, European hunters negotiated with their colonial officials and obtained exclusive hunting rights. In some British southern African colonies for example, colonial hunters almost wiped out wildlife through hunting rights exclusively given to them (Ramutsindela 2004). Besides differential access to resources in economic terms, aspects of exclusion were parts of mental colonization of the colonial subjects aimed at producing a mental set up of generations who would accept the established 'regime of truth'. The institutionalization of hunting right particularly in British settler colonies was adopted from the old tradition of hunting privileges reserved to the Crown that marked the king's political and social status in Great Britain (Gissibl et al. 2012).

Regarding representation of Africa as 'wild' and 'natural' in Western media, academics and public discourses, I would like to make the following arguments. first, these representations were not innocent descriptions of African people, territories and natural resources, especially the wildlife but they were/are used as psychological and ideological tools of expressing the 'civilized' status of Europe vis-à-vis 'traditional' Africa. In addition to landscape consumption, I would argue that the representation of Africa as 'natural' and 'wild' had psychological dimension of empowering and maintaining Europe's 'modernity' vis-à-vis Africa's 'traditionalism'. By establishing protected areas from which they either evicted or restricted natives from drawing resources for their economic and cultural purposes, the colonial authorities used national parks as markers of identity between themselves and the native people.

4.1.2.4 The 'Environmental Degradation' Thesis

Environmental issues were the fourth major arguments behind the exportation of national park ideals to Africa. Colonial and post-colonial states have brought discourses of environmental degradation, climate change and biodiversity depletion into policy agendas (Anderson and Grove 1987). It has become evident that international donors, environmentalists, scholars, politicians, civil society organizations and local people in Africa are all worried of changing environmental conditions and increase in adverse impacts of climate change (Adams and McShane 1996; Brockington et al. 2008). In Africa, as elsewhere across the globe, environmental problems are now evident and campaign to address these issues, although the approaches differ, are considered legitimate. However, one might wonder if the establishment of national parks during the early colonial period was genuinely based on environmental concerns. Positing the motivations of colonial conservation programs within the argument of environmental concerns or 'environmental degradation' thesis, Grove (1990: 17) argues that:

> One can now begin to expose the extraordinary vigor with which conservation programmes, in particular, were pursued after the early nineteenth century, and to understand the technical agendas and powerful motivations behind these programmes. Principal among these motives was a deep insecurity about the prospects for the long-term survival of the colonial state and a deep anxiety about conse-

quences of climatic change and environmental deterioration.

Although I partially agree with Grove's argument that deep insecurity among Europeans about the prospects for the survival of the colonial state could be a motivation for territorial control under the guise of conservation, the notion that Europeans were compelled by consequences of climatic change and environmental deterioration should be scrutinized. It is true that colonial powers imported their practices of private property ownership and management such as land, forest, pasture and water to Africa, and attempted to designate communal lands into private or state controlled system during the colonial period upon which post-colonial states effectively modeled (Grove 1990). Nonetheless, resource scarcity and environmental deterioration were not as such major problems in the nineteenth century. As Mitchell (2012) argues, resource scarcity was not the major source of conflict between the Khoisan indigenous people and white settlers in South Africa in the eighteenth and nineteenth centuries. Rather, according to Mitchell, the conflict was mainly a struggle over meanings of landscape and differences in understanding of property ownership as the settlers imposed private property regimes that were alien to the indigenous peoples. I also argue that most states and non-state actors used conservation as rhetorical tool to justify and legitimize state stewardship of the environment.

In the context of asymmetrical power relations that had already been constructed following the onset of colonial conquest, state conservation discourses were put in place by relegating customary rights. In doing so, colonial states were able to delegitimize indigenous knowledge of resource management and instituted scientific and 'expert-based' environmental knowledge as inalienable form of knowledge. To justify the rationale for the new form of conservation, they emphasized environmental problems and inadequacy or inefficiency of traditional African resource management systems. On the other hand, the colonial powers sought to assess and use indigenous knowledge and law by establishing dual regulatory system, even if this dual system was believed to enhance domination of the colonial forms of administration. The central point to note at this point is that it was colonialism by itself that intensified environmental degradation through commercial plantation, timber production, mining, hunting and infrastructure construction, all of which were part of colonial exploitation and suppression (Asebe 2011).

If fact, it should be acknowledged that Africans took part in practices that were environmentally destructive such as the aforementioned activities mainly introduced by the colonial system. For example, industrial revolution in Europe had increased the demand for raw materials, which made Africa the ideal source of such resources. In the process, Africans were engaged in trade, plantation, mining and logging within the system established by the colonial system. In this regard, the increased demand for wildlife products in Europe and the United States exacerbated wildlife hunting on the side of African and non-African hunters (Rich 2012). Nevertheless, 'conservationists' had put the blame on African hunters and traders, although the system that increased the scale of hunting and trade in wildlife was masked under the stereotypical representation of Africans as poachers.

Nature conservation in Africa was also part of Western representation of the continent as "Eden" that demands human stewardship. As Rich (2012) discusses in her study about Chimpanzee trade in Gabon, conservation discourses in central Africa were basically framed on Western imagination of the region in which apes and monkeys are central in the construction of 'wild Africa'. Throughout the colonial and post-colonial periods, debates on conservation of wildlife in the region hinge on symbolic representation of these animals as nature that needs human stewardship. Since the late nineteenth century, local Gabonese hunters and merchants were engaged in primate trades with European merchants along the Western coast of Africa. Although the local people were encouraged by their economic returns obtained from the chimpanzee and gorilla trade, Europeans later began to label the hunters as cruel, despite the trade being hitherto considered 'legitimate'. Some argue that 'rational' men began feeling paternalistic attitudes to protect the innocent animals from 'backward' and 'cruel' African treatment of the animals (Rich 2012). Wildlife conservation in the colonial context thus reflects discourses of European paternalism versus African 'irrationality' towards wildlife. As Rich (2012: 37) remarks: "Discourses of kind but powerless primates who required European masters matched the paternalism voiced by administrators and missionaries" in contrast to stereotypical representation of African hunters and traders as cruel and inhuman to the animals.

My argument is that notions of nature conservation in colonial and post-colonial Africa should be conceptualized within the complex economic, political and cultural matrix in play. Sense of insecurity to environmental deterioration, threat to wildlife through African 'poaching' practices, 'unwise' use of resources, 'inappropriateness' of indigenous knowledge to modern conservation visions and so forth were used by colonial officials to legitimize the dominant conservation practices that were alien to Africans. It has been portrayed in the broader conservation discourses that conservation of the African environment is an entirely apolitical affair. However, due to the history of conservation trajectories in the continent, which has been full of agonies and tragedies of eviction, the criminalization of customary rights, denigration of indigenous knowledge systems and restriction of cultural and economic spaces of African people, backed by colonial and post-colonial state policies, "naïve and idealized apolitical conservation in Africa is absurd" (Grove 1987: 6). Rather, environmental issues were brought to the front lines of colonial policies to mask the economic, political and cultural motives and legitimize the imposition of scientific knowledge and notions of private ownership and state control on indigenous knowledge and communal resource ownership of Africans, respectively.

In this section, I have attempted to see the trajectories in the exportation of national park ideals to Africa during the colonial period. The countryside ideal that developed in Britain and later spread to North America could be understood from both landscape consumption and identity reconstruction. By offering opportunities for the middle class people to pass their leisure time in the countryside, the countryside ideal was believed to have put them 'back to home'/nature, albeit it was difficult to claim that the countryside was 'natural' as it was presumed. Likewise, the middle class people also used the countryside ideal in their pursuit to construct spatial and social/class boundaries against the working class. When similar notions were exported to North America, they were first used by European settlers as liberation from European social hierarchy, political suppression and despoiled life in industrial cities. However, the conquest of the American West was later transformed into a national iconography of victory of science, technology and capitalism over wilderness and 'savage' people (Brockington et al 2008).

Despite diversities in arguments about the motives of Europeans in introducing the ideals of the national park to Africa, I would argue that national parks were forms of hegemonic control over African people and their cultural, economic and political spaces. It was a form of imposing colonial authority, scientific environmental knowledge and notions of private and state control of land, which were in sharp contrast to indigenous cosmologies of people-nature relations. The irony with conservation in the form of national parks in Africa is that despite its multidimensional social consequences in the form of eviction, restriction of customary rights and denigration of indigenous knowledge systems, post-colonial states modeled and even strengthened the protectionist conservation approach. The questions I will briefly discuss in the following few paragraphs are: why and how do post-colonial states maintained the dominant colonial discourse of nature conservation that excluded Africans from their customary land rights?

4.2 Colonial Legacies and Post-colonial Dynamics in Conservation Discourses

Nature conservation in post-colonial Africa is subject to different kinds of arguments and debates. Due to the complex historical trajectories most African states underwent during and after the colonial era, it has become difficult to discern underlying forces that compel African states to build upon the colonial approaches of national park ideals. While some scholars position the modern forms of nature conservation as colonial and neocolonial impositions (Anderson and Grove 1987; Neumann 1998), others position particularly wildlife conservation in Africa beyond the colonial and neocolonial influences. In this regard, Garland 2008: 52) argues that conservation discourses and practices in Africa should be viewed, "as a productive process, a means of appropriating the value of African nature, and of transforming it into capital with the capacity to circulate and generate further value at the global level." Indeed, Garland (2008) argues that despite rich academic works on conservation discourses in Africa from colonial and post-colonial experiences, much of them emphasize two aspects of conservation. These include the impacts of Western conservation approaches on local livelihoods and the sensibilities and ambitions of Western conservationists, and to a lesser extent on Western conceptions of nature and its nexus with national park ideals in Africa.

However, according to Garland, emphasis on both these dimensions of conservation blurs our understandings of the interplay between Western notions of nature and local cosmologies in Africa, as well as how the two negotiate, contest and perhaps appropriate different values from one another. This is why Garland (2008: 61) observes in the following concerning the works of Western conservationists on the impacts of conservation and ambitions:

> Left out are the shifts in desires and worldviews, new social identities and relationships, and forms of discourse and practice that conservation processes themselves bring into being, not only at the relatively localized level of rural communities displaced by parks, but also in the broader frames of reference represented by nations and the world system. Thinking of conservation as a colonialist assault on the cultures or subsistence strategies of local people cannot, for example, explain the enthusiasm of many independence-era African nationalists for protecting wildlife, or the appeal of work in the fields of wildlife conservation and tourism for both rural and urban youth in many African countries today. Nor can such a framework address the kinds of global power dynamics.

I share Garland's views, particularly concerning the need to move beyond the idea that considers national parks as colonial and neocolonial imposition to conceptualization of the perspective as a complex phenomenon that constitutes new forms of practice and discourse that emerge from the interplay between European and African notions of nature. However, I further argue that conservation in post-colonial Africa should be analyzed within three broader contexts: colonial trends, post-colonial dynamics, and the subsequent results of both contexts. Therefore, local and global forces such as legacies of colonialism, issues of nation-building projects, influences of neoliberal economy, global environmental movements and emerging value systems among Africans, are some of the fundamental driving forces behind the continuation of colonial approaches of national parks in post-colonial periods.

4.2.1 National Parks as Colonial Legacies

Given that forms of colonial legacies continued to shape postcolonial economic and political orders in Africa, conservation discourses can be positioned within the broader institutional and ideological frameworks the newly independent African states inherited from the ex-colonial powers in the 1960s and 1970s. African elites did not rebuild the states as per the pre-colonial socio-political, geographical and cultural settings. Rather, they forged the state on the active ruins of colonial institutions, ideas and worldviews. In the process of crafting 'nation-states' in the postcolonial eras African elites who were one way or another shaped by the very colonial system and ideas they fought had played central roles. Despite African nationalism that flared up against injustices perpetuated by the colonial system, independence failed to dismantle colonial institutions that sustained the system. By introducing notions of scientific knowledge of conservation that legitimated European views of nature-culture relations on the one hand, as well as denigrating indigenous cosmologies and knowledge of resource management on the other, colonial system established 'regimes of truth' that were gradually institutionalized through scientific research, colonial policies and bureaucracy, public imaginings, narratives and rhetoric. As with many other institutions and ideas, post-colonial elites inherited these 'regimes of truth' and in fact used them within the new political and economic settings and interests. National parks and the ideals behind their establishment were part of colonial institutions and ideas that were passed-over to the newly 'independent' states and the elites that took state power in Africa.

I would also like to establish the nexus between colonization of Africans' minds and the subsequent adoption of the colonial discourses of conservation by the postcolonial elites. In contrast to Fanon's (1952 [1986]) aspiration of decolonization of the mind along with political decolonization, the political decolonization did not significantly result in the decolonization of the mind among most African elites, despite many attempts towards the re-traditionalization of African institutions (Mazrui 2002). The colonization of the mind entails institutionalization of different forms of knowledge, values, beliefs and ideologies of the colonizer into the cognitive set up of the colonized. In the process of colonization of the mind, colonizers used different forms of power to transmit their values, norms and beliefs to the colonized by denigrating the knowledge and worldviews of the latter or accompanied by socio-economic rewarding or punishment. However, punishment and coercion alone would not successfully engender acceptance of the colonizers' knowledge or ways of thinking. Rather, subtle but persuasive mechanisms had to accompany the coercive mechanisms. For example, one might ask why African nationalists emulated British and French styles of parliamentarian and presidential systems of government upon independence, despite having fought against the institutions set up by these colonial masters. Some might respond that the nationalists were not given any chance or freedom to choose between different options as the departing colonial powers imposed some conditions, such as the continuation of 'partnership'. However, this argument is not convincing because history reminds us about some nationalist leaders and heads of states who defected from the so-called post-colonial 'partnership' and adopted socialist system.

Rather, the argument might go to the direction that most nationalists were products of colonial education and were brought up under the colonization process that controlled the mind settings of the people at large. They were educated either in Europe, USA or even in Africa under education systems crafted upon colonial curriculum. The education system was designed in a way that Africans would be socialized to accept the presumption of considering Western democracy and social ordering as utmost superior level of civilization than indigenous African governance and practices. It also represented European development trajectories as the right paths to be transplanted to developing countries and in general Western scientific knowledge as the ultimate 'truth'. Do we expect these colonial 'products'/elites while drafting the new constitutions, to sit together and revisit how the Ashanti, the Kikuyu and the Zulu, for example, governed their people before the white men arrived? Mazrui (2002: 19) describes this situation stating that, "Institutions were inaugurated without reference to cultural compatibilities, while new processes were introduced without respect for continuities. Ancestral standards of property, propriety and legitimacy were ignored." Likewise, postcolonial elites institutionalized colonial forms of resource management without looking back to their ancestral ecological wisdom. While this does not completely nullify the agency of African elites in articulating ideas, policies and ideologies that suited their interests, it underpins the conjuring role that colonial legacies played in the elites' understanding of the complex political, economic and cultural circumstances in the continent during the transition from colonialism to independence.

The issue of nation-building project is closely related to the colonial legacies, albeit with some forms of new internal dynamics that has been shaping discourses of conservation in post-colonial era. Like their colonial predecessors, the leaders of the newly independent African states had to struggle to gain legitimacy, authority and state power during the 1960s and 1970s (Herbst 2000). Resistance against the way in which colonial states were crafted and inherited, influences of ex-colonial masters, decaying social and economic systems, imported political and economic systems and so forth were among few of the challenges African states and leaders have been facing since independence. In order to assert control over the people and territories, as well as gaining legitimacy, governments continued similar notions of discrediting local resource use systems and institutionalizing the imported approaches of resource management as the ultimate conservation practices. In this regard, post-colonial governments not only adopted and modeled the notions of national parks but also continued with massive expansion of national parks in most countries since the colonial era (Neumann 1998; Brockington et al. 2008). Apart from its economic reasons, to which I will shortly return, the establishment of national parks was associated with nation-building projects in which nation states embarked on subjugating the homelands of indigenous peoples as a form of asserting state control.

It should be noted that most African states have rejected the call for recognition of indigenous peoples and their claim for land ownership, cultural distinctiveness, political autonomy and economic empowerment (Saugestad 2001; Asebe 2011). While questions of indigenous peoples for recognition, self-government and access to and ownership of resources were portrayed by the nation states as threats to national integrity and sovereignty of the states, actions against these groups were justified as part of nation-building projects (Stewart-Harawira 2005).

4.2.2 Global Influences

The second post-colonial pattern in the analysis of why national parks continued beyond the life span of colonial rule is associated to the influence of global economic institutions and ideologies. Since the 1980s, the influences of neoliberal economic paradigm in Africa were evident particularly in the form of Structural Adjustment Programs (SAP) through which powerful financial institutions such as the World Bank and International Monetary Fund exerted influence on economic and political policies in Africa. As part of SAP, these institutions prompted African states to diversify their economy presenting tourism as one of the major areas of possible investment. In this regard, while the 'Old World' countries rely on heritage, African countries were thought to invest in nature for their tourism economy. Investment in nature for tourism economy entails the establishment of national parks, which often excludes people from their homelands and dispossesses them from their customary resource rights. Despite Africa being a home of charismatic animals, which would indeed boost the tourism industry, the approach used by global institutions such as the IUCN, World Bank and UNESCO in encouraging African states to invest on national parks was paternalistic.

The globalization of environmental movements and the role of conservationist NGOs, individuals and institutions in influencing international and national policy making processes have also been paramount in the post-1960s periods. Like the role of science in constructing 'truth' about nature-culture relations and conservation practices, these major international actors such as IUCN and UNESCO were able to mobilize resources and invest in discursive representations and narratives of 'danger' related to environmental change in their quest to craft international policies and programs along their own approaches of conservation. Owing to their strong economic positions to sponsor academic works and lobby governmental and other policymaking institutions, the influence of these actors in shaping global imaginations and policies has been strong (Garland 2008). For instance, the National Geographic Society and the World Conservation Society played fundamental roles in convincing the Gabonese President, Omar Bongo, to establish thirteen national parks in Gabon in 2002 and in the process lobbied the United States to donate a significant amount of money for the same purpose (Garland 2008).

Likewise, Wilshusen et al. (2002) argue that global conservationist organizations have begun working against the community-based natural resource conservation paradigm in developing countries, because such institutions criticize these approaches for failing to achieve the goals of biodiversity conservation. As the case in Ethiopia demonstrates, UNESCO sponsored the establishment of several national parks in the 1960s and 1970s along the classical protectionist approaches of protected areas governance. Similarly, the World Bank, IMF, IUCN and other international foundations working on 'development' and 'conservation' invested millions of dollars on the development of tourism industry in Tanzania in the post-1960s periods (Neumann 1995). The global political and economic ideologies of market liberalization that imposed different sorts of conditionality on African states had pleased international conservation NGOs by compelling the states to liberalize the market, including the privatization of land.

4.2.3 Changes in Value Systems among African Elites

In the preceding paragraphs, I have discussed how the bogus ambitions of 'nation-building' project, neoliberal economy and global environmental movements have contributed to the continuity and even expansion of national parks in Africa. However, these three arguments seem to have given the agency to external forces and actors but portray African states and the people as passive recipients of external ideas and interventions. Nevertheless, African states, elites and ordinary people do play significant roles in conservation practices in Africa (Garland 2008). Garland observes that behind any successful conservation story in Africa are hundreds if not thousands of Africans who work day and night within the conventional conservation approaches, yet whose contribution towards the outcome has been neglected or consciously made invisible.

Indeed, Africa is a home of different charismatic animals and rich diversities of wildlife–lion and leopard, gorilla and chimpanzee, elephant, rhino, hippo, ostrich, zebra, giraffe, and more (McShane 1996; Garland 2008). The question will be who sustained the African wildlife? This dimension of the question is often ignored from most media coverage and academic discussions. The answer actually depends on different views. While some might applaud Western conservationists and donors who shower millions of dollars for conservation purposes, others might give credit to African elites who have enacted conservation policies and practices in line with global approaches. I doubt if mainstream academics and environmentalists recognize the role of ordinary African people and their knowledge in maintaining biodiversity in the continent. Nevertheless, it has been documented that most African societies had different aspects of ecological wisdom and systems of environmental governance that enabled them for centuries to maintain human-environmental coexistence (Igoe 2004). Conversely, different local epistemologies in Africa influenced the dominant protectionist discourse to undergo some fundamental paradigm shifts–a recognition and adoption of community-based conservation approaches during the 1980s and 1990s. Although debates on this kind of knowledge are significant, my argument in this section is different. It has now become clear that colonial encounter–for better or worse–introduced shifts in value system among African elites and ordinary citizens, in respect of their values to wildlife.

In the case of changes in value system among African elites, it is worth mentioning that wildlife began to possess economic values like minerals, oil and other natural resources (Garland 2008). The following case will further substantiate the argument that colonial system and post-colonial dynamics have introduced some changes among African elites regarding their views to conservation discourses. In 1961, the first Tanzanian president, Julius Nyerere, was invited to deliver a speech on a symposium organized in Arusha, Tanganyika, by the IUCN. On the symposium, Nyerere's speech attracted the attention of conservationists because not only he declared his government's readiness to implement conservation programs, but he also confirmed to the audience that conservation would contribute to the well-being of the future generation. The president remarked, "In accepting the trusteeship of our wildlife we solemnly declare that we will do everything in our power to make sure that our children's grandchildren will be able to enjoy this rich and precious inheritance" (Quoted in Neumann 1998: 40). Nyerere said, "The conservation of wildlife and wild places calls for *specialist knowledge, trained manpower and money and we look to other nations to cooperate in this important task*" (Neumann 1998:40, emphasis added). As a brainchild of colonial education, being the first Tanganyika to receive education in British University, it is not difficult to imagine that Nyerere was influenced by the 'truth regimes' constructed and implanted during the colonial era that made scientific knowledge, expert skills and Western values the superior forms of knowledge and legitimate approaches of conservation. Thus, colonial the encounter had impelled African elites to go through value changes in perceiving wildlife and nature in general as commodities.

It is again imperative to bring how Nyerere labeled wildlife as commodities putting them into similar economic packages of exchange like diamond and sisal when he noted, "I personally am not very interested in animals. I do not want to spend my holidays watching crocodiles. Nevertheless, I am entirely in favor of their survival. *I believe that after diamonds and sisal, wild animals will provide Tanganyika with its greatest source of income.* Thousands of Americans and Europeans have the strange urge to see these animals" (quoted in Neumann 1998: 44, emphasis added). In this context, Nyerere was in favor of the survival of wild animals for economic purposes in the form of tourism. This is consistent with Garland (2008) arguing that wildlife conservation practices in post-colonial Africa should be conceptualized as a productive process–a shift in value system whereby wildlife is seen as economic capital. Whether it is African elite with political position or those employed in conservation organizations, at least the economic values attached to wildlife have now being accepted in Africa. One would argue that Africans' engagement with external forces and actors have transformed their value system and perception of wildlife and conservation practices over time.

In addition to its effect in transforming the value system among Africans towards wildlife, colonial encounter and post-colonial political circumstances in Africa have shaped the elites, particularly governments, to associate state control over resources with sovereignty. This means that while colonial governments used national parks as symbolic markers of identity, control, power/knowledge and practical fields of economic exploitation, post-colonial governments linked the authority of setting aside tracts of land as signs of sovereignty, which is often contested for several reasons. In other words, post-colonial leaders appropriated the national park ideals and systematically used it within post-colonial political circumstances. Besides their economic attraction, national parks in most African countries have been established in peripheral regions or along borderlands as security zones. For example, the Ethiopian governments established most of the national parks along contested inter-ethnic boundaries mostly in pastoralist areas and used such territorialization to control the movement of people from place to place.

I would like to move further what Garland (2008: 62) describes, "The recent efforts of many African rural communities to assert their right to exploit wildlife through community-based conservation and tourism projects, often in the face of opposition from governments claiming state ownership of wildlife, may be understood as struggles over resource control in precisely these terms". I argue that the struggle between local communities and states over territories and resources are manifestations of the formers' resistance against the legitimacy of the state in asserting control over resources and the state's claim and struggle to gain dominance over the society. Like other kinds of natural resources on which the state claims control such as oil, gas and precious minerals, wildlife in Africa have also become areas of state interest. This means that state initiatives of wildlife conservation are also economic and political. This is why political ecologists consider the apolitical conservation rhetoric as naïve and call for greater scrutiny because conservation in practice is a form of environmental control, which entails exercise of different forms of power among actors operating at different levels (Robbins 2004).

Overall, I argue for positioning the establishment and development of protected areas in postcolonial Africa within the complex and interconnected political and historical trajectories in the continent since the late nineteenth century. While most African states inherited national parks and notions behind these arrangements from departing colonial powers, global influences and internal demands have also necessitated and at times compelled African elites to expand protected areas. The expansion of tourism industry as alternative investment option, the support it received from global institutions and ideological frameworks like the 'free market economy' and increased involvement of African states in global environmental conservation forums have all contributed for the expansion of national parks in the post-colonial period. As mentioned earlier, Africa inherited not only the physical enclosure as national parks but they also adopted and widely emulated the models behind these enclosures. Although the resistance from local people and failure of the protectionist approaches compelled states and non-state actors to adopt the so-called 'community-based conservation' approach in the 1980s and 1990s, the dominant approach continued. As Adams and Hutton (2007) noted, despite some remarkable changes in terms of approaches and structure, the dominant discourse (i.e. protectionist conservation) continues to shape the management of national parks in Africa.

4.3 Center-Periphery Relations in Ethiopia and the Establishment of Nech Sar National Park

4.3.1 The Center-Periphery Relations in Ethiopia

In modern Ethiopian political history, state-society relations are more characterized by aspects of domination and subordination between the Abyssinian core (highland, Amhara-Tigrean, Orthodox Christians and Semitic language speakers) and the 'periphery', which encompasses all ethnic groups that were incorporated into the empire in the last quarter of the nineteenth century (Donham 1986; Markakis 2011). Beyond geographical dimension, the center-periphery dichotomy has cultural, political, economic and historical connotations that mark the boundaries, although it is often difficult to draw clear lines between the two. As Markakis (2011: 7) noted, "What distinguishes the centre from the periphery is not simply geography, although this is often a salient feature. The locus of power is the most significant indicator." In this context, what defines the periphery is its marginal position in power relations with the state, and how it is represented in economic, social, political and cultural schemes of the state. In the context of Ethiopia, the center-periphery relations should be understood as a continuum of power relations, in which elites from the Abyssinian core, mainly the Amhara, were in the core center, with their religious institutions/clergies, feudal lords, nobilities and peasants occupying the next space in the continuum of the center (Dereje 2013).

On the extreme pole of the continuum were those groups who were incorporated into the empire, with pastoral ways of life in the lowland areas in the south, east and west of the present day Ethiopian state. Religion, language and culture were also other markers of the dichotomy because most of the people in these regions were not in conformity with the Abyssinian narratives of "Great Tradition" (Donham 1986; Clapham 2002; Markakis 2011; Dereje 2013). For better understanding of the center-periphery relations and its relevance to the topic under discussion, it is important to give brief discussion on the cultural and ideological self-imagery of this in order to get an insight into how this group perceives the 'self' vis-à-vis the 'other' (people and territories in the periphery). By doing so, I will seek to establish analogies between the Abyssinian imaginations of the periphery and European perceptions of Africa during the colonial encounters and how such perceptions shaped processes of establishment of national parks in both contexts.

Three broad markers of identity are identified as distinguishing features of the Abyssinian core groups that also marked symbolic representation of the Ethiopian state. These were feudal modes of governance, property ownership and economy; ethnicity, religion and language; and the myth of 'Great Tradition'. To begin with the first set of state representation, until the 1974 revolution, the Ethiopian empire was characterized by a long history of feudalism, which descended from the ancient core regions of the empire to the newly incorporated peripheries. Long history of hierarchical power structure marked relations between lords and peasants in which access to the ladders of power were mainly hereditary, both as right to land ownership, the expropriation of resources and assuming administrative power. Nobilities, feudal lords and their army had privileged social status with distinctive markers like the type of food, clothing, place of residence, number of servants they possessed, amount of land under their control and number of tenants they governed representing their social boundaries (Donham 1986; Levin 2000). These social strata represented typical practices of domination, discipline and constraint, vanquishing, governing, expropriating and inflicting pain on the lower social strata, namely the tenants.

In terms of property ownership, particularly the right to land, the Abyssinians had different structures that guarantee a person the right to land but commonly, private and hereditary land tenure system was common among these societies. In principle, all adult men were entitled to inherit plots of land from their parents, although the nobility and military officials held administrative rights over the land in extracting tributes in different kinds (Bahru 2002). In Abyssinia and still in modern Ethiopia, agriculture remained the central feature of economy, contributing nearly eighty-five percent to the national GDP (Markakis 2011). With their settlement in high plateau regions in today's Amhara and Tigray regional states, the Abyssinians practiced the ancient plough agriculture, which sharply contrasted them with the livelihood styles of the people in the peripheries (Donham 1986; Turton 2011). They considered plough agriculture as stage of development and thus compared themselves in contrast to either the pastoralists or hoe cultivators inhabiting the southern half of the empire.

The second and perhaps strongest marker of Abyssinians' self-image that still lingers among Amhara nationalists was the strong marriage between Amhara ethnicity, Orthodox Christianity, Amharic language and the Abyssinian state, which became the foundation for the architects of the modern Ethiopian empire to forge the nation-state along these lines. There is no doubt that Christianity entered the Abyssinian state in the 4th century A.D., making this part of Africa one of the ancient Christian states in the continent. However, this history has been used by ordinary people and state rulers to project their 'distinctiveness' and also as signs of civilization in contrast to their Muslim rivals in the peripheries or those groups who practiced different forms of indigenous religion, whom the Abyssinians referred as pagans (Donham 1986). Donham (1986: 20) observes the dichotomy and Abyssinian representation of the people in the peripheries as analogous to European representation of Africa as 'wild' stating that, "According to Abyssinian ideology, the lowlands were to be avoided–they were wild and dangerous, infested with disease and *inhabited by savages who did not acknowledge God*" (emphasis added). The Abyssinians held competing perspectives about the peripheries: wild, savage, dangerous, desolate and hostile on the one hand, as well as reservoir for slave-taking, refuge for defeated lords and as economic paradise to support the pursuit of regional lords to the ladder of power, as happened during king Menelik's campaign against these regions in his pursuit to accumulate wealth.

The long history of conflict between the Christian highland kingdom and the lowland Muslim states played a paramount role in shaping the self-representation of Abyssinians and the state itself as Christian island surrounded by Muslim states both locally and externally. Ethnicity, religion and language were the three central symbolic ascriptions forming the core constituents of the Ethiopian state until 1974 and implicitly continued to date. In this regard, the Amhara, sometimes partnered with their Tigrean junior partners, backed by Orthodox Christianity and making Amharic language and its writing script as signs of advancement, controlled state machinery with which they were identified. For instance, although 'Amharization' and conversion of peoples into Orthodox Christianity were major state policies until recently "'assimilated' individuals were barely eligible to enter the political, religious and social systems of the state" (Markakis 1994: 225). Markakis (1994: 225) further argues that:

> The propagation of what might be called 'state nationalism' in the guise of 'national integration' was one political initiative taken by ruling groups in attempt to reinforce the shallow foundations of the state. Understandably, their [Amhara ruling

> elites] perception of national identity was the mirror image of their ethnic and cultural ego. Thus, the language of Amhara and Christianity became the salient features of Ethiopian nationalism and the Arab language and Islam of Sudanese nationalism. Consequently, integration was premised on assimilation into what was presented as the superior culture of the ruling ethnic group. In Ethiopia, no other indigenous language was allowed to be printed, broadcast or spoken in public functions, and attempts to study the culture and history of other groups were decidedly discouraged.

Narratives often support nation-building, nationalist projects and discourses that discredit counter-narratives of others in the pursuit of projecting a particular group's vision of how the nation-state is ought to be crafted. In this regard, Abyssinians built their narratives upon a legend of 'Great Tradition' which chronicles of kings, church teachings, oral narratives and modern 'historians' constructed, reconstructed and passed down from generation to generation as 'commonplace truth'. The myth of 'Great Tradition' is an aspect of Ethiopian historiography that glorifies ancient Abyssinia and its Amhara and Tigrean peoples as civilized Christians with many more links to the Judeo-Christian culture than black Africa (Markakis 2011). It was used as a diplomatic tool in winning the support of Christian states in medieval Europe against their war with Muslim sultanates during the sixteenth century but later it carried cultural nationalism in internal power struggle between the Abyssinian core and the peripheries. It considered the Abyssinian core as the 'true' Ethiopians and carriers of historical civilization (Ullendorff 1965). They used the myth as a political and psychological mechanism of exerting domination and legitimacy of statehood over others.

By subscribing to the notions of 'civilization' and 'modernization' and tracing their descent to king Solomon of Israel, the Abyssinians worked hard to maintain their position in the center by repelling others further to the periphery. Among several titles of emperors, the common one with religious, psychological and ideological undertones was *"Conquering Lion of the Tribe of Judah, Elect of God, King of kings, Emperor of Ethiopia…"* Beginning from the first king in pre-Axumite period up to the last emperor, Haile Selassie (1930-1974), apart from a few periods of interruption, all emperors and kings in Abyssinian proper subscribed to this legend of the Solomonic dynasty–hereditary link to king Solomon of Israel through the legend of Queen Sheba of Abyssinia.

The Abyssinians' image of the people in the peripheries was constructed in contrast to their own self-representation as 'civilized' people with plough agriculture, centralized governance, Orthodox Christianity, 'developed' writing system and more importantly the legend of 'Great Tradition' (Turton 2011). In contrast, pastoralists, hoe cultivators, Muslims and followers of indigenous religions, and non-Semitic groups were categorized as 'savages', 'uncivilized' and slaves, whereas the regions where these people inhabited were considered as wild (Ullendorff 1995; Donham 1986). Although dichotomizations between the peoples of the highland Christian kingdom and lowland states persisted for long before the birth of the modern empire, the conquest and subsequent asymmetrical power relations between Abyssinian rulers and peoples in the subjugated regions added to the divergence. Indeed, the nature of state-society relations varied within the subjugated regions based on the trends of the conquest (military conquest or diplomatic submission), religion and ways of life. For example, while Muslim communities and highland peasants were the most oppressed and exploited parts of the subjugated regions, pastoral communities in the peripheries were marginal to state interest and intervention.

Discussions on the birth of the modern Ethiopian state in the late nineteenth century that was crafted through relentless conquest under empire-building project of the Abyssinian kingdom (Donham 1986; Vaughan 2003) is not the central objective of this research. However, suffice to mention some underlying historical antecedents and their role in strengthening north-south or center-periphery polarizations. For the last over one thousand years in history, there has been southward expansion of Abyssinian kingdoms in different periods until the present day international border of Ethiopia was demarcated around 1900 with diplomatic agreements with European colonial powers (Donham 1986; Bahru 2002; Vaughan 2003). Prominent among these historical antecedents were first, the aggressive expansions of the highland Christian kingdom from the 14th to the sixteenth centuries mainly as competition against the westward expansion of the Muslim sultanates, and second, the late nineteenth century empire-building conquest launched in a process analogous to the European scramble for Africa during the same period. As Markakis (2011: 3) rightly argues, "While rebuffing imperialism in the north, Ethiopia successfully managed to practice it in the south". Mainly motivated by lucrative resources in the south and partly intrigued by nationalist missions of colonial establishment in competition with European colonial powers and equipped with modern firearms imported from Europe, the Abyssinian kingdom launched its conquest from its administrative center in northern Shewa.

Scholars from different disciplines have discussed in depth the process of the conquest, which I will not repeat in this section (see Bahru 2002; Mekuria 2011; Donham 1986; Teshale 1995). What I would rather discuss is the forms of relations between Amhara feudal lords and the local people in the subjugated regions after the conquest. Based on the nature of reaction of the hitherto autonomous states in the southern, western and eastern part of the country, the imperial government had put in place different administrative approaches. For example, in regions where the local people encountered the conquering army with armed resistance and were eventually subdued by Menelik's military force, governors were brought from the north and implemented harsh administrative measures, including the confiscation of land, abolition of traditional governance and forceful Amharization in the form of conversion to Orthodox Christianity (Mekuria 1996). On the other hand, in areas where incorporation was undertaken through submission of traditional leaders without armed resistance, there were some sorts of granting nominal 'autonomy' to the traditional chiefs, whereby elites were co-opted to facilitate resource expropriation and administration on behalf of the central government. This second aspect continued until 1930s when Haile Selassie's centralization policy brought the remaining traditional rulers under full control of the emperor (Bahru 2002). This form of differentiated incorporation into the state, has led to significant differences in state-society relations, where the state was more predatory and oppressive in areas it forcefully subjugated.

In the Guji and Koore cases for example, two different forms of administration were put in place because while the Guji were subdued by force, the Koore chief negotiated with the invading army and later continued to serve the imperial regime as a local governor. Although state institutions of coercion (the army, police, security, court, prison, etc.) did not entrench well in pastoral and agro-pastoral parts of the Guji, the highland areas were put under heavy feudal administration, exploitation and repression like other groups in the conquered regions. Like settler colonies in other parts of Africa, massive influx of people who sought fortunes in the newly incorporated resource rich regions took place following the conquest and the trend persists due to land degradation in the northern highlands. Although many of the people who moved to the south were fortune seekers, they established higher social status in the new places partly because they had the backing of the state that used them as tax collectors, administrators and religious leaders, and partly nurtured by their self-image and pride as descendants of kingmakers. As Markakis (2011) observes, in the new power constellation that was created between the center and periphery after the conquest, which still continues regardless of regime changes, Amhara peasants identified themselves and were more close to the center than any elite from the periphery. In short, the Ethiopian nation-state has been built upon the cultural and ideological foundations of its constituent subjects, namely the Amhara ethnic group with their Orthodox Christianity, Amharic language, 'great Tradition' and sedentarizing principles (Turton 2011).

In Ethiopia, the north-south dichotomy or asymmetrical center-periphery relationship has been built upon questions of power and cultural assumptions that inhibited patterns of social relationship and smooth upward mobility of the people in the periphery to the center in terms of assuming political power and socio-cultural integration. As Donham (2002: 11) observes, "although individuals from these regions [peripheries] could fairly readily associate themselves with the state tradition through the adoption of Orthodox Christianity, the Amharic language and Ge'ez/Amharic names–and a great many did so–their cultures and societies were condemned to subordination by the nature of the state itself." As a result, the Ethiopian state and its constituent subjects continued to represent the center while other diversified nations and nationalities in the country occupied peripheral positions, although there were different levels of the peripheries within peripheries and centers within the center.

4.3.2 The 'Natural' South and the Establishment of Nech Sar National Park

4.3.2.1 The 'North-South' Dichotomy

In Ethiopian studies, it has been a common trend that historians studied the northern societies focusing on their state structure, wars, church history and plough agriculture, among others, whereas anthropologists studied the culture (rituals, customs, kinship, marriage, initiation rites, etc.) of the people in the south (Donham 1986; Clapham 2002). This is why some scholars tend to argue that while the Amharas and Tigreans had their history, others have their anthropology (Ullendorff 1965; Donham 1986). The myth that civilization originated in the north and drifted to the south (if at all it was there) had shaped the nature of inter-societal relations between people from the north and south, and influenced how the state perceives the south. Describing the competing images of landscape representation in the lower Omo valley among different actors, particularly on how the state perceived the territory, Turton (2011: 164) summarizes as follows:

> The physical, climatic and cultural conditions of the Abyssinian center, with its mountain ranges, heavy rainfall, Christian tradition and plough cultivating peasantry, contrast markedly with those of its hot, dry, lowland periphery, inhabited mainly by pastoralists. This contrast has dominated the process of Ethiopian state building to this day. Viewed from the centre, the lowland periphery was seen as a wilderness, not in the sense given to that word by European Romanticism but in the earlier sense of "a waste or desolate region of any kind" or '"a tract of solitude and savageness"' (Oxford English Dictionary). The Abyssinian soldier-settlers who followed the armies of the Emperor Menelik II into the Omo Lowlands, saw them-

> selves faced, like the seventeenth-century Puritan settlers in New England, with a "wild, untamed landscape, a savage howling wilderness", where no recognizable imprint of civilization had been left by the local inhabitants. They saw it above all as dangerous and threatening a chaotic, disease ridden and unproductive tract of land, inhabited by anarchic and violence-prone "nomads".

From the above excerpt, I would like to draw two arguments that I will later link to the internal 'exportation' of national park ideals to the peripheries. Firstly, there were cultural and ideological assumptions that shaped the popular imagination of the center about the peripheries. It was a self-centric imagery of the 'Other' through which the people from the center contrasted the peripheries. Secondly, center-periphery relation was/is not a matter of geographical location as such, although distance from the geographical center also contributes to the further marginalization of the peripheries. Rather, it is more of an aspect of power positions of both the center (a group who dominated the political, economic, cultural and social positions of the state) and the periphery in the state. The asymmetrical power relations particularly under circumstances where the center subscribes to cultural and ideological domination in addition to political power are analogous to European representation of Africa as wild, untamed, pristine and 'Edenic'. This analogy links the European representation of Africa to Abyssinian representation of the south as natural, wild, savages, wasteland and uninhabited (Turton 2011; Donham 1986; Markakis 2011). It was within such context of north-south dichotomy that discourses of national parks were imported by the Ethiopian state in the 1960s and were systematically 'exported' to the peripheries, mostly located in the south. With this understanding, I will posit the historical trajectories of the establishment of the Nech Sar National Park within the north-south dichotomies that were developed following the conquest. More specifically, the argument is that the Ethiopian state considered the landscape in the south as 'natural' to be converted into consumable commodity through national parks so that it would render economic benefit through tourism.

The history of many national parks in Ethiopia was part of the country's projection of its self-image to the outside world, particularly to the West as part of the global movement for environmental conservation. Since the Second World War, Ethiopia increased its alignment to the global North in areas of diplomacy, development cooperation, education and environmental conservation schemes (Clapham 2006a). The periods 1941-1950 and 1950-1974 were commonly termed as the British and American eras in Ethiopia, respectively, when the two global powers influenced the socio-political and economic dynamics in the country (Bahru 2002). During these periods, Ethiopia became strategically important for the USA and its allies. From the part of the Ethiopian government during the period of Haile Selassie, ambitious modernization programs were designed to attract Western aid and diplomatic supports. Despite its feudal orientations, the imperial regime appropriated some aspects of capitalist development models in areas of agriculture by giving large tracts of land for private investors and introduced state farms, which the communist regime later intensified (Clapham 2006a).

The disillusioned feudal regime began amalgamating capitalist notions of development (such as granting land to multinational corporations for agricultural investment) with a feudal ethos, neither of which ultimately benefited the society as a whole. Rather, the emulated development model served the interest of the aristocrats, landowners and the nobilities while exacerbating the destitute living conditions of the poor. For instance, the government established the first large-scale commercial farms such as Awash Valley project and the Chilalo Agricultural Development Unit (CADU) in the 1960s. While the Awash Valley project led to the displacement of pastoralists blocking their seasonal grazing land, CADU brought massive resettlement and displacement of peasants among the Arsi Oromo (Clapham 2006a). Such 'modernist' approaches of development, which valued sedentary life style over pastoralism and state farm over peasant agriculture, denigrated the local livelihood conditions of the people and neglected their views of development.

4.3.2.2 Ethiopia's Partnership with UNESCO

Ethiopia's importation of Western models, ideas and practices was also evident in areas of conservation, particularly the establishment of national parks. Before the mid-1960s, the country had only one officially designated national park–the Managasha national park that was officially established in 1958 (Huxley et al. 1963). However, in the 1960s and 1970s the government established all the major national parks largely based on the Yellowstone national park model–the exclusionist and protectionist approach. Before proceeding to the central theme of this section–namely, the establishment of the Nech Sar National Park–I will briefly discuss some historical antecedents that made the 1960s unique in the country's history of national parks. In 1962, the Ethiopian government sent a delegation to attend the UNESCO 12th conference held in Paris. The Ethiopian delegation led by the then Minister of Agriculture, Akaleworld Habtewold, finally submitted a request to UNESCO Secretary General seeking support in the process of surveying, planning and establishing national parks in Ethiopia. During its 12th session (9 November to 26 December 1962), the UNESCO General Conference adopted two important Resolutions, which I will discuss in this section. These were the resolutions concerning "*Economic Development and Con-*

servation of Natural Resources, Fauna and Flora", and "*Recommendation Concerning the Safeguarding of the Beauty and Character of Landscapes and Sites*".

The first Resolution emphasized the economic values of natural resources (flora and fauna) for economic development, particularly in developing countries. It announced this emphasis as:

> The General Conference, conscious of the extent to which economic development, especially in developing countries, requires that attention be paid to their renewable natural resources, particularly flora and fauna, in some cases, may be irreplaceable if such development is pursued without due attention to their conservation, restoration and enrichment, and to their increased productivity, [...] Calls upon the Secretary-General, the specialized agencies and other interested international and national organizations, in support of the above mentioned-resolution of the United Nations Educational, Scientific and Cultural Organization, to continue to give the fullest co-operation and to provide technical assistance to the developing countries, at their request, in the conservation and restoration of their natural resources and their flora and fauna (Huxley et al. 1963: 41).

I draw some interpretations from the above statements. Although many African countries had gained independence by the time, UNESCO continued advocating for the commoditization of nature in Africa regardless of the decolonization rhetoric. More importantly, the resolution gears towards encouraging developing countries to invest in nature (flora, fauna and landscapes) in their attempts to gain economic development through tourism. It is also important to critically investigate the technical assistance the UNESCO promised to provide to developing countries in the process of conservation and restoration of their natural resources. What kind of technical support would UNESCO provide? Whose knowledge would be central in the process of surveying areas for national parks? As has been practically evidenced in the case of Ethiopia in 1963 and subsequent years, the team from UNESCO undertook the survey and proposed to the national government and other organizations on areas to be protected as national parks or game reserves based on their own conceptions of protected areas. UNESCO's commitment to provide technical assistance to developing countries shows the paternalistic views ex-colonial powers and their institutions continued to project upon the developing countries.

The second Resolution was specifically "*Recommendation Concerning the Safeguarding of the Beauty and Character of Landscapes and Sites*" and aimed at encouraging member states to take corrective measures against practices, which, according to the conference, destroy the beauty and character of landscapes and sites (Huxley et al. 1963: 6). As a point of departure, I will focus on the following recommendations of the general conference that stated:

> Considering that at all periods men have sometimes subjected the beauty and character of landscapes and sites forming part of their natural environment to damage which has impoverished the cultural, aesthetic and even vital heritage of the whole regions in all parts of the world. [...] (recommends among other things that), when conditions are suitable, Member States should incorporate in the zones and sites to be protected, national parks intended for the education and recreation of the public, or natural reserves, strict or special. Such natural reserves and national parks should form a group of experimental zones intended also for research into the formation and restoration of the landscape and the protection of nature (Huxley et al. 1963: 5-6).

Since the mid twentieth century, the impact of capitalist encroachment on the environment had been pronounced partly due to the post-War economic rehabilitation programs and partly as a response to global political and economic forces. From this angle, UNESCO's conference was likely influenced by the environmental movements of the period, although it sought to homogenize actors responsible for the destruction of the beauty and character of landscapes and sites. In a critical analysis of the content, however, one can easily understand the continuity of the notions of national park ideals that presume landscapes in Africa as pristine, aesthetic and untamed wilderness. Since the late nineteenth century, this notion dominated the advocacies for national parks. Besides its recommendation for the establishment of national parks and reserves, the UNESCO General Conference adopted what it termed as "protective measures" in which it suggested the application of traditional "fence and fines" approaches stating that:

> The fundamental norms and principles governing the protection of landscapes and sites in each Member States should have the force of law, and the measures for their application should be entrusted to the responsible authorities within the framework of the powers conferred on them by law. [...] Violation of the rules governing the protection of landscapes and sites should involve payment of damages or the obligation to restore the site to its former condition, as far as possible. Administrative or criminal prosecution should be provided for in the case of deliberate damage to protected landscapes and sites (Huxley et al. 1963: 48-49).

As I discussed earlier, scientific knowledge and expert based conservation approaches were modeled as the ultimate 'regimes of truth' in conservation discourses during the colonial period. In 1962, UNESCO advocated for inculcation of Western principles of conservation into education systems and educating the public about conservation values in developing countries. It was stated that

"Educational action should be taken in school and out of school with a view to arousing and developing public respect for landscapes and sites and publicizing the regulations laid down to ensure their protection" (Huxley et al. 1963: 49).

At this conference, the Ethiopian delegation presented its request to the Director-General seeking assistance in the field of conservation and establishment of protected areas. As Huxley et al (1963: 6) stated in their report after their UNESCO mission to Ethiopia:

> The Ethiopian Delegation to the General Conference of the UNESCO, who had throughout the discussion given their fullest support to these motions, submitted to the Director-General upon their adoption, through H.E. Mr. Akalework Habtewold, Minister of Agriculture and Head of the Delegation, a request for assistance in this field. In his letter, the Head of the Delegation pointed out to the Director-General that "it is our wish to manage and develop (national parks and wild life reserves) in such a way as to secure the preservation of their flora and fauna, provide centers of biological and ecological research, and contribute to the growth of the national economy, especially the development of tourism and game cropping".

UNESCO accepted the request from the Ethiopian Delegation and sent a team of high profile professionals to Ethiopia in 1963 to survey potential areas for national parks and wildlife reserves. As the report indicates, the government of Ethiopia was keen to 'modernize' its wild life conservation practices to increase solidarity with international conservation organizations and particularly to increase the economic benefit from the resources. The team was invited by the Emperor, who, according to Huxley et al. (1963) provided it with all necessary facilities including transportation, security and field guides during their extensive tour in different parts of the country. After the fieldwork in Ethiopia, the team produced a forty pages document and submitted to UNESCO and the Ethiopian government in 1963. Later, UNESCO adopted the document as "*The Conservation of Nature and Natural Resources in Ethiopia*" (Huxley et al. 1963). Given that the document shows how Ethiopia imported Western discourses of national parks and later internally 'exported' it to its peripheries, I will analyze few points from the resolution and the report that the UNESCO team compiled in 1963.

In the document, the team emphasized that Ethiopia has had immense resources (wild nature and landscapes, as well as cultural and historical sites) but the country's natural resources were under series threat from human and livestock pressure. This is the 'environmental degradation theses that conservationists often emphasize while projecting their interests in Africa. By classifying conservation under three categories as conservation of wild nature, cultural conservation and general conservation principles, Huxley et al. (1963) underscored the urgent need for action in Ethiopia in these three categories to capitalize on Ethiopia's wealth of wild life, wild areas, landscape beauty and forests and cultural, historical and archaeological sites. Apart from the general underlying question behind the motives of Western conservationists in conservation programs in Africa, it can be learned from the document that scientific research (zoological, botanical and archaeological), economic interest (in the form of tourism) and aesthetic (recreational) purposes of the areas were mentioned. To this end, the team proposed seven areas as potential protected areas in the form of national parks and wildlife sanctuaries, one of which was identified between Lakes Abaya and Chamo–that means the current Nech Sar National Park. However, in contrast to the representation of the area by successive regimes as 'uninhabited' territory, the team noted its witness of heavy pressure from agriculture and pastoral ways of life:

> The first site proposed for the field station was on Lake Awassa but a section of opinion now favors some place further south on Lake Abaya or perhaps between Abaya and Chamo. This latter area is said to contain a fair surviving assemblage of large mammals and its aquatic biology is interesting in that it contains the main ingredients of the Nilotic fish fauna which does not exist further north in the Rift Valley. However, the Lake Abaya area is said now to be under heavy pressure from both cultivators and domestic stock so that the best choice for the Park may ultimately be further north, in the more accessible area between, and if possible actually containing, Lakes Awassa and Shalla (Huxley et al. 1963: 25).

Like other conservationist predecessors and colonial conservation policy makers in Africa, the team not only proposed the establishment of protected areas, but also recommended the enforcement of fines and fences policy, as well as the eviction of indigenous peoples to keep the 'natural beauty and character of landscapes'. For example, regarding the pastoral community around a proposed Matahara national park that later became Awash national park in 1966, the team stated that:

> The question of nomadic Matahara people within a future National Park is a difficult one. At present they have a reputation for keeping other people away and this may be of some advantage to the surviving wild animals. Experience from many other parts of Africa indicates, however, that the chances of the present small numbers of wild life multiplying and becoming sufficiently tame to be a major attraction will be remote *unless these people and their stock can be induced to leave the National Park area, and also unless the cultivation which is taking place in certain places is stopped* (Huxley et al. 1963: 22, emphasis added).

Two central issues that later became the major problems of national parks management in Ethiopia, yet the UNESCO team did not recognize, were related to customary rights of local people and issues related to the recognition of their indigenous knowledge of resource management. This entails participation in planning, management and benefit from national parks. As common elsewhere in protectionist conservation approach, the report paid due attention to scientific knowledge and 'educating' the local people on conservation ethics and significances of wildlife, landscapes and other cultural/archaeological sites. Nonetheless, how did these resources survive long before the advent of the state? Does it mean that local people do not have ecological wisdom? In the previous section, I argue that behind the spectacular imagination of Africa as "Eden", there are African people who contributed through their indigenous knowledge for the preservation of the biodiversity. Although I do not subscribe to the stereotypical representation of indigenous peoples as 'noble savages', I would rather argue that most indigenous peoples have had culturally and spiritually embedded ecological wisdom. The Guji Oromo are the good case in point, in that the people's relation with nature is mediated through different aspects of rituals, prayers and blessings enacted by elders (Baxter 1991; Asebe 2012c).

So far, I have discussed the involvement of Ethiopia in the UNESCO General Conference on nature conservation and its request for technical assistance in the process of establishing national parks in the country. The UNESCO resolutions and the report of UNESCO team laid grounds in terms of which national parks in Ethiopia were established. It laid the foundation for the management approach by introducing the protectionist conservation approach, placing the state and scientific knowledge as the 'right' paths to lead conservation. Based on the survey results and recommendations of the UNESCO team, Ethiopia established the major national parks such as Awash national park, Semien Mountains national park, Omo national park, Nech Sar National Park and Bale mountains national park in the 1960s and early-1970s (Dessalegn 2004).

With this in mind, I will discuss the practical establishment of the Nech Sar National Park and along my discussions, I will link its management approaches with the 'conventional' conservation discourse and how this eventually affected both the people and the wildlife alike. In addition to the UNESCO team who surveyed several places in the country and suggested the area that later became the Nech Sar National Park among other areas to be designated as national parks, few other British and American conservation experts suggested the area as a national park between 1965 to 1970. For instance, Blower, a British biologist had carried out a survey in the region about human habitations and the status of wildlife in 1966-67. Regarding human settlement, he recorded that:

> There were a number of small temporary encampments of cattle owning Guji peo-

ple throughout the Nech Sar area. No permanent habitations were seen though it was understood that there were permanent villages in the hill to the east and signs of cultivation could be seen on slopes of Amaro Mountains. There were several thousand cattle in the area belonging to the Guji, though for the most part these animals did not appear in very good condition and it was reported that mortality due to unknown disease was heavy (Blower 1967 cited in Getachew 2007: 48).

4.3.2.3 The Management Approaches of the Nech Sar National Park (1974-1991)

As has been discussed in chapters one and three, the area constitutes diverse fauna and flora. In fact, the big game animals such as buffalo, elephant, giraffe and rhino that Smith (1897 [2008]) mentioned as spectacular features of the territory shortly before the advent of the Ethiopian empire in the region were no more in existence in the 1960s. Above all, the territory was noted for the presence of the endemic Swayne's hartebeest, whose number has dramatically reducing over time. According to some sources, the major objective of the establishment of the national park was conservation of this animal with tourism development as another motive behind the government project (Dessalegn 2004; Freeman 2006). As some argue, the scenic landscapes of the territory that includes the two lakes and different ecological niches were also among the major factors behind the establishment of the national park in connection to its importance for attracting tourism (Markin et al. 1974; Briggs 2006).

Depending on the recommendation from foreign advisors who surveyed the area in the 1960s, the Ethiopian government began the enclosure of the area before it formally designating it as a national park in 1974. For example, a document compiled by the British Foreign and Commonwealth Office Overseas Development Administration states that the entire lowland areas around Abaya and Chamo Lakes, Arba Minch town and Nech Sar areas were proposed as potential development corridor for different sectors (Markin et al. 1974). Regarding the Nech Sar, Markin et al. (1974) suggested the establishment of the territory as a legally constituted national park, by prohibiting all cultivation and permanent settlements and restricting grazing in the enclosure. The team rather recommended a kind of community participation in the revenue sharing from the tourism industry.

According to Guji informants whose ancestors lived in the area for several generations, the restriction on grazing land in the Nech Sar plains was initiated during the last years of the imperial regime. Guji elders recall the early state intervention in the form of wildlife conservation when the government deployed police force, who were said to have been in charge of keeping the 'security' of the people and the animals from "criminals"–hunters and robbers in the Bonke hill forests. A Guji elder, Chari Galgalo who had burial sites of his ancestors in the Bonke forests, narrated the beginning of the government intervention as follows:

> Some five or six years before the coming to power of Mengistu Haile Mariam's regime [in 1974] the governor of Gamo Gofa province sent to our villages his lower administrators and police force. They told us that the government wants to secure the area so that people would safely travel to and from Arba Minch town. Actually, there were sporadic conflicts and some people used to hide inside the forest and killed people they found along the way. It was either retaliation or motivated by looting. The government built three police stations in the center of what later became the national park. People did not oppose the establishment of police stations because the government did not tell us it was for the park. After they finished the construction of police stations and staffed it with police forces, road construction was started from both Gamo Gofa and Sidamo provinces directing to our homes. We were told that the roads would connect Dilla town [some 60km to the East] with Arba Minch town [some 20km to the west from the center of the park]. The road destroyed our ancestors' burial sites and sacred spaces but it was not completed. Towards the downfall of the king [Emperor Haile Selassie], they stopped the construction of the road but began telling people not to graze and hunt inside what they called wildlife reservation area. It was not all part of the park but included all the forest area and portions of the grassland. People who had settlements inside the restricted area were told to move out. That was started during Haile Selassie period. The Derg came and strengthened the restriction. It actually took strong measures and evicted our people at gunpoint in 1982. We first thought government does not lie. But they deceived us first by telling it was for our security, then telling us they were building roads for us and later they told us that we would get benefit from tourism income. Still they are telling us this for now over one generation. They pushed us gradually and we have now nowhere to go except to perish (informant: Chari Galgalo, April 2011)

Restrictions and criminalization of customary practices of the local people should also be contextually seen within broader ideological and political changes in the country in the post-1974 period. Soon after it assumed state power, the military junta introduced a national policy regarding land. After the 1975 Land Reform Proclamation[9] that transferred land ownership right to the state, Ethio-

[9] In 1975, the military regime promulgated a land reform proclamation, which usurped the rights of land once again from the Ethiopian people and bestowed it upon the state. Although the central motto of the Ethiopian Student Movement of the 1960s and early-1970s that contributed to the collapse of the imperial regime was "Land to the Tiller", the military regime that usurped the revolution paralyzed the tenets of the revolution and once again brought Ethiopia to seventeen

pian peasants faced once again the century-old dispossession of their customary rights over the land (Markakis 1981). This was rooted in the military regime's socialist ideological orientations that reinforced state command of the economy and sources of production. As Dereje (2011) argues, the socialist military regime in Ethiopia had crashed traditional values, customs and practices and worked hard enough to abolish them. In doing so, the government nationalized rural land and introduced scientific and expert-based resource management systems.

In the case of the Nech Sar National Park, the territory was administered by the central government and was included under the then Sidamo Province. Although it first strived to reveal its legitimacy as security provider and development agent, literally the state occupied the territory and asserted its control in the form of deploying hard power–deployment of police force and territorialization of the space. The military regime (1974-1991) introduced strict protectionist approach of park management by evicting local people at gunpoint in 1982 (Taddesse 2004; Dowie 2009). During the eviction, over 2000 people from both Guji and Koore communities were removed from the park based on the 'modernist' discourse of the socialist regime that considered traditional practices, customs, values and religious beliefs as obstacles to development (Dessalegn 2004). Through the incident, houses, crops and properties were burnt to ashes. Many cattle died due to a shortage of water and pasture en-route to new settlement areas. Because the state did not prepare any resettlement area for the displaced people, they were prompted to compete over resources with other neighboring communities such as the Konso and Burji ethnic groups. This led to prolonged inter-ethnic conflicts that further destabilized the region and impoverished the people. It also restrained the local people from their economic and cultural spaces (sacred sites, ritual places, burial sites, extraction of medicinal plants, etc.).

The Guji people consider the eviction as a major social crisis because it exposed them to famine, diseases and inter-group conflict in the new settlements where they were seen as aliens. According to one informant, some clans perished due to famine and disease:

> That period [the period of eviction] was horrific. They [government forces] came with heavy arms and gave us only two days to move out. If we refused to move, they would destroy our livestock and crops and kill us by tanks and bombs. That was what they told us. We responded that we had nowhere to go. We were born here; our ancestors lived here and were buried here. Where could we go? Our cattle

---

years of dictatorship (see Markakis 1981; Teshale 1995; Bahru 2002).

> could not adapt to other areas. But they were not ready to listen to us. On the third day, they came fully armed. Immediately, they began destroying our houses, burning all what they came across. They chopped down crops and burnt to ashes. They shot at people. There is a woman who lost one of her eyes during that incident. Then we moved to Oddoo Darba [some 15km to the east]. However, this new place was a disease infested area that killed thousands of cattle and many people. There was no road to medical centers. Even some clans were cut off their roots [perished]. Then we moved to Abulo Alfacho [to the south of the park]. Because it was adjoined by many other ethnic groups such as the Burji, Koore, Konso and Gidole, Abulo-Alfacho was not good either. There were conflicts over resources. We lost many people in the conflict. Then we returned to our home [the park] following the downfall of the military regime [in 1991] (informant: Dhugo Waqayyo, April 2012).

While the state claims control over the territory for conservation and economic purposes, the local Guji and Koore communities consider state intervention as a threat to their livelihood and cultural survival. As will be detailed later in this research, the Guji challenge the legitimacy of state approaches of conservation building their argument upon their local knowledge of conservation that is embedded in their cosmological views. For instance, a Guji elder argues that: "these park people came few years back and began telling us what to do and what not to do regarding wildlife. But we lived in harmony with the animals for centuries. Our fathers and grandfathers and our ancestors in general lived here with the animals. We care for the animals not because these people told us what to do but we do so because we fear our creator, Waaqa" (Informant: Gagasa Dokale, May 2012).

The Koore people had also similar stories because the 1982 eviction displaced them as well from their farmland in the Tsalke valley. As a Koore woman from Tifate village recalled the incident:

> Our people used to move to Tsalke for rituals in ancient times. It was called Tobbe ritual to get rain, good harvest and for other blessings. Later, people started going to Tsalke during dry seasons to cultivate cotton and maize through irrigation along Sermale River. But the park people came and first chased the Guji from the desert [Nech Sar plains] and then they evicted us. Koore people initially refused to move out of their land but the government used force. I was young by the time. My parents had plots of land here in Tifate and Tsalke. The Derg government took our land in Tsalke. Many people returned to the highland. Some people went back to Tsalke when the new government took power. We did not go there since then. There was conflict with the Guji and the park people. We wanted just to cultivate what we have here (informant Werkinesh Ware, February 2012).

It is also important to position the state intervention in the form of setting aside land as a national park from which people were excluded within the broader state conceptions of development vis-à-vis the livelihood styles of pastoralist or agro-pastoralist communities. Despite differences in political ideologies and economic policies, the last three successive regimes in Ethiopia share some common features in their views and policies towards pastoralist groups. While the feudal and the military regimes used to consider the pastoral territories as threats for national security but insignificant for the national economy, the current regime considers the pastoral territories as viable sources for investment in different sectors (ecotourism, agriculture, mining, dam construction etc.). However, in all cases, pastoral communities remained victims of state 'development' projects, resettlement programs and displacements (Hagmann and Alemmaya 2008). The military regime under Mengistu Haile Mariam introduced the villagization program in the 1980s, aimed at grouping together scattered farming and pastoral communities throughout the country into small village clusters to enhance land redistribution, resource conservation and provision of social services. The Guji, despite their agro-pastoralist life style, were forced to comply with villagization programs that entirely disrupted their socio-cultural structure and livelihood conditions (Taddesse 1995). In the post-1991 period as well, despite the government's rhetoric of local autonomy, intervention from the federal state into the life worlds of people in the peripheries under the guise of development projects (dam construction, national parks and large-scale agricultural investment schemes) has been evident (Markakis 2011).

Summing up

In this chapter, I have discussed how and why the notions of national park were exported to Africa. To understand the motives behind the establishment of national parks beyond the 'conventional' arguments for biodiversity conservation, I have endeavored to observe the nexus between the post industrial revolution 'countryside ideal' in Europe and the national park ideal in Africa. The central argument is that the countryside ideal in Europe was informed by three developments related to industrial revolution. First, it served as a retreat from the despoiled industrial cities. Because of urban pollution, population pressure and stressful work ethics following industrial revolution, the countryside was considered an ideal place of 'salvation' where humans were in harmony with nature. Second, industrial revolution had created a social hierarchy whereby the middle class used landscape appreciation, countryside tour, mountain hiking, summer camping and in general landscape consumption as a form of drawing class boundary between themselves and the working class who was unable to afford for leisure time in the countryside. Therefore, countryside ideal was a form of identity/class boundary marking both spatial as well as social differentiations between members of different classes in post-industrial revolution Europe, particularly England. The third dimension of countryside ideal was part of the continuation of renaissance and enlightenment thoughts that encouraged human dominion over nature, which later complemented the developments after industrial revolution where people began to domesticate 'nature' for human aesthetic consumptions.

When early settlers and conquerors arrived in the Americas and other parts of the world, the notion of subduing 'wilderness', which actually constituted wild nature and indigenous peoples, became another marker of distinctiveness and was used as a triumph of the 'civilized' over the 'primitive' culture (Merchant 2003). Particularly in the United States, the conquest of wilderness and setting aside territories in the form of national parks became signs of national identity. In the long run, the Anglo-American imagination of 'wilderness' and its conservation was geared toward a nationalist and identity project through which human victory over nature as well as rights of some people to 'privileged' spaces were deemed to indicate racial as well as class boundaries. During the colonial encounter, Europeans brought with them the myth of 'wild' Africa, which was represented in ambivalent perspectives–the continent represented as wilderness, deserted, dangerous, ridden by diseases and death on the one hand, and as an "Edenic" garden on the earth, on the other (Adams and McShane 1996). Both perspectives complemented each other in the process of conquest because explorers, colonial invaders and missionaries were longing to reach the "Edenic" garden, along the way demonstrating their mastery over what they considered as wilderness.

Although there is no single and conclusive factor behind the motives of Europeans to establish national parks in colonial Africa, I argue that national parks were used as arenas of colonial control, as markers of racial and class boundaries and as forms of resource expropriation in the colonial system. During the long periods of encounter between colonial discourses of nature conservation and different local ecological epistemologies among African communities–that in fact constitute human actors in the process of the interaction–there were apparent contestations, negotiations and appropriations of perspectives and practices on both sides. However, as discourses and their translation into practice largely depend on power positions of actors, the dominant exclusionist notions of nature conservation served as 'regimes of truth' during the colonial periods. This is partly why most of the post-colonial states maintained and even strengthened national parks for economic, political and environmental reasons.

Today, the environmental/conservation discourse is used mainly to draw financial aid from international donors under the pretext of 'biodiversity crises'. Nevertheless, protected areas conservation in Africa adopted the so-called community-based conservation programs since 1980s because of two major reasons. First, it was owing to local communities' resistance against the dominant exclusionist conservation practices and the failure of the 'fortress conservation' approach to serve the missions of biodiversity conservation and local development. Second, different actors began to recognize some success cases of indigenous resource management approaches in contrast to the scientific approaches. Nevertheless, as I have argued earlier, the rhetoric of community-based conservation programs have often served as mechanisms of creating environmental subjects and as forms of establishing local regulatory community where local elites would get privileged positions to get access to environmental resources.

Within the local context in Ethiopia, the establishment of national parks in the historically subjugated regions of the South shares much in common with similar phenomenon in colonial Africa. In this regard, the cultural and socio-political identity markers of the Ethiopian state (derived from its core centers)–Amhara-Tigrean ethnicity, Amharic language, Orthodox Christianity, and sedentary and plough agriculture–were fundamental elements of boundary making in the north-south dichotomy. Thus, the south (the periphery) was perceived as a 'wild' awaiting the hands of the center to be converted into civilization (Markakis 2011; Turton 2011). During the 1960s, the Ethiopian government adopted the dominant dualist notion of conservation and systematically exported it to its peripheries, which it already categorized as 'natural', 'wasteland' 'desolate' and 'uninhabited'. Thus, the management approaches of the Nech Sar National Park were partly informed by the global discourses successive Ethiopian regimes adopted, and partly by the perpetual representation of pastoralist and agro-pastoralist life styles as 'backward' that would justify state interventions as legitimate steps for development.

CHAPTER FIVE: ETHNIC FEDERALISM AND CONTESTED SPACES IN ETHIOPIA: SITUATING THE NECH SAR NATIONAL PARK IN THE POLITICAL DISCOURSE

Introduction

Within the century-long history of modern Ethiopia, the country embarked on a remarkable reconfiguration of political order in 1991, marking a major departure from its political tradition at least in principle (Dereje 2010; Markakis 2011). The reconfiguration of the political order was both cultural and political in nature. Culturally, it deconstructed and questioned the national identity of the state, which was associated with one ethnic group–the Amhara–and its language, religion and culture. Politically, it challenged the century-old hegemony of the center and centralization of state power. In chapter four, I have discussed how center-periphery relations were constructed and maintained in the country's political, economic and cultural patterns, as well as mythical imaginations. The Abyssinian state that later became the modern Ethiopian empire was crafted in conformity with Amhara and Tigrean cultural ethos and religious understandings (Donham 1986; Markakis 2011). Since the mid nineteenth century, a highly centralized and hegemonic state was established, whereby the core Abyssinian constituents controlled the state's political, economic and cultural affairs. In the process of creating such a hegemonic state, the peoples in the periphery were excluded from the political participation and socio-economic benefits that the state is supposed to provide its citizens.

The new political order of the post-1991 period can be seen as an attempt to reverse both the cultural and political foundations of the Ethiopian state by institutionalizing ethnic federalism. I would argue that despite conspicuous debates between its proponents, who consider it a solution to redress the century-long Amhara domination, and its detractors, who perceive it as a divisive political system that threatens national integrity, ethnic federalism demonstrated some levels of progress in the country's political culture.

In this chapter, I will briefly discuss the ideological tenets of ethnic federalism focusing on how the remapping of the political order has changed the center-periphery relations. By doing so, I will probe into the lived reality of the new political order by focusing on community voices, representation and participation, particularly regarding the planning and management of the Nech Sar National Park. This specific case acts as a mirror to gain an insight into the broader political and economic dispositions in the country and informs us about the nature of state-society relations at large. I argue that the national park is one among many contested arenas in which the national political discourse encounters local realities. Therefore, by taking the experience from the Nech Sar National Park within the context of the post-1991 political and economic orders, I will investigate what the new political order means for the local communities who, despite the apparent new forms of empowerment, continue to face different forms of intervention from the federal and regional states.

5.1 The Reconfiguration of the Periphery within the New Political Order

In the mid twentieth century, the quest for deconstructing the hegemony of the center flared up among different sections of the Ethiopian population, especially in those regions incorporated by force into the state through the empire-building project. However, the major challenge to the bogus ambition of nation-building came from one of the historical centers, Tigray, which was the center of Abyssinian state formation. In any case, social dissatisfaction from different corners of the population first invoked mass movement under different banners that questioned the century-long ambitions of the Abyssinians in building the nation according to ethnic, religious and cultural sentiments of the dominant group during the 1960s (Markakis 1994). Questions of nationalities were invoked for the first time as potent political agenda, serving as instruments of mass mobilization during the period of student movement in the 1960s and early-1970s in which different ethno-nationalist movements, including the current architects of ethnic federalism, participated (Balsvik 1985; Merera 2003).

Initially, the activists in the Ethiopian Student Movement (ESM) were inspired by the Marxist-Leninist ideology of national oppression and prescribed nations and nationalities' right to self-determination, including the right of secession as a solution to cure the problem (Merera 2003; Asebe 2012a). By doing so, university students challenged the state hegemony and domination. For example, an article written by Wallelign Mekonnen, then student at Addis Ababa University (formerly Haile Selassie I University), "sparked a political bombshell for the regime by explicitly addressing ethnicity and exposing the Amhara dominance and oppression to the public" (Asebe 2012a: 6). A portion of his article reads as follows:

> Is it [Ethiopian national identity] not simply Amhara and to a certain extent Amha-

> ra-Tigre supremacy? Ask anybody what Ethiopian culture is. Ask anybody what Ethiopian language is. Ask anybody what Ethiopian religion is. Ask anybody what the national dress is. It is either Amhara or Amhara-Tigray!! To be a 'genuine Ethiopian' one has to speak Amharic, to listen to Amharic music, to accept the Amhara-Tigre religion, Orthodox Christianity, and to wear the Amhara-Tigre shama in international conferences. In some cases to be an 'Ethiopian', you will even have to change your name. In short, to be an Ethiopian, you will have to wear an Amhara mask (Quoted in Balsvik 1985: 277).

Wallelign's article and other similar works by the student circle questioned the bogus ambition of the Abyssinian nation-building project, which equated Ethiopian national identity to Amhara ethnic identity, with many dozens of nations and nationalities being considered as subjects rather than equal citizens of the country (Markakis 1994). The students encompassed different ethnic backgrounds.

During the 1970s, different ethno-nationalist movements were established, embracing ethnicity as a primary instrument of mass mobilization to redress the ethnic-based dominations and exploitation that had engulfed the country since the late nineteenth century. One of the ethno-nationalist movements is the Tigray People's Liberation Front (TPLF), established in 1974 as a guerrilla movement (Vaughan 2003). I briefly elaborate on the political ideology of the TPLF because upon creating other ethnic-based movements and aligning with already existing ones, TPLF has continued as the central architect of ethnic federalism and in general the mastermind of the new political order in the country. Despite its representation as the orbit of ancient Abyssinian 'civilization' and state formation, Tigray lost its political role in the country's power game, particularly following the death of emperor Yohannes IV (1872-1889) in a war with the Mahdist Sudan in 1889 at the battle of Matama. The shift of political center from Tigray to Shewa was both a change in geopolitical center and constituted ethnicity. Beginning from the mid eighteenth century, power contention between Amhara and Tigrean regional lords changed the relation between the two into inter-ethnic competition. This rivalry continued until the formation of the Ethiopian People's Revolutionary Democratic Front (EPRDF) in 1989 as a coalition of political parties from Tigray, Amhara, Oromo and Southern Ethiopian Nations, Nationalities and Peoples. Nonetheless, this does not mean that the political 'marriage' has settled the historical tension between the two, given that the opposition parties are still organized along ethnic lines, contesting the TPLF/EPRDF regime.

I would now like to return to my initial point of TPLF's political struggle. After emperor Menelik II assumed power in 1889, ethnic-based political domination, economic exploitation, cultural marginalization and generally state hegemony under the Amhara supremacy became the common feature in modern Ethiopian history (Markakis 1994; Vaughan 2003). Although the nature of state repression and exploitation differs between provinces, with the extreme repressive system imposed on the subjugated south, Tigreans also felt the pains of domination (Vaughan 2003). Later in the 1960s and 1970s, ethno-nationalists from different ethnic groups, including those in Tigray, articulated the economic impoverishment and political underrepresentation of their people into ethnic questions. Thus, informed by the Marxist-Leninist ideology of national oppression thesis, TPLF also mobilized the Tigrean mass to emancipate the people from Amhara oppression, whether by establishing independent state or within the larger Ethiopian framework (Vaughan 2003; Aregawi 2009). Thus, TPLF embraced ethnic federalism and its principles of the right of self-determination of nationalities, with the historical center turned into a periphery, at least in terms of economic, political and social representations, if not in terms of its cultural representation in the state. Through the umbrella party–the EPRDF–the TPLF was able to entrench to the broader regions in Ethiopia beyond its original base in Tigray.

After taking state power by deposing the military regime in May 1991, EPRDF formally institutionalized ethnicity by restructuring the administrative map along ethno-linguistic lines (see map 2 in chapter one). More importantly, the 1995 Constitution declared the establishment of the Federal Democratic Republic of Ethiopia (FDRE), embracing the approach of federal arrangement that in principle claims to be 'coming together' of member states on a voluntary basis (Dereje 2006; Turton 2006; Berhanu 2009). In principle, the constitution introduced a new chapter in the process of nation-building in the country, with a remarkable departure from the past, by reversing both the identity and political foundation of the state. As Markakis (2011: 237) maintains:

> [T]he 1995 constitution was a remarkable departure from Ethiopian political tradition. It rejected the unitary model of the state and the centralized form of government that previous regimes had struggled to forge. It accepted cultural pluralism and rejected the link between culture, nationality and citizenship inherent in the model. It recognized the voluntary nature of membership in the state, as well as the right to renounce it. All in all, it seemed a reversal of direction followed in the century-old process of Ethiopian state building.

From the onset, I would like to make two important points clear prior to moving to the discussion. First, it is not the primary objective of this chapter to discuss in detail about the party ideology of the EPRDF government–the architect of ethnic federalism and the mastermind of policies, strategies and programs in the country for the last twenty-three years. I will rather focus on how the new political order has reconfigured center-periphery relations, in both principle and practice, as well as what such a reconfiguration entails in state-society relations. Second, it is not my intention to analyze the general discourses of ethnic federalism in terms of its practices and consequences. Since 1991, these issues have become major topics of debates among Ethiopian and foreign scholars, with several books, articles and dissertations produced on the topics (see Aalen and Tronvoll 2009; Aalen 2008; Abbink 1997, 2009; Asebe 2010; 2012a; Assefa 2006; Clapham 2002, 2006, 2009; Dereje 2006, 2010; Merera 2003; Turton 2006; Vaughan 2003, among others). My aim is rather to highlight some of the promises of ethnic federalism, particularly concerning the historical peripheries, as well as investigating its lived realities.

The new political order was thought to introduce two fundamental changes. First, an attempt to deconstruct the historical hegemony of the center and the centralization of the state was promised in the constitution by introducing the devolution of state power to constituent members of the federation that were built along ethnic lines. Second, the new political order ostensibly seemed to be a radical departure from the past by divorcing the century-old marriage between Amhara ethnicity, religion, culture and state identity and Abyssinian/Ethiopian state identity. This divorce was intended by recognizing all nations, nationalities and peoples of Ethiopia as important contributors in building a 'New Ethiopia' (Assefa 2006; Berhanu 2009). Markakis (2011: 279) summarized how both the decentralization of state power and the recognition of cultural pluralism were meant to contribute to build a sense of belongingness of the peoples in the peripheries, noting that:

> [T]he introduction of a decentralized federal system of government promised to end the centre's historical monopoly of ruling power, while the reformulation of Ethiopia's national identity on the basis of cultural pluralism lifted the burden of cultural inferiority from the periphery and the threat of forced assimilation. This was fundamental shift that opened new and hopeful perspectives for the country's future.

In principle, decentralization means the devolution of state power to lower administrative structures and the empowerment of local level decision makers to decide on economic, political, cultural and other social affairs of their communities. Nevertheless, as will be discussed later in this chapter, the success of this unfinished political project should be evaluated based upon how the political discourse is translated into practice.

5.1.1 Recognition of Cultural Pluralism

The preamble of the 1995 constitution begins by subscribing to the notions of common belongingness and embracing cultural diversity as follows:

> We, the Nations, Nationalities and Peoples of Ethiopia: Strongly committed, in full and free exercise of our right to self-determination, to building a political community founded on the rule of law and capable of ensuring a lasting peace, guaranteeing a democratic order, and advancing our economic and social development; [f]irmly convinced that the fulfillment of this objective requires full respect of individual and people's fundamental freedoms and rights, to live together on the basis of equality and without any sexual, religious or cultural discrimination (FDRE Constitution 1995, Preamble, emphasis in Original).

In contrast to the historical marginalization of the people in the periphery in terms of cultural, political, economic and social values and practices, the recognition of 'nations, nationalities and peoples of Ethiopia' as visible constituents of the federation marked a remarkable reversal of the past political order. During the imperial period under Emperor Haile Selassie I (1930-1974), the emperor was celebrated as 'elect of God' and the law vested absolute power upon him. The emperor and members of his royal family, as well as loyal nobilities, not only monopolized state power but also the identity of the state, with which members of the core center identified themselves.

Although constitutional analysis is not the central objective of this section, an overview of the 1955 revised constitution of the imperial regime provides us a glimpse into the state-society relations of the period. The proclamation that promulgated the adoption of the revised constitution reads as follows:

> WHEREAS, twenty-four years ago, at the beginning of Our Reign, We granted to Our faithful subjects and proclaimed a Constitution for the Empire of Ethiopia; and WHEREAS, Almighty God, the source of all benefits, has strengthened and inspired us to lead Our beloved People, during Our Reign, through the greatest of trials and hardships, to an era of great progress in all fields; and WHEREAS, being desirous of consolidating the progress achieved and of laying a solid basis for the happiness and prosperity of the present and future generations of Our People, We have prepared a Revised Constitution for Our Empire after many years of searching study and reflection; and WHEREAS, Our Parliament, after due examination and deliberation has submitted to Us its approval of this Revised Constitution; Now therefore, We, Haile Selassie I, Emperor of Ethiopia, do, on the occasion of the Twenty-Fifth Anniversary of Our Coronation, hereby proclaim and place into force and effect as of to-day, the Revised Constitution of the Empire of Ethiopia

> for the benefit, welfare and progress of Our beloved People. Given in Our Imperial Capital, on this the 24th day of Tekemt [October], 1948 (Ethiopian Calendar), and on the Twenty-Fifth Anniversary of Our Coronation.[10]

Moreover, articles 2 and 4 of the revised constitution clearly attached state identity to ethnic lines, which was built upon the legend of hereditary linkage between the emperor and king Solomon of Jerusalem, as can be understood from the following statements in the constitution:

> The Imperial dignity shall remain perpetually attached to the line of Haile Selassie I, descendant of King Sahle Selassie, whose line descends without interruption from the dynasty of Menelik I, son of the Queen of Ethiopia, the Queen of Sheba, and King Solomon of Jerusalem (Article 2). By virtue of His Imperial Blood, as well as by the anointing which He has received, the person of the Emperor is sacred, His dignity is inviolable and His Powers indisputable. He is consequently entitled to all the honors due to Him in accordance with tradition and the present Constitution. Anyone so bold as to seek to injure the Emperor will be punished (Articles 2 & 4 of the 1955 Revised Constitution).[11]

The proclamation and the above articles in the constitution clearly indicate the invisibility of diversities of the Ethiopian peoples and the exclusion of the non-center from the making of the state. The pronoun 'We' in the two constitutions has different meanings. The preamble of the 1995 constitution begins with the pronoun 'We', as in "We the nations, nationalities and peoples of Ethiopia", while the proclamation for the 1955 revised constitution begins with the pronoun 'We', as in "We granted to our faithful subjects and proclaimed a Constitution for the Empire of Ethiopia". While the pronoun "We" was used to represent the emperor rather than the peoples of Ethiopia in the 1955 constitution, the 1995 constitution considers the peoples as "authors" of the constitution itself. However, practical issues regarding the translation of the constitutional provisions into practice have reflected points of contention in both cases.

Under the past regimes, different ethnic groups in the country were denied their dignity of existence as distinctive groups and were forced to comply with alien language, religion and culture in their pursuit to participate in the country's socio-political affairs. As many scholars have indicated regarding the assimilation policy against the diverse nationalities in the conquered regions under the previous regimes, the Ethiopian empire worked hard to relinquish their

---

10 http://www.himchurch.org/Constitution/Ethiopia/The_Revised_Constitution_of_The_Empire_of_Ethiopia-Chap_I.html

11 http://psokolovsky.tripod.com/constitution.html

identity by demoting their language, culture, religion and indigenous system of governance (Markakis 1994; Mekuria 1997; Hamesso 2001). The military regime's modernist dream worsened the situation of cultural practices by denouncing them as 'backward' and being at odds with the government's development and modernization programs (Dereje 2011). In the post-1991 period, constitutional recognition and formal celebration of traditional practices have been evident, although Amhara nationalists still find such recognition too bitter to swallow.

It can be argued that the recognition of cultural pluralism within the new political order has been anticipated to boost the psychological mindset of the peoples in the periphery by erasing their cultural stigmatization, historical distortion and, more importantly, by empowering every ethnic group (nations, nationalities and peoples). In terms of language and culture rights for example, the constitution opens the space for all ethnic groups, "to speak, to write and to develop its own language; to express, to develop and to promote its culture; and to preserve its history" (Article 39.2, FDRE Constitution 1995). Despite practical limitations that I will discuss in the next section, it can be considered as a breakthrough to gain the constitutional recognition of self-government for the peoples in the extreme geographical and social peripheries, such as the Anywaa and Nuer in Gambella, the Gumuz in Benishangul Gumuz, the Mursi, Surma and Nyangatom in South Omo, the Somali and Afar, to mention a few (Clapham 2009). In areas of resource ownership and utilization, the new political order presumably grants different ethnic groups the right of administering, utilizing and managing resources on territories they historically inhabited. Accordingly, it is against this backdrop of departure from the historical past that contradictions between the lived realities of the political order and community expectations will be discussed later in this chapter.

5.1.2 Decentralization of State Power

The decentralization of state power was another major area intended to deconstruct the center hegemony and centralize the government system. The two constitutions during the Haile Selassie's imperial regime (1930-1974) and military rule (1974-1991) had enacted a firm centralized system of government by restricting the autonomy of administrative regions, let alone granting self-government to ethnic groups. For example, according to the 1955 revised constitution, "[T]he Sovereignty of the Empire is vested in the Emperor and the supreme authority over all the affairs of the Empire is exercised by Him as the Head of the State, in the manner provided for in the present Constitution" (1955

Revised Constitution, article 26)[12]. Far beyond the concentration of power at the center, it is fair to argue that power was monopolized in the hands of the emperor, as clearly stipulated in article 27 in the constitution. It reads, "[T]he Emperor determines the organization, powers and duties of all Ministries, executive departments and the administration of the Government and appoints, promotes, transfers, suspends and dismisses the officials of the same" (1955 Revised Constitution, Article 27)[13].

In the post-1991 period, decentralization did not merely mean the structural devolution of power from the center to subsequent lower administrative units (regions, zones and districts); rather, it entailed reconfiguration of the state structure from a unitary system to a federal arrangement. The Ethiopian ethnic federalism has been portrayed as if it constituted a voluntary membership of constituent groups into the federation, which subscribes to the notion of federalism called 'coming together' in contrast to the top-down imposition of federal arrangement, termed 'holding together' (Berhanu 2009). In fact, the federal system maintained the state that was built through conquest and thus it best fits the 'holding together' federal arrangement rather than 'coming together'. In any case, the 1995 constitution went very far in the country's history of hegemonic center in including a controversial article that grants nations, nationalities and peoples of the country the unconditional right to self-determination, up to and including secession (Berhanu 2009). Article 39 sub-articles 1 and 3 of the constitution stipulate the following:

> Every Nation, Nationality and People in Ethiopia has an unconditional right to Self-determination, including the right to secession. [...] Every Nation, Nationality and People in Ethiopia has the right to a full measure of self-government, which includes the right to establish institutions of government in the territory that it inhabits, and to equitable representation in state and Federal governments (Article 39 Sub articles 1 &3, FDRE Constitution 1995).

Some scholars argue that issues of ethnic-based domination under the Amhara supremacy during the imperial regime and the heavy hand of the military rule necessitated the provenance of carrying ethnicity as a rallying political instrument during the heydays of ethno-nationalist movements in the 1960s and 1970s. The architects of ethnic federalism learned from their home region in Tigray that the ethnicity card could be played within the larger Ethiopian context both to win legitimacy from non-Tigrean mass as well as experimenting with new forms of democratization along Western notions of multination feder-

---

[12]http://www.angelfire.com/ny/ethiocrown/Constitution.html

[13] http://www.angelfire.com/ny/ethiocrown/Constitution.html

alism (Vaughan 2003). Whatever the objective behind such political machination, there is no doubt that it generated a sense of recognition in the mindset of people in the peripheries, for whom issues of self-government had previously been unimaginable (Markakis 2011). According to the federal arrangement, administrative regions and their respective lower units established along ethno-linguistic lines were believed to be autonomous with respect to the right to administer their own affairs and govern the people and their resources. In this context, the federal arrangement can be considered as a process of restoring the autonomy the peripheral groups lost during the late nineteenth century Abyssinian conquest.

On paper, the constitution defines the role of the federal and regional states by limiting the former to major national issues such as defense, foreign policy, currency, inter-regional affairs and national economic policies and strategies (Berhanu 2009; Markakis 2011). It was assumed that the federal arrangement would boost the economic capacities of the regions by empowering them to administer resources, plan and extract revenues and administer investment within their respective areas of jurisdiction. Nevertheless, the regional states' autonomy and right to decide upon investment projects particularly related to land was retracted in 2010, with the federal government since being formally entitled to control, monitor and administer processes of transferring land in the regions to foreign or domestic investors (Assefa 2013). Although the question of whether the regional states voluntarily gave up their right over land or whether it was political decision from 'above' needs further research, it clearly shows how regional states are gradually becoming nominal administrative units with limited power over key aspects such as land, budget, investment, taxation, education and other major social service sectors. Moreover, the lack of fiscal devolution and the regions' increased dependency on the federal government for financial subsidies are among the major weaknesses of the federal arrangement, indirectly paralyzing the economic and political autonomy of the regions.

From a different angle, apart from the psychological aspect of empowering the historically marginalized groups, some scholars argue that decentralization could be a right approach to enhance the economic development and democratization in the country by ensuring the equal participation and equitable distribution of resources (Taye and Tegegne 2007). The discourse presumably seeks to redress the past asymmetrical power relationship between state and society and presumes to ensure equal opportunities for the citizens (Mohamed and Markakis 1998). Mohammed and Markakis (1998: 8) further argue that:

> Decentralization in Ethiopia is not seen merely as a device for the satisfaction of ethnic political demands, but also as the path leading to democratization through

devolution of decision making in a manner that enables more people to influence the political process. Furthermore, since decentralization and democratization are regarded as requisite to development, the empowerment of ethnicity is intended to harness ethnicity to the purposes of development.

In principle, ethnic federalism is supposed to serve as a path to democratization and development, harnessing ethnicity for such purposes. This is supposed to be achieved by granting ethnic groups the right for self-determination and empowering them to freely exercise their culture, knowledge, language and religion towards the same vision (Asebe 2012a). However, Mohammed and Markakis (1998: 8) critically warn us that the success of this unfinished project will depend on "whether the formal, i.e., constitutional provisions of decentralization and democratization are realized in practice." Indeed, I will discuss this critical dimension of the political order in the next section.

5.2 The Lived Reality of the New Political Order (mixed results)

In the previous section, I outlined some of the promises of the new political order and some aspects that marked it as a remarkable departure from the past. The devolution of state power, rights to self-government and autonomy, recognition of cultural pluralism and, more importantly, the reversal of center-periphery asymmetrical power relations and deconstruction of the identity of the state that was attached to a dominant ethnic group were among the remarkable features of the new political order, at least in principle. This is why the early-1990s was considered as the 'honeymoon' period for the proponents of ethnic federalism as a dawn of democratization, peace, stability and development in the country. Such enthusiasm was more evident among the people in the periphery, because the new system promised them something that they had not experienced for the past century or more. In this section, I will discuss the lived reality of the new political order, with particular emphasis on the center-periphery relations.

Ethnic federalism has generated mixed results. On the one hand, it has created a government system that has remarkably reversed the century-old centralized system and granted rights of self-government to administrative regions. Besides the opening of some political spaces for the exercise of multiparty democracy in the early years, the government has delivered in some areas of social services, such as education and healthcare. On the other hand, ethnic federalism has ethnicized every aspects of life among the peoples of Ethiopia. As Abbink (1997: 160) argues, "In contemporary Ethiopia, the discourse of ethnicity has become strongly politicized, more so than before", and it influences every form of social interaction, policy formulation, citizenship, education policy, budget allocation and inter-group relations. As a result, inter-ethnic conflicts have become apparently frequent during the post-1991 period, despite the relative peace and security in terms of major wars. Moreover, issues of the further marginalization of the periphery have become evident owing to the government's policy of "Land to Investors" by reversing the motto of the 1960s Student Movement, which was based on "Land to the Tiller" (Markakis 2011). In the following few paragraphs, I will discuss both dimensions of the mixed results of ethnic federalism, to contextualize the discussions on the Nech Sar National Park within the macro-political arena.

At a macro level analysis, EPRDF government's model of ethnic federalism can be marked as a successful experiment, at least in some key areas, especially during the model's heydays in the 1990s. Despite continuous intervention and regulation from the federal government, which undermines the autonomy of regional governments (Assefa 2006), the establishment of regional states and lower administrative units along ethnic lines has increased a sense of recognition, identity and autonomy among the people in the periphery. This is thought to reconfigure their peripheral status and is believed to have brought them closer to the center.

Until the 2005 contested election, there were relatively open political spaces in which multiparty politics was exercised. Opposition parties, mainly organized along ethnic lines, won some seats in the past three subsequent elections (from 1995 to 2005). Besides the constitutional provision of freedom of expression, Ethiopians began to enjoy the right of free press until the government closed down almost all private newspapers in the post-2005 election, allegedly accusing them of instigating conflicts (Markakis 2011). Some scholars argue that the EPRDF government's economic policies were also partly translated into economic schemes, enabling the country to record significant economic growth (Clapham 2009; Dereje 2010). For such a group of scholars, the government's emphasis on rural agriculture and to some extent the 'free market' economy was contributory factors for the so-called economic growth. However, issues of fair distribution of the economy and resources remain points of contention. In areas of decentralization, the distribution of primary education and healthcare services to peripheral regions was enhanced because regional and other local administrative units were given the autonomy to decide upon these matters. Moreover, as mentioned earlier, the coming to power of EPRDF ended the long and large-scale wars that the country suffered during the military regime and installed relative peace and stability (Hizkias 1996; Abbink 2009).

Another important result of ethnic federalism worth mentioning is the establishment of 'autonomous' regional states and their respective lower level administrative units. The following statement by Clapham (2009: 187) indicates some aspects concerning the decentralization of power and the continuation of federal state intervention:

> To have a government in Konso or Welayta that was visibly run by Konsos or Welaytas was a revolutionary change in Ethiopian politics, and one, which, like the land reform of 1975, established a new, and for many purposes irreversible transformation in the relationship between people and the state. That policies were still to a very large extent made in Addis Ababa for local implementation, and that the constitutional guarantee of a right of secession was more apparent than real, did not at least initially matter anything like so much as the change in the visible appearance of government on the ground.

From regional states to local administrative units, administrative councils were established to deliberate on issues pertaining to their specific areas. Regardless how the autonomy is contested, there are clear cases in which local governments can decide on local budget and other projects, such as in the areas of cultural revitalization, language choice for administration and education, designing education curriculum for primary level and the extraction of taxation from small business enterprises.

However, the relationship between the federal government and regional states has remained contentious throughout the last two decades, with latent frictions between the federal state, regional states and different administrative units along ethnic lines apparently testing the federation in many instances. This includes claims for regional status, like the Sidama[14] question, as well as claims for autonomous zonal status, particularly in the multiethnic Southern Ethiopian nations, nationalities and peoples (SNNP) regional state (Merera 2003; Vaughan 2003). As Assefa (2006) argues, continued intervention from the federal state in the affairs of the regional states has threatened the latter's autonomy to the extent that the country has once again returned to centralization. Although it has granted some form of self-government to historically marginalized groups in principle, the EPRDF has continued to play a centralizing role by reducing

[14] The Sidama are the largest ethnic group in the SNNP regional state with a population number of nearly three million people according to the 2007 population census of Ethiopia. From 1991-1992 following the coming to power of EPRDF, Sidama was organized as 'autonomous' region but it was reduced to a status of zone since then. Following such rearrangement, different Sidama political actors both in the government and in opposition were raising questions of regional status and at times, it instigated conflict between these actors and the regional/federal authorities (Aalen 2011).

regional and local administrative divisions to the status of implementers of policies made in Addis Ababa (Assefa 2006; Clapham 2009). This ambivalent role of the federal state and contrasts between the constitutional provision and practical realities are part of latent tensions between regions and the federal state, as highlighted in the case of management of resources such as protected areas, to which I will return shortly in this chapter.

Among the drawbacks of ethnic federalism, I will focus on two fundamental areas directly relevant to the current study, namely issues of ethnicization of resource ownership and their role in instigating inter-ethnic conflicts and aspects of strong presence of the center in the peripheries. Issues of one-party domination in the country's political landscape, continued intervention of the federal state in the affairs of the regions and the narrowing of political spaces and limited freedom of press, including the collapse of civil societies in recent years, have become topics of debate among politicians, academics and human rights organizations (Assefa 2006; Abbink 2009; Markakis 2011). Of course, it is not my objective to repeat the same debate in this research. Despite the right for self-government being granted to ethnic groups to the extent that they could negotiate aspects of federation up to secession, ethnic federalism has increased inter-group divergences rather than building on similarities. In fact, such creation of inter-group tension is not necessarily a problem that can be attributed to the notions of ethnic federalism, but rather due to weaknesses in implementation and the way in which it has been translated into practice (Asebe 2010, 2012d). As a result, common resource areas happened to attract ethnic labels of ownership from which members of different groups were restricted. Thus, territorial claims, competition over natural resources and questions of autonomy have triggered conflicts.[15]

As I have discussed elsewhere on the Guji-Burji conflict, contestations over resources have assumed ethnic dimensions through the involvement of local ethnic entrepreneurs reactivating old memories of antagonism within the current ethnic based divisions (Asebe 2012d). Conflicts between pastoral communities or between pastoralists and agriculturalists over resources (farmland, grazing land, water sources, hunting grounds, etc.) are common occurrences elsewhere in Africa (Zelessa 2008). However, conflicts in post-1991 Ethiopia have mainly adopted ethnic dimensions for resource claims or in questions for self-government and autonomy. One of the early tests to the implementation of

[15] For different inter-ethnic conflicts in the post-1991 period such as the cases between Borana and the Somali, Guji and Burji, Guji and Gedeo, Afar and Karayyu, Guji and Sidama, Nuer and Anywaa, Oromo and Amhara settlers in East Wallaga and so forth see Asebe 2010, 2012d; Fekadu 2011; Tesfaye 2009; Dereje 2011).

the federal arrangement came from the Guji Oromo and Gedeo, who entered into two rounds of conflicts in 1995 and 1998 concerning questions of self-government (Hussien 2002; Asebe 2010).

Given that decentralization meant to open opportunities for local elites in administering their respective districts, the first decade in the experiment of the political model experienced a proliferation of questions concerning the establishment of distinctive self-governing units based upon any possible markers of differences. In the SNNPR, there are over fifty ethnic groups merged together, albeit with different levels of demands for autonomy (Vaughan 2003; Cohen 2006; Markakis 2011). Whether these consequences are fundamental drawbacks of ethnic federalism or weaknesses of the government in translating its discourses into practice is a matter of controversy. In my view, ethnic federalism could have been a viable system of democratization and development process, harnessing ethnicity for the purposes of establishing harmonious relationships between the state and society through genuine empowerment and participation. However, EPRDF has partly failed to genuinely translate its policy rhetoric into practice and partly decentralized conflicts from the center as a mechanism of consolidating its power and building legitimacy as an arbiter of inter-regional or inter-group tensions.

The second area of discussion, which is not a direct consequence of ethnic federalism, but rather seems to have reversed EPRDF's assertion of decentralization and self-government of regional states and more importantly enhanced a process of re-marginalization the periphery, is the center's intrusion into the periphery through its development programs. For many decades, the lowland parts of the country remained marginal to the state's 'development' interests, partly due to the state's failure to 'transform' these regions into economic resources, as well as owing to the long-established representation of these regions by the mainstream society as 'wasteland' and being inhabited by violent pastoral communities. Despite the incorporation of the lowland peripheries into the Ethiopian empire in the late nineteenth century, state presence was minimal during the past regimes (Hagmann and Alemmaya 2008; Markakis 2011). On the other hand, the EPRDF government has successfully reached the periphery, both through its ethnic-based administrative structure and in the form of 'development' projects that were aimed at taming the 'wilderness' and converting it into an economic resource. It departed from its predecessors in reversing the old assumption about the periphery as a violent prone source of trans-border insecurity and marginal in its contribution to the national economy. The periphery has become viable a source of state economic interest in the form of providing ample arable land for foreign and local investors and government projects, such as a hydroelectric dam, large-scale agricultural projects and tourism.

State intrusion into the periphery during the post-1991 period is evident in two different yet interdependent areas of development interventions. While the first is related to the government's economic policy of liberalization through the adoption of market liberalism and privatization, attracting foreign and local investors, particularly in areas of agriculture, the second is linked to the government's ambition to diversify the economy to foster development. To begin with the privatization policy and how it has enhanced the state intrusion into the peripheries, it is imperative to highlight on the ideological shift the TPLF/EPRDF made along the process of assuming state power in 1991 and during the years that followed. TPLF/EPRDF dropped its Marxist ideology soon after it assumed state power and embraced a capitalist ethos of market liberalism and multiparty democracy (Markakis 2011). In the process, privatization and the opening of investment schemes for domestic and foreign investors were envisaged to boost the economy. According to a policy document of the government, granting land to foreign and domestic investors was thought to contribute to the country's ambition to reduce poverty and ensure sustainable development (SDPRP 2003). To meet its ambitious plans of generating foreign income from agricultural investment sectors, leasing out land in the peripheries to investors has become the government's major target (Markakis 2011).

The government's transfer of large tracts of land to investors is a major departure from the 1960s "Land to the Tiller" slogan of the student movement in which the architects of ethnic federalism were part. This new approach of "Land to Investors" sharply contrasts with the notion of self-government of nations and nationalities and their autonomy to decide on their resources. For example, according to the 1995 constitution: "Ethiopian peasants have right to obtain land without payment and the protection against eviction from their possession. Ethiopian pastoralists have the right to free land for grazing and cultivation as well as the right not to be displaced from their own lands" (FDRE 1995 Constitution, Article 40, sub articles 4 and5).

By contrast, the constitution has left a loophole for government appropriation of land, given that it reads, "Without prejudice to the right of Ethiopian Nations, Nationalities, and Peoples to the ownership of land, government shall ensure the right of private investors to the use of land on the basis of payment arrangements established by law. Particulars shall be determined by law" (Article 40, sub article 6). The central point is that resettlement schemes are still underway in Ethiopia to create 'empty' land for the investors in the peripheries for the sake of different investment programs. The resettlement program should be assessed in terms of issues related to participation, socio-cultural and economic impacts and customary rights of local communities. For example, agricultural investment programs in Gambella regional state and South Omo in the SNNP regional state and large-scale flower farms in central Oromia have resulted in the resettlement of hundreds of thousands of local people, with serious ecological and social consequences (Markakis 2011, see map below).

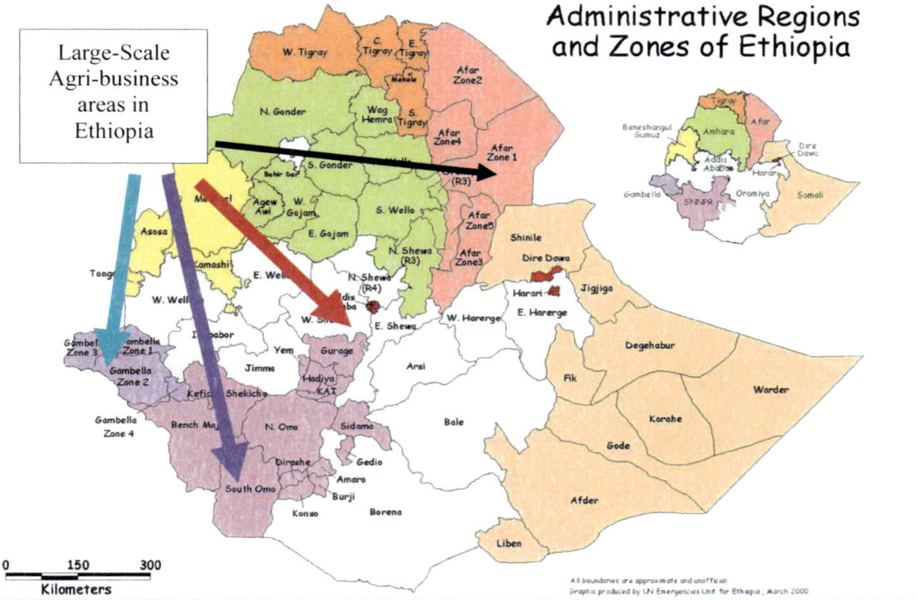

Figure 11: A Few Areas of Large-Scale Agri-business in Ethiopia (Adapted from discussions in Markakis 2011)

The second dimension of government meddling in the peripheries has been through its own development projects, including hydroelectric dam construction, sugar cane plantation, sugar factory projects and national parks. The government builds hydroelectric dams to generate electric power, both for national consumption and as means of generating foreign currency by selling electricity to neighboring countries. Besides hydroelectric power generation, it builds dams establish large-scale irrigation agricultural projects from the water in the reservoirs, as is the case of the Gilgel Gibe III hydroelectric dam under construction on the Omo River. The government's engagements in large projects such as ten sugar cane plantation and sugar factories currently under construction (six of them in South Omo) reflect just a few cases in which the peripheries have become arenas of state interest and intervention. Different media coverage has speculated that the construction of Gilgel Gibe III hydroelectric dam and the sugar factories will have disastrous ecological and social implications, potentially leading to the displacement of over 500,000 indigenous peoples from the region.[16]

From the part of the government, the intervention has been justified as a process of modernizing and transforming the peoples in the region. The late Prime Minister Meles Zenawi once stated that:

> The Federal government is working hard to bring a permanent solution [to problems pertaining to pastoralists]. The Gilgel Gibe 3 dam is developing rapidly and when it is finished the flood, which has been a huge problem for years in this region, will end forever. It will then be possible to create a big irrigation system in this wide and fertile area of South Omo. Following the good results we have achieved in the Afar region, the government is planning, and working hard to establish, a 150,000-hectare sugarcane development in this area starting this year. When this development work is done, we believe that it will transform the entire basis of the area. This will benefit the people of this area and hundreds of thousands of other Ethiopians, by creating employment. The pastoralists who live around this area will be given some fertile land from this irrigation system, which can be used for their own cultivation. There will be support for the pastoralists to combine agriculture with modern cattle herding.[17]

By presenting the federal government as a guardian of the people and a legitimate development actor, the then Prime Minister brought the government rhetoric of development down to the pastoralists, whose agency in deciding on their ways of life has been usurped through a top-down development plan by the federal government. According to the government's version of development, "big irrigation projects, commercial farming as in the case of sugar cane development, creating employment opportunities and introducing combination of agriculture with modern cattle herding have been presented as the basis of transforming the pastoralist lives towards development" (Asebe 2012c: 40).

The establishment of protected areas in the form of national parks is another aspect in which state control of territories in the peripheries has been exercised. Although many national parks were established during the previous regimes, land appropriation for conservation purposes and subsequent resettlement programs have also been underway during the post-1991 period (Turton 2011). While some of the national parks have been established since 1991, the existing parks have been consolidated, gazetted and the resettlement of local people has been made within the current political system. In short, the EPRDF

---

[16] http://www.businessweek.com/news/2011-09-13/ethiopia-building-10-new-sugar-plants-to-become-leading-exporter.html

[17] http://www.mursi.org/pdf/Meles%20Jinka%20speech.pdf

government has also maintained the status quo that made the historical South a home of national parks, with over eighty percent of the national parks in Ethiopia located in this part of the country.

5.3 Ethnic Federalism and Community Voices in the Periphery

In principle, the new political order is supposed to grant ethnic groups the unconditional right to participate in the making of the new federal state. By contrast, the government's increased interventionist policy seems to have resulted in further suppressing the participation the people in the peripheries in economic and political spheres. In this section, I will discuss the interplay between ethnic federalism and issues of representation and participation of the periphery in the decision-making processes on development projects that affect their lives. More specifically, I will focus on the promises of ethnic federalism, which grants ethnic groups the right to self-determination on the one hand, as well as EPRDF's shift to a 'developmental state' paradigm and how it suppresses notions of participation and representation on the other. In the previous section, I have discussed that the state has penetrated into the periphery through 'development' projects. Thus, state intrusion into the periphery changed the center-periphery relationship once again to a form of state predation over the periphery concerning economic interests and political control (Markakis 2011).

In administrative aspects, the EPRDF government has extended its areas of influence far beyond its predecessors by reaching the periphery. One of the major challenges for the TPLF in the late-1980s and early-1990s was the question of transforming its ethno-nationalist program to an Ethiopian framework and thereby gaining legitimacy beyond Tigray. To tackle this challenge, it allied with a few existing ethno-nationalist movements and forged new ones, through which it could reach their respective populace. This was partly successful following the establishment of EPRDF in 1989 (Vaughan 2003; Clapham 2009). However, the establishment of the EPRDF did not solve the historical challenge of successive regimes in the country to their efforts to exert effective control over peripheral regions, particularly the pastoral communities. Thus, in its pursuit of reaching the people in the peripheries, EPRDF has managed to co-opt local elites (primary school teachers, secondary school dropouts and people with very low political or educational competency), training them in Civil Service University College in Addis Ababa, where they were socialized into the political ideologies and policies of the EPRDF. Upon the completion of their education, these government cadres returned to their home constituencies and assumed administrative positions at different levels, where they served as loyal subjects in implementing the policies and strategies designed in Addis Ababa (Vaughan 2003; Markakis 2011).

Therefore, before converting the territories the state historically represented as 'wasteland', 'desolate', 'wild' and 'violent' into economically viable spaces, the EPRDF government has managed to secure it under its administrative control. Through such nominal and symbolic empowerment of the periphery, EPRDF has developed notions of self-government, decentralization and sense of ethnic identities (i.e. sense of pride in one's markers of distinctive group representations) in the periphery, in both principle and practice. As Dereje (2011) argues, during the post-1991 period, the Anywaa in Gambella regional state have resorted to aspects of self-pride in being distinct from the highlanders, who historically used to demote them under the previous regimes.

However, the honeymoon in the periphery did not last long. After consolidating its administrative grip by establishing satellite parties, the EPRDF government entered into a new chapter of intrusion to the peripheries through the 'domestication' of the 'wilderness' as an economic resource. Contradictions in the EPRDF's relations with the periphery emerged from two angles. Firstly, there are some contrasts between constitutional provisions and its practices. It has been stated that, "Nationals have the right to participate in national development and, in particular, to be consulted with respect to policies and projects affecting their community" (FDRE Constitution 1995, Article 43 sub article 2). Nevertheless, because the government already established a dichotomous relations between the center and the peripheries by labeling the latter as 'backward', 'developing' and 'emerging' regions, it rarely recognizes their right to be consulted and participated in development plans. Second, there is an apparent contradiction between the EPRDF's political rhetoric and its economic interests. Politically, the notions of ethnic federalism envisage the promotion of the self-government of nations, nationalities and peoples in their respective regions and/or units with autonomy. This would entail the right of local administrative units to administer resources or their traditional land according to local norms, values and systems of governance.

Economically, it has devised ambitious plans to transform the economy of the country and the lifestyles of peoples in the country, particularly in the periphery, through large-scale development projects. In the process, there have been apparent clashes between promises of ethnic federalism and development programs of the government (Markakis 2011). In a different context, Dereje (2006, 2013) contends that despite experiencing some senses of political and psychological empowerment, the people in the peripheries practically felt the pains of the new political order as it enhanced their marginalization in terms of economic benefits, political representation and autonomy, and because it has become a factor for the escalation of existing conflicts.

The glaring drawback in the EPRDF's administration is that it failed to effectively translate its political rhetoric into practice in many cases. For instance, in contrast to the much spoken participatory approach of development, hierarchical structures persist in the state-peasant relationship in the process of implementing development programs. Rather than providing the political spaces for the citizens to contest and negotiate discourses and practices of development, it has become a technocratic and top-down process channeled down to society for implementation (Zerihun 2004). In other words, despite the rhetoric, the new political order has demonstrated a retreat to the hegemony of the center, with strong state intervention into local affairs specifically in the resource-rich peripheries, particularly since the early-2000s. As one former administrator of Amaro district recalled concerning the resettlement process of the Koore in 2004: "the higher officials decided everything without any consultation with us. Neither local people nor the district administration was asked. We were only given orders to resettle the people according to the decision" (interview with former Amaro district administrator, April 2012). This experience of the district administrator was widely shared among Koore local communities in the villages adjoining the national park, who blame the government for ignoring them in the planning processes of the resettlement.

Moreover, both local government officials and local communities on the side of the Guji reflected their discontent concerning the government and the park administration for not considering them as important contributors for the management of the resources. One Guji elder once remarked, "The people from the government visited only the animals but they did not talk to us. We came to know that they care for the animals than for people. When we go to government meetings, they tell us we have right to talk. But when we talk, they detain us. What type of government is this?" (Informant: Aga Guracha, May 2012). The elder clearly noted the mismatches between the government's rhetoric and practice in areas of participation and empowerment of the local communities. While the government officials encouraged the local people to be engaged in debates concerning issues that are important to their communities as signals of democratization, the actions of detaining those who are critical in their speech during local meetings can be seen as a strategy of suppressing critical voices. As I will discuss in chapter eight, Galana district administration detained seven Guji elders in April 2012 for appealing to the higher government structure against the destruction of sixty-four houses in Gode village by the district and park administration.

At this point, the basic question to ponder is, how does the government justify its compromise of political rhetoric through development interventions? Although one might give different possible answers to this question, the most common explanation on the part of the government is that the EPRDF government prioritizes economic development over democratic rights. The government intervention into the autonomy of the periphery and its apparent retreat from rhetoric of democracy, good governance, multiparty politics and market liberalism to 'development' can be understood as an ideological shift within the EPRDF government, from a neoliberal ethos to a 'developmental state' paradigm. After a decade of adopting a neoliberal paradigm mainly along its shift in ideological orientations from Marxist-Leninist philosophy to neoliberal political and economic discourses, the EPRDF has once again began to emulate experiments from the East. Unlike when it was a rebel group, the ideological emulation that the EPRDF has introduced since the early years of the 21st century was not from Russia but rather from the Southeast Asian countries. According to an article posted on a pro-EPRDF website called *Aiga Forum*, the ideological shift that EPRDF made from a neoliberal political and economic paradigm to a developmental state philosophy has been in response to the failure of the neoliberal experiments during the Structural Adjustment Programs implemented from 1991-2001.[18]

As the central figure in the country's political and economic ideologies in the last two decades, the late Prime Minister Meles Zenawi was an advocate of a developmental state as an appropriate model for the renaissance of African development. In his monograph published on the *Initiative for Policy Dialogue (IPD)* website in 2006, Meles described neoliberal paradigms as 'dead ends' and developmental state paradigms as 'new beginnings' in Africa's quest for development (Meles 2006). The late Prime Minister argued that the neoliberal paradigm has taken Africa on the wrong path, in both political and development dimensions, in terms of understanding the underlying problems and the solutions it prescribes. In his opening speech at the AU Conference of Ministers of Economy and Finance held in Addis Ababa (24-29 March 2011), Meles urged member states, academic institutions and political parties to further the discussion on a developmental state paradigm and practically institutionalize it in their respective countries as a right path to development in the continent.[19]

Apart from the bulky debates on theoretical underpinnings of neoliberal

18 http://aigaforum.com/articles/meles-development-paradigm.pdf

19 file:///C:/Documents%20and%20Settings/user/My%20Documents/Books/Ethiopian%20PM%20bashes%20neoliberalism%20_%20AU%20Ministers%20call%20transforming%20rentseeking%20political%20systems%20_%20Danielberhane%27s%20Blog.htm

versus developmental state perspectives, a developmental state paradigm gives the state the opportunity to intervene into economic schemes while also limiting the participation of citizens in political and economic decision-making processes. A developmental state paradigm is a political and economic school of thought that favors a strong state or one that intervenes and regulates the economy, in contrast to the neoliberal development doctrine, which encourages a night watchman state. In other words, by providing the state the space to control the political and economic schemes, the developmental state paradigm allows the government to justify its resource expropriation, as well as the restriction of local people's participation. In his latest chapter of the book "*Good Growth and Governance in Africa*" (Noman et al. 2012, eds.), Meles argues that "developmental state can be semi-democratic and semi-parliamentarian at best" (Meles 2012: 167). This clearly shows how Ethiopia has experimented with being a developmental state at the expense of democratic rights, but rather in favor of strong state intervention in the economy. The idea that the first demand of the poor is 'bread' before they ask for other sorts of rights is central among the proponents of a developmental state paradigm, including the EPRDF.

However, the argument that attempts to justify the EPRDF's shift to a developmental state political and economic discourse in response to the failure of the neoliberal policies is questionable from many angles. In the first place, according to Markakis (2011), the EPRDF has never dropped its Marxist ideology of popular democracy, even during the heydays of its neoliberal experiments in the 1990s. Rather, liberal democratic principles were juxtaposed with a Marxist-Leninist ideology of popular democracy that was largely embedded in the party ideology of the TPLF. Secondly, the assertion that the EPRDF moved to developmental state in response to the failure of neoliberalism masks the internal political and economic motives of the party. The EPRDF government relatively opened spaces for political parties, civil societies, media and NGOs to engage in political debates in the country during the early years. However, it was challenged along the way, from both within and outside the party. Particularly following the split in TPLF in 2001, which had finally shaken all member parties in the EPRDF, the party has entered into a new chapter of power consolidation in both political and economic schemes. From the outside, pressure from opposition parties, civil societies, the media, international donor countries and organizations for more democratization processes have moved the party in a different direction (Markakis 2011).

The government used developmental state paradigm as a justification of the intervention on the one hand, and as a way of silencing pro-democracy critics on the other. As Meles (2012) argues, to bring sustainable development, the state should first take the leading role in designing, implementing and managing development projects that are autonomous from the society and private sectors. By allowing the interventionist policies of the state, the developmental state proves to be hegemonic at times, in that "the developmental agenda must be hegemonic if successful development is to take place and the developmental state is to be established" (Meles 2012: 168).

5.4 The Experience from the Nech Sar National Park

The Nech Sar National Park should be seen not only as a physical space with valuable environmental resources, but also as a contested political, cultural and symbolic space within which multilayer actors contest, negotiate and at times conflict on the meaning, utilization, management and ownership of the territory. With the articulation of the political order along ethno-linguistic lines in the post-1991 period, resource spaces have become political spaces with ethnic connotations. Thus, apart from the involvement of many actors with different interests, the new political order has also created new forms of representation of the space as a territory that some ethnic groups link to their identity, while others associate it to their history of settlement. One fundamental absurdity in the political model is the absence of an alternative approach to accommodate non-ethnic spaces and territories such as common resources, which are difficult to clearly delineate along ethnic lines. In this regard, the principles of ethnic federalism do not explicitly provide a guideline concerning how to administer protected areas that happened to fall along inter-ethnic borders such as the Awash, Nech Sar, Omo and Mago National Parks. Putting such areas under the federal government would undermine local autonomy, whereas designating them under specific regional state inevitably raises inter-ethnic and inter-regional competition for the ownership and administration of the territories.

For example, in the case of the Nech Sar National Park, different strategies of administration have been attempted, although none of them has been successful to date. From 1991-2004, it was been decentralized to the southern nations, nationalities and peoples (SNNP) regional state, although this strategy instigated discontent from the Oromia regional state, which also claims ownership and administrative right over the territory. Although it was assumed to neutralize the inter-regional contestation by transferring the management of the park to an international company in 2004 through a twenty-five year contract, the company terminated the contract in 2007, mainly due to a lack of inter-regional collaboration between the two regions concerning the issue of the park. Since the withdrawal of the African Parks Foundation (APF), the federal government has administered the park through the Ethiopian Wildlife Conservation Authority

(EWCA). In this section, I will treat these different trajectories in the management of the national park, hinting at the national discourses of ethnic federalism and whether or not has fostered or hindered access to common resources and the participation of local communities in issues that affect their lives.

5.4.1 Ethnic Federalism, Decentralization and the National Park

Following the reconfiguration of the administrative map of the country along ethno-linguistic lines, the Nech Sar National Park has been included under the SNNP regional state within the Gamo Gofa Zone. Shortly before the new arrangement, the Guji and Koore people who were evicted during the military regime returned to their former home (inside the park). According to some sources, the people reestablished their living in their previous settlement areas using the opportunities of the power vacuum during the transition following the collapse of the military regime (Dessalegn 2004; Freeman 2006). After people returned to the park territory, a hostile relationship was created between the people and the park (and the wildlife) because the people began to associate the park and its resources with the state. In this context, the state has been seen as a coercive, predatory, exclusionist and suppressive entity. It is evident that due to the nature of the state and the experiences and people's memories of problems related to state induced resettlement and displacements, society has considered the state project and institutions as something to be avoided and resisted. As a result, they have used different strategies to dismantle the national park, the details of which will be the subject of chapters seven and eight. The regime change in 1991 also created a new form of competition between Oromia and SNNP regional states concerning questions of administrative rights over the park and added a different form of claim between the Guji and Koore, which took ethnic lines rather than purely a struggle over environmental resources.

According to claims from Guji elders and political elites from Oromia regional state, this arrangement was in sharp contrast with the principles of ethnic federalism. One government official from Borana zone (Oromia region) argues that:

> Our people [the Guji Oromo] have been living in Ergansa [Nech Sar] for many centuries before the park was established. No Koyra [Koore], no Gamo and no other people lived there with the Guji. Our ancestors actually used the current Arba Minch town as their grazing areas and all areas around Abulo were Guji customary lands. The previous government evicted the people from their home but now the dictatorial regime has gone. However, when the new administrative regions were established, the park was included into the Southern region. In reality, it was supposed to be included into our region [Oromia] because the very principle of ethnic

> federalism is to reconfigure administrative units according to ethno-linguistic aspects of belonging. The people in Gamo Gofa simply contended that the Guji were not legal inhabitants of the area. Unless they deny historical facts, the Guji are rightful inhabitants and thus the area belongs to them. This has created unresolved but latent tension between the two regions (former administrator of Galana-Abaya District, April 2012).

On the other hand, according to views from the part of officials in the SNNP regional state, particularly in the Gamo Gofa zone, the national park's inclusion into their region was based on historical records of administration. During the imperial and military regimes, the park was partly administered under the Gamo Gofa and partly under Sidamo provinces, which are now largely under the SNNP regional state. According to members of the Gamo ethnic group, the park is a resource space and not an ethnic space. Nevertheless, one can easily refute this argument from the very notion of ethnic federalism that ethnic groups and their territorial units are the central focus in the reconfiguration. Therefore, if the Gamo or Sidama did not inhabit the present day Nech Sar National Park, how can its inclusion into the region be justified? This contestation is important to mention because it hints at the politicization of resource spaces along ethnic configuration, as well as some inconsistencies in the translation of the federal arrangement into practice. The national park has become an arena of contestation between the two regional states, among local communities, between the park and the local communities, as well as between regional states and the federal government. It has also been a field in which different local and external non-state actors contest different missions, including environmental, economic, human rights, etc.

As part of its decentralization endeavors in other aspects of administration, the government has initially attempted to devolve the administration of national parks to their respective regions, with the exception of few contested national parks such as the Awash National Park (Jacobs and Schloeder 2001). Later, many other national parks were centralized under the federal government due to tensions between regions or different adjoining ethnic groups. For example, Awash, Gambella, Semien Mountains, Nech Sar, Omo and Abijata Shalla National Parks are now under the Ethiopian Wildlife Conservation Authority (Young 2012). In the 1990s, during the time in which it administered the Nech Sar National Park, the SNNP regional government began an ambitious plan to rehabilitate three major parks (Nech Sar, Omo and Mago National Parks) , with five-year financing from the European Development Fund. The preliminary phase of the project began in 1994 for two years, during which it intended to accomplish gazettement of the three parks, resettlement of people who lived within the boundaries of the parks and strengthening the capacity of the then Ethiopian Wildlife Conservation Organization (EWCO) in areas of legal, infrastructural and institutional capacities (Turton 2002).

According to the interim report of the preliminary phase of the project, the resettlement of about 3,220 Guji and 3,580 Koore individuals who were living inside the park was proposed as an agenda not to be compromised (Technical Report No.4, 1995). This relocation plan was arranged to meet the donor's precondition for further large-scale funding, as stated in the report, "Concerning Nechisar National Park, the main precondition to any further development inputs is the successful relocation of the people living in the park according to a programme that satisfies all parties" (Technical Report No. 4, 1995: 3). One of the glaring weaknesses of the project implementation process was the park administration's rush to meet the prescribed precondition of resettling the people who lived in and around the park, without due consideration of their customary rights. As the technical report indicated that the park administration considered the people as illegal intruders, it should be noted that all resettlement plans, socio-economic surveys, discussions with the people and plans for establishment of community services (if any) were to be conducted within this asymmetrical power relation, between a 'legitimate' government body and 'illegal' settlers. It was stated that, "thirty years ago, there were no people resident in the park. At various times since then, Koyera and Guji people have settled in the park to cultivate and graze cattle so that today over 7,000 people and 4,000 cattle occupy the park–considerably outnumbering the large wild animals" (Technical Report No. 4, 1995, Annex 1: 2).

According to a socio-economic survey in 1995, over 460 Guji and 550 Koore households were reported to have been engaged in agro-pastoralism and farming in the park, respectively (Getachew 2007). It should be noted that any discussion or negotiation between the state and the local communities concerning the future of the national park would be difficult because both actors held competing narratives about the history of human settlement in the area. Eventually, the government planned the resettlement program with serious drawbacks. In the preparation for the implementation of the project, the participation of grassroots communities was not recognized, apart from discussions held with few local elites whose legitimacy as representatives was contested, given that most of them were handpicked by the state based upon their loyalty. The project was planned without genuine participation of the local communities. As a result, some argue that the government sought to impose its notions of biodiversity conservation in a military style of top-down approach (Freeman 2006).

As can be learned from the attendees at the final workshop held in Arba

Minch town in May 1995, only nine community elders were represented from ten Guji and Koore villages, while the remaining thirty-one participants were government officials from park administration, district offices and regional bureaus (Technical Report No. 4, 1995). Here, I would argue that the government resorted to the nominal type of 'participation' to pursue its predetermined conservation agenda. Although participation might take different forms, such as nominal, passive and interactive participation, the only way for development projects to achieve success is through genuine interactive participation, whereby local communities participate in the planning, management and benefit of the project (Turton 2002). The essential feature of interactive participation is that local people are involved in the design, planning and implementation of a project from the inception to the end. As Turton (2002: 110) argues: "It must surely be accepted by now that, unless people have real power to influence the way a project is designed and managed, they will not feel that it is 'theirs', whatever (often temporary) benefits they derive from it."

Thus, the management approach of the Nech Sar National Park has largely been exclusionary of the local people, except the involvement of a few local elites whose participation has been considered by their members as the state's strategy of controlling and gaining access to the communities. According to Halake Chari, a Guji elder from Mi'o, who I interviewed in Darba Manana in February 2012, "The government people always know how to inspect what we think about the park. They pick one or two young men and give them money or I do not know what they do but when they come back to our villages, these individuals are completely changed. They pass our secrets to the government. That is how the government controls our activities."

One can imagine the nature of discussion that would take place between participants within such power-laden contexts. According to key informants, the chairman of the workshop delegitimized the customary right of both groups by presenting them as illegal settlers and attempted to persuade the local community representatives to accept the resettlement plan without further precondition. However, the representatives, particularly of the Guji Oromo, finally pledged to discuss with their community members because they believed that they had to discuss on the issue at community level rather than deciding based upon the views of a few individuals.

Another drawback in the process of the resettlement plan was that beginning from the inception throughout the planning period, local communities did not get the chance to contribute their views. Government technocrats designed and brought the plan to the workshop without the consent of the people. Accordingly, it ignored the cultural, spiritual and livelihood connectedness of the people to the environment. Rather, it only focused on economic assessments such as fruit trees as major variables for compensation and for provisions of other social services in the new settlement areas. For example, although it recognizes the Guji as a 'nomadic' group, the new resettlement plan was meant to completely shift the Guji lifestyle into sedentary farming. Apart from the cultural repercussions of such a shift in livelihood style, the difficulty it would pose upon the people to adjust to new environment with a new way of life should have at least been taken into consideration.

The EU funding for the Nech Sar National Park phased out at the preliminary phase because the Ethiopian government failed to fulfill the preconditions of resettling the Guji and Koore households that was set in the financing agreement. Freeman (2006) stated that the resettlement program failed because local communities, and particularly the Guji, resisted the resettlement plans. A former warden of the park who later served as head of the Culture and Tourism Bureau in the SNNP regional state pointed out in one of my interviews that Oromia regional state was responsible for the failure of the EU project in the mid-1990s and later for the withdrawal of the African Parks Foundation in 2007. According to the informant, government officials from Oromia regional state "were discontented from the very beginning with the park's inclusion into the south, and thus they were encouraging the Guji not to agree to the resettlement plans" (personal communication with former Warden, March 2012). Based upon fierce resistance from the Guji, who were indirectly backed by the Oromia regional state, the resettlement plan did not succeed. The plan to resettle the Koore families was also suspended because both the people and government officials from Amaro district insisted that the decision should also be effected on the Guji.

The government did not proceed to demarcating and gazetting the park, given that it raised discontent and resistance from local communities, for whom the space represents economic, cultural, spiritual and historical meanings. For example, from Arba Minch side, low-income sections of the town dwellers, mainly women and youth, depend on the park for firewood, charcoal and construction materials from the forest, as well as fish from the lakes. The dependence of the Guji and Koore on the park for grazing, farming and water has been discussed earlier. In addition, for the Guji Oromo, giving away the territory is considered as a compromise to their history and detachment from their *Waaqa* (God) in spiritual aspects because they had sacred spaces, ritual sites and burial places of their deceased forefathers, while the resettlement would also radically transform their pastoral way of life, which is cultural and economic.

Despite the phasing out of the EU funding for rehabilitation of the three parks in southern Ethiopia, the government continued its policy of promoting tourism by strengthening the existing protected areas and establishing new ones. It is evident that the government considers tourism as an important sector to contribute to the country's economic development and thus entrusted the federal agency and regional bodies with the tasks of strengthening existing protected areas and surveying potential areas for similar objectives (FDRE Draft Wildlife Policy 1997). Although the policy document claims to have been completely new, national park management approaches in the country remained top-down and protectionist in practice. One important aspect of the policy document, despite its failure to be translated into practice, was its commendation of revenue distribution to local communities whom the conservation process affects. It was stated that "[t]he federal and regional governments shall receive and distribute revenues generated by wildlife exclusively for the administration and management of wildlife and for the benefit of communities associated with or participating in the conservation of wildlife resources" (Draft Wildlife Policy 1997: 3).

Despite such policy rhetoric for community participation and benefit sharing, the park management resorted to the classical protectionist conservation approach by fully restricting local communities from access to the resources and excluding them from the management aspect of the park. According to Freeman (2006), the Guji were entirely excluded from any affairs of the national park, including the planning, management and benefit of the park. The author states that, "the current stalemate means that the Guji are not involved in ongoing discussions about the park and natural resource management in the area. This situation is to no-one's benefit" (Freeman 2006: 19). After the EU project phased out, relations between the park and the Guji became tenser, with the latter blamed for resisting the resettlement plans, which subsequently resulted in the termination of the EU project before its planned project span. In inter-ethnic relations as well, Guji's refusal was interpreted as ethnic and regional issue in that it generated sense of common concern to the national park among the Gamo and Koore elites, who constructed boundaries of "us" and "them" against the Guji within the ongoing ethnic politics in the country. One government official from the Amaro agricultural department described the situation as follows:

> From our childhood period, we have grown up being told that the park belongs to the Koore on the eastern side and the Gamo on the other side. Both of us had close historical relations. There were trade relations across Abaya and Chamo Lakes to Chencha and then Arba Minch. Later, these nomads came in and began wandering from place to place within and around the park. After the military government, our region has got full control over the park and those Guji nomads were supposed to live either in Oromia region or elsewhere in our region but they do not have the

> right to live inside the park (informant Kassahun Ayele, April 2011).

Within the context of ethnic politics, whereby local cadre members often used to foment inter-group antagonism rather than for a genuine democratization process, comments like the one above by the government cadre can be well understood as a manifestation of a latent inter-regional tension concerning questions of ownership and administration of the national park. From the Gamo Gofa side, the park is more symbolized with identity of both the people and the Arba Minch town. In one of my discussions with a young employee in the Culture and Tourism office of Gamo Gofa zone, I was astonished by the emotionally loaded reflections of this young university graduate, who emphatically described the national park as a strong identity marker of his group, the Gamo, as well as the town of Arba Minch. The informant describes the relationship between the Nech Sar, Gamo identity and Arba Minch town as follows:

> This park is a national treasure. It is for all Ethiopians. It is particularly an identity for our people and our town. We are proud of conserving this national heritage and when Nech Sar is mentioned, nothing comes first than Arba Minch and our people. Nech Sar is our heritage, our pride and shows how our people to conserve the wildlife. Tourists flock to this part of the country because of it. If this park is registered as a UNESCO heritage like what the Konso did for their terracing, we will be internationally known. However, there are still challenges from another side. The pastoral people do not care for this national treasure. They claim it belongs to them but they did not have a single hut inside the park as they were always on move. The Koore are relatively better in their understanding of the resources but we should cooperate and bring effective management of the park. Another problem is intervention from the federal government. After the federal government took it some three years ago, our people have lost the initial sense of belongingness. I fear that if it continues to be administered by the federal government, people would not care for it and there will be severe damage to the resources (informant Molla Alemayehu, May 2012).

Within such polarized images of the park, the SNNP regional state administered it until 2004, before it was transformed to the Dutch Multinational Company called the African Parks Foundation (APF). As I mentioned earlier, Guji's exclusion from the park in terms of gaining any benefit from tourism and access to customary resources has antagonized their relation with the park. For instance, out of twenty-six park scouts before 2004, only two were from the Guji, while the remainders were from Koore and Gamo ethnic groups. According to some park authorities, the Guji have been given limited job positions in the park because they did not have people who went to 'formal' education (informant:

former park Warden, February 2011). However, during my conversation with Guji elders and youth in Ergansa village, they questioned the government's excuses, arguing that they have more rich knowledge of wild animals' behavior, management and its relations with the people than those who have modern education. In the context of ethnic federalism, resettlement programs, employment opportunities, restriction to local communities' customary rights and administrative measures on the part of the park administration were mainly part of the existing ethnic politics. However, in terms of the administration of the national park, the initial administrative decentralization developed a sense of ownership among the Koore and Gamo political elites, as well as generating discontent and alienation on the part of those in Oromia regional state.

5.4.2 Privatization of Nature Conservation–the African Parks Foundation

The political and economic conditions within the new political order after 1991 have introduced resource privatization under the name of investment. Despite its categorical rejection of the neoliberal development paradigm in recent years[20], and particularly following the contested 2005 national election, the EPRDF government has selectively subscribed to neoliberalism mainly to comply with some conditions of donors and generate income by leasing land to investors (Aimé 2008). As a policy discourse, one of the markedly core tenets of neoliberalism is the privatization of the "hitherto unoccupied, state-owned, or communally owned aspects of the social, cultural and/or natural worlds" (Castree 2010: 10). In the Ethiopian context, this aspect of neoliberalism remains in practice, such as in leasing arable lands, mining areas, protected areas, factories and hotels to foreign and domestic investors. In this sub-heading, I will focus on the privatization of nature conservation that brought a Dutch company (the African Parks Foundation) to take over the management of three national parks in southern Ethiopia–the Nech Sar, Omo and Mago National Parks–in 2004 and 2005 (Markakis 2011; Turton 2011).

The African Parks Foundation (APF) took over the management of the parks for both business and conservation purposes under a twenty-five year contract (Freeman 2006). The agreement was reached between the Ethiopian government represented through the SNNP regional state and the Ministry of Agriculture on the one hand, as well as the African Parks Foundation on the other. The agreement had a precondition to come to a full-fledged operation only if the Ethiopian government relocated all the people residing inside the park. Accord-

---

[20] http://danielberhane.com/2011/03/31/ethiopian-pm-bashes-neo-liberal-paradigm-african-ministers-call-transforming-rent-seeking-political-systems/

ing to the agreement, "the Government, in accordance with its intention that the Park shall meet the IUCN criteria for Category II Protected Areas and for other reasons, undertakes to relocate all people living within the boundaries of the Park, their livestock and possessions, and to deal with all matters of compensation that may arise."[21] Moreover, the government took the responsibility of demarcating the boundaries, while the company retained the responsibility of fencing and patrolling the park.

The discussion concerning the transfer of the park to the APF and the challenges the government and the company faced in the management process will help us to understand disparities between local resource rights and the implementation of ethnic federalism. It is not my objective to describe all the activities of the APF in this discussion; rather, I will focus on resettlement programs that the government attempted to implement, local people's responses to the resettlement plans, the conservation approaches introduced by the company, as well as the company's withdrawal from Ethiopia. The responsibility that the government reserved to relocate the people who were vaguely described in the agreement as "squatters" later became a significant test of inter-regional relations and challenged the federal government's capacity to handle the tension. In the process of transferring the management of the park to the APF, the SNNP regional state pursued plans of relocating the Koore and Guji people who lived in and adjacent to the park boundary.

Many meetings were held with government officials from Amaro district, Gamo Gofa zone, park administration, the SNNP regional Bureau of Agriculture and Natural Resource Development, as well as later with representatives from Galana Abaya district (Oromia regional state). However, people from the local communities did not participate in this long process of planning for the resettlement, because, according to one government official, "it was thought that consulting them through meetings and conferences was time consuming and also would instigate conflict because the people do not envision broader development policy of the government. Instead, local communities always want to satisfy their immediate needs" (informant: Ayele Chane, May 2011). For technocratic government politicians, notions of community participation, consultation and engagement are considered as wastage of resources because local communities are not capable of contributing substantial input to the project, according to their views.

Rather, community participation is often used in Ethiopia for media and political consumptions, as well as fulfilling donor requirements in the case of

[21] Article 8.3 of the Agreement Document between the Ethiopian government and African Parks Foundation (2004) (see the full document in appendix).

implementing foreign-funded projects. For example, within this specific resettlement plan, only few representatives from the Koore and Guji people attended the final meeting arranged as a "Community Conference", held in Arba Minch town from July 12-13, 2003. During the conference, an expert from the regional Bureau of Agriculture and Natural Resources Development presented a research paper titled "*Relationships between Local Communities and Natural Resources*".[22]. As one would expect, the conference paper concluded that the local people were illegal settlers and should be resettled somewhere else to create a better home for the wildlife and enhance the development of the tourism industry in the national park, which would benefit the people in turn. In addition to such a self-evident conclusion, the fundamental point to question is whether the conference was organized to accommodate views of the people or whether it was simply used as a forum to inform community representatives about the government decision.

As became evident in the later years, the resettlement plan and other forms of resource management approaches were top-down and exclusionist. The government's orientation to the top-down approach, dominance of scientific knowledge production and negligence of local peoples' voices in development projects was evident in such cases where the 'experts' presented 'academic research' papers to local people who had not attended modern school. While the paper was presented in Amharic, only a few members of the local communities understand a few words, let alone have the ability to comprehend the concepts discussed in the research paper. One Guji elder recalled that:

> The government officials called us for a meeting in Arba Minch. They told us that we illegally came to the park. They also told us it was a government policy to protect the animals. They warned us that the government would take tough measures unless we voluntarily move out of the park. Because we rarely speak and understand the language [Amharic], we could not express our ideas well. The authorities were also not ready to listen to our ideas. They already finished [decided] the issue before bringing it to us. That was why we discussed among ourselves and decided to take the case back to our people (informant: Danbala Badacha, April 2011).

Another Koore elder particularly focused on the influence of local government cadres before and during the conference, saying that: "We had a meeting in Kelle town before the conference in Arba Minch and decided with our officials not to resist the resettlement but to negotiate terms of resettlement such as compensation and other provisions. Although we wanted to stay in our home, our officials warned us not to oppose the government policy" (informant: Shibiru

[22] Minute of a Meeting dated 14/09/2003

Ayele, February 2012).

A taskforce organized to produce an implementation plan for the relocation of these people held a meeting on 18 December 2003[23], during which neither the local representatives of communities nor the government officials from Oromia regional state attended. The team underscored in the discussion that the Guji were considered as having intruded into the park boundary during the transition period, while their current settlement was considered illegal. Likewise, the team considered the Koore as if they had expanded to the lowland slope of their district in search of farmland around the same time. Eventually, the taskforce proposed to resettle the Guji in Oromia region, but since the team labeled them as illegal settlers, there was no proposal for material or monetary compensation for costs incurred during the relocation process. On the other hand, the taskforce recommended that the SNNP regional state should take the responsibility of resettling the Koore, with necessary compensation. In any case, the Amaro district officials 'agreed' and received an order from higher officials to convince their people about the resettlement plan without any condition (interview with former Amaro district administrator, May 2012). With financial assistance from the government and the APF as compensation and initial establishment, which included two hectares of land, a school, a clinic, and food aid for eight months, around 1,089 Koore people were relocated to Abulo and Alfacho villages some fifteen kilometers to the south of the park in 2004 (Dessalegn 2004; Freeman 2006; Dowie 2009). One Koore elder described the process as deception by the government. He bitterly recalled the incident as follows:

> Our government officials deceived us by telling us that 'if you do not agree to the plan, the government will bomb you and your properties'. We asked them what about the Guji? And, they replied to us saying that 'you will see what will happen to them. The government decided to take a severe measure against them if they would continue maintaining their rigid stand against its policies'. We believed them because it reminded us of the attacks the military regime took some years ago. By that time, houses were burned to ashes and crops were destroyed. Later, the Guji did not move. Our politicians cheated us. We suffered in Abulo and Alfacho because there was no water, no fertile land and no road to the market (informant: Shibiru Ayele, February 2012).

Indeed, some local government officials also shared the same idea. During some

[23] Minutes of a meeting held to discuss "Planning the Implementation of Relocating Illegal Settlers in Nech Sar National Park", SNNP Culture and Tourism Bureau Archive, dated 18/12/2003.

of our discussions, they implicated that higher officials ordered them to persuade the people. A former administrator of Amaro district who ordered the resettlement of the Koore people expressed the situation with regret, given that opposing the decision was beyond his administrative mandate at the time. He described the situation as follows:

> We actually discussed the issue of resettlement but it was more of a dead deal. The regional and federal governments had decided the issue and brought it down to us just as information. During the discussions, government officials told us that if we oppose the resettlement, the government would consider us that we were resisting its policy. The government already agreed with the African Parks on the relocation of the people. We had to accept it. You know things in our country. Sometimes you accept decisions looking at an existing scenario. But, now after some eight years, when I heard the Guji people did not move out and our people suffered from lack of good agricultural land in Abulo and Alfacho, I regret for leading the resettlement. But, even if I opposed it, others would have accomplished it (informant: former district Administrator, May 2012).

Unlike the Koore case, whose local government officials were said to have been 'co-opted' to the government project and forced their people to accept the plan, both the Guji people and their government officials covertly joined hands and resisted the resettlement plan from the very beginning. However, despite the refusal on the part of the people and the Oromia regional state, the Guji were evicted at gunpoint by the police force from the SNNP regional state in November 2004, before retreating to the eastern outskirt of the park (Refugee International April 19, 2005; Abiyot 2009; Dowie 2009). In contrast to the claims from the SNNP regional state officials who deny Guji's customary right to the area on the grounds that the people only 'intruded' the park during the transitional period in the early-1990s, the Guji opposed any government program that would dismantle them from their homes. A Guji elder said:

> The government people came and told us that the land belongs to the government. They evicted us from our home. That was first during the Mengistu era [the military regime]. Under Meles Zenawi [EPRDF government] they came again and told us the same thing. But this time we decided not to be fooled again. We know this is our land, a Guji land. Our ancestors lived here and died here. You can see their burial sites inside the park. Unlike in the past, we have got our people in the government. They speak our language and understand our problem. We appealed to them at district and regional level in Oromia and they helped us to stay here (informant: Galgalo Edema, March 2011).

In the previous chapters, I discussed Guji's connectedness to the land from religious, cultural, economic and historical dimensions. What makes the 2004 resistance against the resettlement program different was the collaboration between the local people and Oromia regional state against the government's resettlement plan. According to one higher government official in the Borana Zone, who requested for anonymity, there was common understanding among some government officials in Oromia region concerning the issue that the people were not supposed to move from their homeland. According to the informant, there were two reasons for the opposition against the program. First, Oromia region still raises questions of ownership right of the park, although they did not formally bring the case to the concerned bodies in the federal government. Second, like the local Guji people, the politicians believe that the Guji were prior inhabitants of the area long before the park's establishment. They held the opinion that displacement would not solve the presumed resource degradation in the park because unless the right of ownership and utilization was recognized, the local people would continue defying government restrictions. According to this view, conservation efforts without the consent and participation of local communities would result in failure. Moreover, local politicians raise issues of the cultural and economic rights of the people, whose livelihood and cultural practices heavily rely on the environment. While negotiations were underway, the police burned 463 Guji houses, destroyed their properties and evicted the people on November 25, 2004 (Refugee International, April 19, 2005; Dowie 2009). Since then, the Guji were pushed to the outskirts of the park on the eastern side, until the two regional governments demarcated some part of the park for human settlement, grazing and access to water in 2008.

The APF initially conformed to the exclusionist national park management approach when it agreed with the Ethiopian government concerning the relocation of the local communities. As discussed in chapter two, environmental conservation and development rhetoric are often used by state and non-state actors in their justifications for establishment national parks, which often compromise the customary rights of local communities. Likewise, APF's collaboration with the Ethiopian government was part of the broader conservation and development discourse that has brought nation states and multinational business/conservation companies together. However, as it faced local resistance and attracted international criticism, the APF was prompted to involve some members of the local communities in some job opportunities in the national park, although their role in the management process was limited. For instance, it organized ninety-eight households who were earlier dependent on the resources from the park for their livelihood and engaged them in different activities within or outside the park (APF Annual Report 2006). The company paid compensation for over one thousand Koore households who were relocated to Abulo and Alfacho villages. Regarding the Guji Oromo agro-pastoral community, the APF increased the number of scouts recruited from the group from three to ten, although the number was far below than the number of scouts from Koore and Gamo ethnic groups. The company made another important concession with the Guji when it came to terms with their demands after some three years of tense relationship.

According to some sources, the APF was prompted to engage the Guji people mainly because it learned that a complete exclusionist approach did not serve its conservation and business missions. Moreover, it was also partly due to external pressures, such as criticisms it faced from Refugee International and Survival International, which championed against the company's exclusion of the people (Personal communication with a former APF staff, March 2012). On the 30th of September 2007, only three months before the final termination of its contract, APF reached on an agreement with the Guji people and allowed them to get access to some portions of the 'park territory' for their settlement, grazing and access to water from Lake Abaya and the bore-salt around *Xabala* hot spring (APF Annual Report 2007). The employment of a few Guji men was also part of the company's recognition that any discontent on the part of the Guji would counterfeit the missions of the organization because they had direct contact with the park territories and the Guji had indirect support from Oromia regional state. However, this concession was utterly opposed by government officials in the Gamo Gofa zone and regional Bureau of Culture and Tourism in the SNNP region.[24]

To elucidate the position of Oromia regional state regarding the Guji's claim of entitlement and the concessions given to the people by the APF in September 2007, the following letter sent from the president of Oromia regional state, Abba Dula Gemeda, to the then Deputy Prime Minister and Minister of Agriculture and Rural Development, Addisu Legesse, holds significance. The letter reads as follows:

> Dear H.E. Ato Addisu Legesse,
>
> Although Nech Sar National Park was established in 1974, studies show that its boundaries were not clearly demarcated like the Semien and Awash Parks. Following the establishment of the park, the Guji who were living scattered in the park had faced serious brutality during the imperial and Derg regimes. Specifically, the Derg regime burned their properties and houses and as a result, they were forced to

[24] Gamo Gofa Zone Office of Chief Administration, Letter sent to the Southern Nations nationalities and Peoples Regional State, Office of the Regional President, dated 10/06/2008, Ref. G.G/Z.A.D/3509/23, Archive of Culture and Tourism Bureau of the SNNP regional state.

> leave their homeland. However, after EPRDF controlled the country, these people returned to their home and began to lead stable life. Nevertheless, following the February 2004 agreement with African Parks (Ethiopia Ltd.) in transferring the administration of Nech Sar National Park, it can be recalled that disputes erupted on issues of demarcating the park boundaries and subsequent relocation of the Guji. We believe that the park is not only a habitat for wildlife but also is supposed to contribute to improvement in living conditions of the surrounding communities and to the national economy at large. Therefore, to settle problems pertaining to the boundaries and administration of the park, the governments of the two regional states (Oromia and South) arrived at the area on April 8, 2008 and observed the overall situation. As we understood from the field observation and discussions with local people, it came to our attention that there are several people and livestock in the area. Thus, to avoid the influence of people and livestock on the park core areas, it is agreed that the park boundary that was agreed upon between the Guji and African Parks shall be maintained and the management and conservation of the park shall be done in close collaboration between the park administration and local people. Therefore, we would like to confirm to you that Oromia regional state accepts the boundary agreement reached between Guji Oromo and African Parks on 30 September 2007[25] (my own translation).

This letter was written after the withdrawal of the APF in December 2007. Nevertheless, it hints at the view of Oromia regional state on Guji's settlement in the area, basing its argument on history. It is evident that some of its contents indicate some controversial issues about the human habitation of the area before the park, disputes concerning resettlement plans, as well as Oromia's acceptance of the boundary agreement between the Guji and the APF. In contrast to the dominant government narratives about the land as 'uninhabited' before the establishment of the park, this view depicts heterogeneity within the government itself and poses a fundamental question about the motives behind these competing notions of representation of the space as 'uninhabited' and 'inhabited', which will be dealt with in chapters six and seven.

Despite its ambitious plan to introduce 'modern' techniques of protected areas management such as an electric fence, airplanes and a well-equipped patrolling system, as well as doubling its income within the first three years, the company came to a premature end in 2007 citing several challenges it faced along the process. According to the APF's Annual Report (2007) and its termi-

[25] A Letter dated 21.04.08 with Ref. No. BM /409/112/Z11 was sent from the then Oromia regional president, Abba Dula Gemeda to the then deputy prime minister and minister of agriculture and rural development on issues related to Nech Sar National Park and the Guji Oromo (my own translation from Amharic).

nation letter, the Ethiopian government failed to accomplish the responsibilities that it was given in the agreement, particularly concerning issues of boundary demarcation and the relocation of the Guji agro-pastoralists. In its statement of terminating the contract dated December 7, 2007, the APF highlighted a lack of cooperation on the part of the Ethiopian authorities to settle the issue of resettlement or recognize the negotiation reached between the Guji and the company. Part of the termination letter stated that:

The resettlement was partially accomplished with the Kore community being resettled to the south of the park. However, 3 years into the project one of the communities still remains in the Park, and the increased use by them and their livestock herds are further threatening the sustainability of the parka. In the first two years of the project, the authorities made little progress with negotiating an acceptable compromise to the mutual benefit of both the community and the park. Therefore this year African Parks decided to make a concerted effort to negotiate, with independent specialist advice, an agreement with the Guji on the limits of use of the park. External organizations were invited to participate in and witness the negotiations. To an extent this process was successful, and a formal agreement was reached on 30 September 2007 with the Guji defining a core area, which would be free from both people and cattle, with use permitted in the remainder of the park. The authorities were requested to recognize this agreement as an acceptable and practical compromise for the benefit of both people and nature. This recognition has not been forthcoming. Therefore, African Parks has decided that it shall terminate all operations in Nech Sar.[26]

Eventually, the APF terminated its contract on the three national parks (Mago, Omo and Nech Sar) in southern Ethiopia, citing complexities of the situations in all three cases, including pressure from international human rights organizations, a lack of common understanding among Ethiopian authorities and the involvement of several ethnic groups with different kinds of claims for ownership rights. Following the withdrawal of the APF, the management of the Nech Sar National Park was reverted to the Ethiopian Wildlife Conservation Authority (EWCA) at the federal level. Overall, the transfer of the national park to APF was part of the country's privatization policy as a strategy of enhancing its development visions. However, because both the internal political and economic systems and the company's business/conservation discourses and practices did not practically address the rights of local communities, the new state-non-state actor partnership invoked local resistance and attracted international criticism.

---

[26] http://www.conservationrefugees.org/APF_withdrawal_statement.html

5.5 Extending the Nech Sar Case beyond Local Contexts

In the preceding sections of this chapter, I discussed the new political reconfiguration in the post-1991 period as a remarkable cultural and political departure from the past. However, the federal arrangement failed to bring the periphery close to the center; rather it enhanced the center's penetration of the periphery through political control and resource extractions. Another important repercussion of the new political order has been the emergence of multiple contested spaces. Although ethnic federalism and decentralization of state power were supposed to have coalesced center-periphery relations, both the center and the periphery have become arenas of contestation for different purposes. In addition, ethnic federalism has brought the ethnicization or politicization of territories, resources, spaces, policies and development programs along ethno-linguistic lines. It has added new meanings to territories and resources by changing latent forms of competition over resource ownership into potent questions of entitlement based upon ethnicity. For instance, common resource areas and territories that were previously located along inter-ethnic borders have been split into different ethnic-based administrative units. As a result, access to resources such as grazing land, water, forest and fishing grounds requires ethnic membership to a group that has given administrative rights over the territories.

The state has also penetrated the periphery through both political and developmental rhetoric and practices by using local elites as channels. For the first time in the history of the modern Ethiopian state, the EPRDF succeeded in effectively controlling the periphery and in turning these regions into viable economic spaces using technological subjugation of the territories in the form of mega dam constructions and large-scale plantation agriculture. In the process, multiple actors have been involved in the 'taming', 'controlling', 'domesticating' and 'commodifying' the periphery. This is why I have situated the contestations over the meaning, ownership right and utilization of the Nech Sar National Park with broader center-periphery relations in the country.

Despite the rhetoric of decentralization, the Ethiopian state has still featured as a dominant center with the vanguard party (the EPRDF) consolidating its strong power at the federal level (Assefa 2006; Turton 2006). In this regard, the center has become a contested political and cultural space, within which different groups struggle for fair representation and the equitable distribution of wealth. Moreover, the periphery has become a contested space where different actors contend to gain access to these regions for political and economic control. Unlike its representation as a 'desolate', 'violent' 'wilderness' and 'deserted' land with limited contribution to the making of the state (economic, cultural and political contributions) in the past, this new political articulation has successfully enhanced state intrusion into the periphery. The contestable dimension of the new dynamics in the periphery mainly reflects the state's attempt to determine, dominate and control the periphery for its economic and political interests. On their part, people in the periphery have become the targets of the federal government's 'carrot-and-stick' approaches. By granting them some degree of autonomy and recognition, and some aspects of socio-economic services, the federal government has presented itself as the 'redeemer' of the people from past injustices. On the other hand, the people in the peripheries have experienced the introduction of massive development projects that have bulldozed their homeland, cultural sites and ritual spaces, as well as generally transforming their territories into consumable commodity in the form of plantation agriculture, hydroelectric dam, mining, etc. On their part, the local communities in the peripheries also struggle to maintain their right over their resources or deter the state and other external actors that push them away from areas they consider as their home. Such interplay between competing perspectives concerning the order of things (development, conservation, modernization, etc.) has at times grappled the people to resort to different forms of negotiations and resistance.

Likewise, it is fair to position the development and conservation programs of the state in the Nech Sar National Park within these broader political, economic and environmental discourses. In this regard, I would seek to place the discussions made in this chapter on conservation enclosure within two broader intertwined developments in the post-1991 political and economic dynamics in the country. While the first scenario is the political landscape in which the conservation enclosure was inherited and how it was adjusted into the new political landscape, the second context is related to the idea of nature marketing, which was appropriated by the Ethiopian state from global environmental discourses.

As discussions in the previous chapters have indicated, the exclusionist conservation approaches, hostile park-people relations and issues of environmental degradation in the park were not new developments in the post-1991 period; rather, the EPRDF government inherited the dominant discourse that masked the history of human settlement in the national park. The unique part of the post-1991 developments is the contradictions between the political rhetoric and practical experiences on the ground.

Despite its applauded rhetoric of decentralization and self-government, the government maintained the old representation of pastoral and agro-pastoral communities in the area as illegal settlers. The justifications from the government side for the resettlement plan were its responsibility to improve the lives of the people and safeguard the environment, including the wildlife, as well as the notion that people and wildlife were not supposed to live together (Turton 2002, 2011). The question of who decides the status of the territory as 'inhabited' or 'uninhabited' and whether humans and the wildlife were not supposed to live together remains contested, because different actors impose their views and narratives upon others based on their power relations. By presenting itself as a steward of nature and the people, the government has tried to broadcast its authority to the periphery. As Herbst (2000) maintains, states in Africa during the pre-colonial, colonial and post-colonial periods have struggled to exert their authority over the mass through control over resource areas on which the people depend for their economic and cultural purposes. As during the previous regimes, the legitimacy of the state has been questioned in the peripheries and thus the EPRDF government has sought to build a new image of the state as 'developmental', 'environmentalist' and more importantly 'designer' of the future generation, passing down a green, developed, prosperous and peaceful nation to the future generation.

Another important point to note regarding the macro-political reconfiguration and the specific developments in the Nech Sar National Park is how the state has decentralized conflict. In the past, most of the conflicts in the country were mainly between the state and society, with the country experiencing major wars whereby different ethnic groups fought the state due to its oppressive political systems, exploitative economic policies and socio-cultural marginalization. Although these forms of conflict have been reduced, there has been a shifting trend from society-state conflict to inter-ethnic conflicts. By decentralizing conflict to the local level, the federal state has reserved the right as an arbitrator in such conflicts. For example, apart from being enclosed as a protected area, the inclusion of the park into the SNNP regional state has created a sense of exclusion on the part of the Guji, while the Gamo and to some extent the Koore people have begun to associate themselves with the park. Thus, conflicts concerning resources and issues of access or restriction from natural resources have adopted ethnic dimensions.

Differences in commitment on the part of the two regional states concerning the implementation of conservation enclosure, including the resettlement plan of the local Guji and Koore people, clearly demonstrates heterogeneities within the government itself. We cannot talk of 'the government' as a homogenous political entity whose policies and programs can be smoothly channeled down to the grassroots without challenge, particularly in contested regions. However, the federal government's failure to settle such internal conflict of interests and claim of entitlement among the regions also signaled a lack/weakness of the institutional setup to handle inter-regional issues such as trans-border resources.

The discussions in this chapter show some inconsistencies and disparities between the constitutional provisions and practical implementation of the political discourse on the ground. However, one would ask how the government negotiates these inconsistencies in terms of translating its policies into practice. In this regard, the inconsistencies are explained in three different ways. Firstly, it should be acknowledged that lack of interest and capacity among government officials at different state structures in terms of understanding, translating and implementing the policies, perspectives and programs have been evident in several sectors (Assefa 2006). Secondly, the EPRDF government is keen to maintain its control over the social, economic, cultural and political spaces, despite the provisions in the constitution. For example, any cultural practice or ritual that seems to strengthen the people's sense of self-consciousness or identity slightly different from the way the government wants to shape the citizens faces government scrutiny. The case of Oromo traditional religious and cultural practice called *irrecha* represents, a good case in point, because the government has placed it under close inspection and control since 2003 (Asebe and Meron 2014). Likewise, the government considers traditional life styles such as pastoralism and indigenous knowledge and practices of resource management in the national park as obstacles to the government's development and transformation visions.

Thirdly, the government seeks to justify its interventions from 'developmental' state perspectives. As discussed earlier, the EPRDF government's developmental state paradigm represents the government embodying the ultimate path of development and thus gives it the absolute right to design and plan development projects and mobilizes the subjects for its implementation. According to this perspective, setting aside land as protected area, resettling people to 'transform' their life style and planning and managing protected areas are considered part of the state mandate. In one of his speeches on Pastoralist Day celebrations, the late prime minister, Meles Zenawi once described the federal state as a rescuer of pastoralist communities from poverty and backwardness, as follows:

> In the coming five years there will be a very big irrigation project and related agricultural development in this zone [South Omo zone]. I promise you that, *even though this area is known as backward in terms of civilization*, it will become an example of rapid development. I also want to assure you that the work we have started in this area on infrastructure and social development will continue stronger than ever. I want to assure you again that all our development work will be in line with protecting the environment and the friends of backwardness and poverty,

> whatever they say or do, can't stop us from the path of development we are taking.[27]

In other context, although the EPRDF government adopted a strategy of nature marketing within the broader neoliberal perspective of privatization of development sectors, the effort of leasing out the park to the APF failed prematurely. The failure was owing to internal political incompatibility, particularly because of inter-regional discontents on issues of ownership and administration of the park. The approach of community-based conservation that the APF strived to introduce after some years of resistance from the local communities against its exclusionist practices also failed because the Ethiopian government authorities were not ready to negotiate the protectionist resource management they inherited from the previous regimes (APF Annual Report 2007). Nonetheless, the government evidently practices the notion of nature commodification and marketing through tourism in the national park.

To summarize, the three successive regimes in Ethiopia share in common many features of nature conservation and representations of the peripheries, despite their differences in the political discourses. The EPRDF government's intrusion into the periphery through 'development and conservation' schemes has further exacerbated the marginal existence of the people in the periphery. The government's rhetoric of granting ethnic groups the rights of self-determination on the one hand, and its developmental state paradigm that neglects these rights on the other, have added to the complexity. As the analysis in the next chapters will show, such competing narratives and practices have generated different forms of resistance, including conflict between different actors, particularly between the state and the local communities, as well as within local communities, along ethnic lines. Therefore, I conclude that the Nech Sar National Park is one of the multiple contested spaces in the political, economic and cultural schemes within the post-1991 political order, which enhanced ethnicization of claims of entitlement to environmental resources.

[27] http://www.mursi.org/pdf/Meles%20Jinka%20speech.pdf (emphasis added)

PART THREE:
THE NECH SAR NATIONAL PARK: AN ARENA OF CONTESTATION/NEGOTIATION ON NATURE-CULTURE RELATIONS

Introduction

In this part, I will analyze how different actors have contested, negotiated and at times appropriated various perspectives and discourses of human-environmental interactions in the process of their struggle over the Nech Sar National Park. Political ecologists discussing human-environmental interaction have focused on the involvement of multilayer actors in environmental degradation and competitions over the utilization, ownership, control and conservation of natural resources. Within this broader context, social actors with different interests, characteristics, practices and power positions interact within different levels of power relations. Thus, in order to understand the interplay between different actors, it is essential to discern their political, economic, cultural and environmental motives in their struggle over a particular environmental resource (Bryant and Bailey 1997; Robbins 2004). As analytical directions, some have focused on specific environmental problems such as soil erosion, deforestation, water and air pollution and land degradation (Blaikie 1985), while others have strived to approach the issues from discourses and ideas used in the politics of environmental problems (Escobar 1995). Ideas, narratives and stories on environmental change shape how people and institutions deal with the environment. In this regard, notions such as the 'tragedy of the commons', 'environmental degradation', 'climate change' and 'biodiversity depletion' are all powerful in shaping how state and non-state conservation institutions deal with the 'commons'. Moreover, because human-environmental relations are intertwined with other forces across spatial and temporal boundaries, some political ecologists have also employed a regional approach of analysis (Blaikie and Brookfield 1987).

However, although human interaction with the environment can be addressed within the context of political ecology in either of the above approaches, including a focus on "socio-economic characteristics such as class, ethnicity or gender" (Bryant and Bailey 1997: 23), an actor-oriented approach offers a comprehensive understanding of the complexity of political and ecological dimensions of the topic. This is because an actor-oriented approach probes into competing notions and practices that social actors employ regarding the utilization, management and ownership of environmental resources, as well as hinting at how they articulate, negotiate, appropriate or compromise different discourses and practices in the process of their interaction.

In this context, I use the concept "actors" to refer to individuals and social groups who deploy their agency in formulating or changing or advancing the course of action in societal, political and economic spheres (Long 1992). In the development and conservation context, an actor-oriented perspective recognizes the agency (knowledgeability and capability) of actors in using resources at their disposal–human, social, capital, political, cultural and knowledge–to turn a certain program or intervention into their advantages or deter the interest of others (Few 2001; Long 2001). Although his work does not directly concern human-environmental interaction, Norman Long's (2001) seminal book titled *Development Sociology: Actor Perspectives* illuminates the advantage of employing this analytical approach, given that it helps to interrogate questions of why different actors respond to the same phenomenon differently.

In the analysis of human-environmental interaction and conflicts that ensue from the interaction between different actors with competing interests, an actor-oriented approach is an appropriate analytical method. It is an important tool to comprehend not only covert and overt interests and actions of the actors, but also to understand notions of power and agency of the actors in their conflict, cooperation and negotiation (Bryant and Bailey 1997; Long 2001). Recognizing the influences, voices, actions and interests of actors encompasses an epistemological paradigm shift in the representation of grassroots social actors in the development and conservation discourses. It demands a change in understanding from considering them as passive recipients of development discourses, ideas and programs to recognizing these actors as "active participants who process information and strategize in their dealings with various local actors as well as with outside institutions and personnel" (Long 2001: 13).

In the three subsequent chapters that follow, I adopt an actor-oriented approach to analyze the interests, characteristics and views of place-based and non-place-based actors in their pursuit of achieving their interests or restricting the interests of other actors over the Nech Sar National Park. The choice of this analytical approach has been informed by the fact that the Nech Sar National Park has become an arena of contestations, at times with negotiations between competing perspectives and practices among various actors. In this case, the chapters address both place-based and non-place-based actors as having their own role in shaping human-environmental interactions and in partaking in the struggle with varying degrees of involvement. Place-based actors refer to those social groups who have had direct interaction with the territory, through either administrative authority or livelihood and socio-cultural engagements, whereas non-place-based actors include external actors whose policies, advocacy works or perspectives have had some influence on the human-environmental relations in and around the park. Because both categories of actors influence the course of action or intervention programs in the national park, I will discuss both as per their level of involvement, influence and power positions.

Based on the above understanding of social actors, I have identified actors that have employed different forms of capability and knowledge to influence the course of human-environmental interaction regarding the national park. Although a comprehensive investigation of all social actors would have been essential to complement an understanding of the complex dynamics in the area, I decided to focus only on major actors whose direct or indirect presence in the area and whose agency in influencing decisions concerning the national park have been evidently visible. In this regard, chapter six deals with the Ethiopian state, its positions in the conservation discourses and practices, as well as how it has channeled down its policies to the grassroots level. In chapter seven, the perspectives, strategies and interests of local communities, a transnational business and conservation organization (i.e. the African Parks Foundation) and international human right advocacy organizations over the Nech Sar National Park will be analyzed vis-à-vis their interaction with the environment, the state or among each other. While chapters six and seven mainly deal with the actors' conceptions of the human-environmental interaction more at the epistemological and ontological levels of conceptualization, chapter eight places much emphasis upon the agency of grassroots actors in resisting or coming to terms with external interventions. Nevertheless, the study still recognizes internal heterogeneities within each actor and strives to bring together inter-actor and intra-actor competing perspectives. Taking each actor or group of actors individually, I will discuss issues related to their characteristics, interests, actions and conceptualizations of the territory within each subsequent chapter.

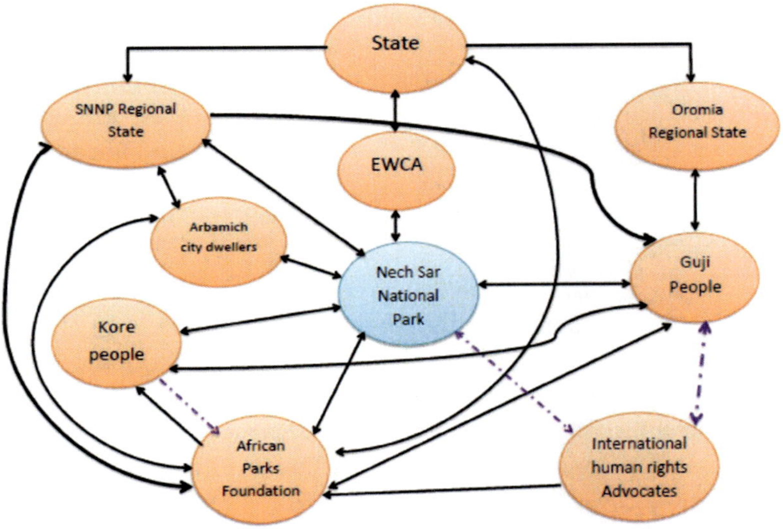

Figure 12: Actors and their Interactions
(Source: My own graphic adapted from the empirical data)[28]

[28] Although this graph might simplify the complexity in the actors' interactions because it does not show contestations and negotiations between different actors, it gives us a preliminary overview of the major actors and some chains of interaction. I would like to acknowledge the contribution of my colleague, Megerssa Tolessa in revising the graphic.

CHAPTER SIX:
THE ETHIOPIAN STATE: CONSERVATIONIST, 'DEVELOPMENTAL' OR ENCROACHER?

6.1 Situating the Conservation-Development Debate within the Identity of the Ethiopian 'State'

In Africa, perhaps as is common elsewhere, the state defines itself and is represented in the mainstream political and development discourses as a legitimate body to govern the people and resources within its jurisdiction and to design development visions (Bryant and Bailey 1997). What differs between countries is the 'how' aspect of the governance. In countries such as Ethiopia, the state uses sovereignty, power and resources in the quest for control over the society and the environment. In other words, the control and expropriation of environmental resources engenders the state with political power, economic wealth and boosts its claims of legitimacy, which in turn replicates the exercise of power in the process of subduing the subjects. Likewise, although political power, legitimacy and economic wealth are supposedly essential to enable the state in gaining the capability to deliver services to the society, the Ethiopian state used it as a source of power to control and dominate its subjects. As Scott (1998: 184) notes, "the logic behind the state's large-scale attempts to redesign rural life and production from above" is the process of 'civilizing' the 'uncivilized', according to the modernist state discourse. In the process of imposing a centrally designed or imagined development or conservation program upon societies in the periphery, the state institutionalizes policies and practices of what Scott calls "domestication, a kind of social gardening devised to make the countryside, its products, and its inhabitants more readily identifiable and accessible to the center" (1998: 184).

My point of departure in this chapter is that under hegemonic government systems , like the case of Ethiopia, where the state's legitimacy at the periphery is questioned and under contexts where it is described as 'extractive' institution, environmental conservation is not an end by itself; rather a means to other goals like economic and political interests. Fletcher (2010: 178) posits conservation debates within discussions on governmentality, concerning the governance of mentality or the art of conducting subjects' conduct'. Within the context of state conservation discourses and practices, governmentality is a process or an art of controlling or governing environments and environmental users as a means to achieve economic, political, environmental, emotional or psychological interests (Fletcher 2010; Peet et al 2011). The state discourses of environmental conservation are often forms of exercise of power over the people and territories they inhabit through different strategies. These strategies include establishing 'environmental subjects' through education, which is described as 'disciplinary environmentality', the 'participation' of local communities in conservation projects that can be termed as 'neoliberal environmentality' and a pure protectionist approach that is placed under 'sovereign environmentality' (Fletcher 2010: 177).

Owing to limited state authority over the society, particularly in peripheral parts of Africa, postcolonial states have resorted in some cases to environmental enclosures "promoting environmental conservation for reasons related to internal security or social control. Conservation initiatives might thus be a means for states to assert their authority over peoples and environments hitherto subject to weak control, thereby strengthening the position of the state in relation to other actors (even if grassroots 'participation' is ostensibly part of the process" (Bryant and Bailey 1997: 62). However, while I concur with the above argument, I would like to move the discussion further, suggesting that it is not an outright nullification of the state's intentions and capabilities of environmental conservation in Africa. Rather, in a quite different scenario from some global discourses of environmental conservation that promote conservation for the sake of inherent salience of preserving biodiversity as the end goal, nature conservation in Africa has been promoted for economic, political, scientific and aesthetic purposes, using conservation discourse as a means to achieve these ends.

In Ethiopia, the state appropriation of land from local communities for the purposes of 'biodiversity conservation' in the form of national parks shares much in common with the colonial legacies that other African states inherited from the colonial powers. Because the state remains the key actor in influencing human-environmental interaction, mainly in developing countries, it is imperative to discuss the nature of the Ethiopian state, its interests in the struggle over environmental resources and its agency in creating environmental subjects. In this regard, I argue that the historical antecedents that ensued after the birth of the modern Ethiopian state have influenced its relations with the society and its claims over environmental resources, particularly in the peripheries. By drawing on the notion that the state deploys different forms of power, narratives and technologies in its quest for environmental conservation, which is often a means to other economic and political ends, I will analyze how the Ethiopian state under different regimes deployed resources at its disposal for the same purpose. As Li (2007: 275) maintains, the fundamental principle of a government is to foster social change and the progress of its citizens, and more specifically to "improve the welfare of the population." Nevertheless, it is a point of great contention as to whether successive Ethiopian regimes designed and implemented conservation projects such as the Nech Sar National Park for the sake of improving the welfare of the population.

To situate the discussion within the identity of the state, I reiterate what I discussed in the two preceding chapters about the political and cultural identities of the Ethiopian state. I have stated earlier that the Ethiopian state was characterized by a strong hegemonic center, with clear asymmetrical power relations between the center and periphery. In terms of cultural identification, the Ethiopian state was built upon the cultural, ethnic and religious manifestations of one dominant group, namely the Amhara. As a result, the worldviews, values, customs, religious outlooks, economic lifestyle and other forms of cultural practices of the peoples in the periphery did not have any representation in the making of the Ethiopian state (Markakis 2011). In other words, the imagination, identity and worldviews of its constituent society shaped how the center related itself to the periphery. This goes in line with Agrawal's (2005) arguments that politics, institutions and identities of the state are the three fundamental technologies of governing the environment and creating environmental subjects. While the political setup prepares the ground for institutional frameworks for the state's environmental policies and practices, identities are elements of values and beliefs that bring together state elites regarding their representation of the state, subjects and territories (Migdal 2001). In this regard, Orthodox Christianity, Amharic language, peasant agriculture, hereditary land tenure system, hierarchical social structure and the myth of 'great tradition' are the central constituting elements of its cultural identity, making the cultural self-identification of the Amhara and the Ethiopian state alike (Markakis 1994, 2011; Turton 2011).

Despite the new political order after 1991, attempting to deconstruct this trend and imagination, there are continuities of the past in most aspects of state-society relations in the country. The EPRDF government is confronted by three mutually contradictory policy paradigms regarding its intrusion into the periphery. Firstly, its policy of decentralization and rhetoric of local autonomy has created ambitious expectations on the part of the people in the periphery, who envisage seeing the policy rhetoric translated into practice. In fact, this local aspiration for autonomy, self-government and recognition to their cultural values and practices can be seen as part of the global phenomenon of the late twentieth century, which sparked global indigenous peoples' movement across the globe. Although grassroots movements in Ethiopia did not have a direct link with global indigenous peoples' movements, it should be noted that there were some global dynamics and changes that encouraged ethno-nationalist movements in the country that linked local grievances with national and international discourses.

Secondly, as is common in most postcolonial African states and in contrast to the constitutional provisions of local autonomies in the post-1991 Ethiopia, the government has strengthened and expanded protected areas in the form of national parks, game and forest reserves, which largely exhibit the protectionist approach of conservation adopted from the Yellowstone National Park model. As elsewhere in the world, where national parks exclude local people from their livelihood, cultural and religious spaces and customary rights, the government's conservation policy restrained the right of the people to decide on their affairs. It is clear that the dominant protectionist approach of nature conservation aims at maintaining a particular environment and its resources unchanged by restricting human interference.

The third scenario again contradicts the previous incongruities in the government policy. In contrast to the political rhetoric of self-government, which ostensibly granted ethnic groups, including those in the periphery, the right to administer resources under their administrative jurisdiction, the federal government seems to have usurped their rights through its 'developmental state' policies. There are some contradictions between the practices of the state, which represented itself as the main development agent on the one hand and as a guardian of the rights of its citizens on the other. This is because by placing itself as the ultimate planner and designer of development, the state denies the local people their rights to decide on matters that affect their lives. In addition, development interventions such as hydroelectric dam construction, large-scale plantation agriculture and leasing land to foreign and domestic investors stand in sharp contrast with conservation discourses that intend to maintain the status of biodiversity unchanged. It is evident that conservation and development projects are often contradictory to one another, although both share similarities in hampering local people's customary rights (Brockington et al. 2008). The modernist developmental discourse of the state encourages sedentary agriculture and denigrates pastoralist livelihood style. Therefore, it is within these broader understandings of the Ethiopian state and the competing policy paradigms, particularly under the EPRDF regime, that this section will attempt to address the aforementioned themes.

6.1.1 The State and Peripheral Representations

To understand how the state represents the territory, it is essential again to recall the nature of the Ethiopian state, its relation to the periphery and some dynamics in the process. This link is important because the Nech Sar National Park was established in the periphery that was incorporated to the empire in the 1890s. The representation of the periphery as 'uninhabited', 'wasteland', as well as the stereotypical representation of the people and their ways of life as 'violent' and 'backward' have persisted among the mainstream Abyssinian core society until present (Turton 2011). This representation is analogous to what Mbembe (2001: 1) refers to as the "negative interpretation" of Africa in the global discourses, where he further contends that "Africa is never seen as possessing things and attributes properly part of "human nature." Or, when it is, it's things and attributes are generally of lesser value, little importance, and poor quality. It is this elementariness and primitiveness that makes Africa the world par excellence of all that is incomplete, mutilated, and unfinished, its history reduced to a series of setbacks of nature in its quest for human kind."

The hegemony of the center that was installed following the birth of the modern Ethiopian empire continued unchallenged despite major changes in the political order of the country particularly since 1991. Following the late nineteenth century conquest of the south, a new form of state-society relations emerged between the center and the periphery (Vaughan 2003; Markakis 2011). The center can be understood as an arena that constitutes three interdependent and complementary elements, "values and beliefs, institutions and elites" that enable the center to exert its authority over the periphery (Migdal 2001: 44). While the values and beliefs constitute the cultural, religious and mythic imaginations and practices considered 'sacred' among the members of the core, institutions are important instruments through which to channel the core values and beliefs of the center. Likewise, the elite category constitutes the core actors (political, business, religious, cultural and academic elites) who maneuver the institutional setups in the quest to exert state authority over the society.

In my previous discussions, I have explicated how the imposition of economic, cultural and political domination by the Amhara ruling elites upon the peoples in the subjugated regions had established a hegemonic system similar to the colonial encounters in other parts of Africa. The core values and beliefs, such as the so-called 'great tradition', Orthodox Christianity, Amharic language and plough agriculture, remained central in the identity of the state and its "Othering" of the periphery. From center-periphery perspective, the modern Ethiopian state should be understood not as an institution accommodating both the center and the periphery, but as a center striving to encroach, control and engulf the periphery. The approaches of such encroachment and engulfment were different across different regimes. For example, while the imperial regime annexed and sought to control the peripheries by installing its hegemonic institutions such as Orthodox Church, state bureaucracy and feudal modes of resource extraction in the form of settler or absentee landlordism, the military regime attempted to broadcast its authority through collectivization and the establishment of peasant associations. On the other hand, the current regime uses two strongly interconnected strategies: the co-optation of elites from the periphery and encroachment through 'developmental' rhetoric and practices.

To contextualize the discussion within the scope of this chapter, I will first briefly discuss how the process of the conquest and the subsequent center-periphery relations that it created subsequently contributed to the dominant state representation of the territories in the periphery as 'wilderness' and 'wasteland'. Two different but interconnected phenomena shaped such representation. Firstly, the conquest produced a victorious-defeated relationship between the center and the periphery, and forced people in the subjugated regions to come to terms with and at times contested their subordinate status. Whether it was represented as 'wasteland' or recognized as inhabited territory, the conquest removed the customary right of the indigenous peoples and vested 'authority' on the conquerors to decide the status of the conquered land (Markakis 1974; Donham 1986). Moreover, the fact that it was established as an Empire through the invasion of several autonomous states that contributed to the establishment of the subsequent colonial type of relationship between the center and the conquered subjects has had tremendous implications in framing the nature of the state and its representations of the lands and the people that it conquered. This is why Donham (1986) describes the Ethiopian Empire and its relations with its subjects in the South as predator-prey relations, whereby these regions were expropriated of their resources through the exploitative feudal system. Upon the incorporation of the resourceful regions into the empire, exploitation of the peasants in the Abyssinian center was significantly reduced, while it became harsher on the conquered subjects.

During the process of the conquest of indigenous peoples' lands elsewhere, the conquering powers used discourses, narratives and myths such as the 'doctrine of terra nullius' and 'wilderness' that were invented to delegitimize indigenous peoples' customary practices and their rights to the lands (Stewart-Harawira 2005; Asebe 2011). I have also discussed in chapter two that the notions of wilderness were used as discursive instruments to nullify the history, culture and customary rights of indigenous peoples. Likewise, similar representations of the periphery were embedded in the identity of the Ethiopian state and in the north-south dichotomy, which were passed down from regime to regime.

Secondly, the conquest and the subsequent asymmetrical power relations between the core and peripheries created dichotomized relations, within which the state and its constituent core members identified themselves as cradles of civilization, whereas they demoted the conquered subjects as 'backward'. Myths and narratives such as the legend of 'great tradition' and hereditary line of connection to king Solomon of Israel (the "Solomonic dynasty") were invented to exclude non-Amhara and non-Tigrean groups from the political center (Markakis 2011). In this regard, the state and its dominant group used religion, language and state institutions to engrave the synonymy between the state and the dominating group. Donham (1986) maintains the view that the core-periphery dichotomy that existed long before the conquest was further strengthened by the post-conquest political, economic and cultural asymmetries in the empire. Prior to the incorporation of the periphery into the Ethiopian empire, the encounter between the two geographical and cultural niches took the form of trade, mainly constituting slave that flew from the periphery to the center and then sold to the outside world. Thus, slavery, slave trade and differences in ecological and economic lifestyles contributed to the establishment of cultural boundaries between the two.

The power-laden relationship between the Amhara ruling elites, the Orthodox Church, the military and their ordinary peasants on the one hand and the subjected people in the peripheries on the other had exacerbated the dichotomy between the two. In other words, the cultural self-representation of the core as the cradle of 'civilization' and its portrayal of the periphery as 'backward' entailed the assertion that cultural and religious practices and economic livelihood other than those of the dominant group were considered 'inferior' (Donham 1986). In short, both the military conquest and the center-periphery dichotomies in cultural and political spheres enhanced the center's justification for its actions in eroding the customary rights of the subjected people in the periphery. As part of the justification to its administrative interventions and assimilation policies in the South, the imperial regime presented the conquest and its interventions as 'civilizing missions' through the homogenization of cultural and religious pluralities into the Amhara/Tigrean culture and religion (Mohammed 1996).

The military regime that took state power from Haile Selassie in 1974 sustained similar notions of representing the territories, livelihood and cultural practices of the people in peripheral regions as backward (Dereje 2011). Although the regime did not officially identify itself with a specific ethnic group, its 'high-modernist' orientations placed the periphery at the margin of the state, whereas its inclination to mainstream cultural practices favored the continuation of the status quo. According to Scott (1998), high-modernist ideology entails uncritical acceptance of the legitimacy of science and technology by powerful actors, such as states who consider 'rational' social order or 'social engineering' and mastery of nature, including human nature as signs and instruments of modernization. For example, state-initiated development and social re-ordering such as "the Great Leap Forward in China, collectivization in Russia, and compulsory villagization in Tanzania, Mozambique, and Ethiopia are among the great human tragedies of the twentieth century, in terms of both lives lost and lives irretrievably disrupted" (Scott 1998: 3). Likewise, the military regime in Ethiopia strengthened the centralization of state power, suppressed multiculturalism and disrupted socio-cultural practices under the guise of modernization (Clapham 2002; Dereje 2011).

I discussed earlier that the Ethiopian state inherited the socio-cultural bias towards sedentary agriculture as a form of modernity, in contrast to pastoralism. For instance, the former prime minister of Ethiopia, Meles Zenawi, once described the pastoralist areas in the southern and southwestern parts of the country as 'backward' in terms of civilization. According to the late prime minister's speech on 'Pastoralist Day Festival' in 2011, the government planned to launch a large-scale sedentarization program to convert the pastoral communities in the Lower Omo valley to sedentary villages, where they would abandon pastoralism and begin a new way of life based upon agriculture and wage labor.[29] As I discussed in the preceding chapters, the dominant idea in nature conservation also presumes the domestication of wilderness, for both biodiversity conservation and utilitarian purposes (sporting, aesthetic consumption, tourism and scientific research). With the exception of few internal differences within the state structure, this view concerning the area has continued among the state authorities throughout the history of the park.

For example, in the case of the Nech Sar National Park, successive Ethiopian regimes described the territory as an empty land and thereby made the people who lived in and around the park 'invisible' through this dominant discourse, which privileged farming agriculture over nomadic pastoralism (Getachew 2007). In his discussions of 'high-modernism'–a condition in which states, elites, individuals and institutions "claim to speak about the improvement of the human condition with the authority of scientific knowledge"–Scott (1998: 93) further argues that many parts of the world experienced tragedies of state intervention. Scott further contends, "all human habits and practices that were inherited and hence not based on scientific reasoning–from the structure of the family and patterns of residence to moral values and forms of production–would

---

[29] On a Pastoralist Day festival held in South Omo zone (Jinka town) on 25 January 2011, the late prime minister was quoted saying "I promise you that, even though this area is known as backward in terms of civilization, it will become an example of rapid development." see http://www.mursi.org/pdf/Meles%20Jinka%20speech.pdf

have to be reexamined and redesigned" (1998: 93). Likewise, despite formally recognizing the traditional values, customs and practices of diverse ethnic groups including those in the periphery, the ambitious developmental plans of the current government resemble the high-modernist orientations of its predecessor.

Nevertheless, the state-society relations in the post-1991 political reconfiguration have been most contentious and ambivalent for some contradictions between the political rhetoric and its lived realities on the ground. Here, it is sound to argue that despite its ostensible departure from the past in terms of deconstructing the Ethiopian state, in terms of both its political foundation and cultural identity, the current regime has inherited much from its predecessors. For instance, in contrast to the constitutional provisions as well as some practical appearances on the ground, there is a clear continuation of centralization of state power as well as hegemony of the center, albeit with a shift from Amhara to Tigrean ascendancy (Merera 2003; Markakis 2011). In this regard, the federal state has often been criticized for interfering with the internal affairs of regional governments by appointing its loyal officials and dismissing others, by assigning cadre members from the TPLF party to oversee the works of regional states and by maintaining authority over the budget (Merera 2003; Vaughan 2003; Assefa 2006).

Moreover, the developmental state paradigm that has been in place since 2002 has remarkably undermined the autonomy of the local governments and the people in the peripheries, because this political and economic discourse gives ultimate right and mandate to the state in planning, designing and implementing development projects and impinging the democratic rights of the citizens. However, unlike the previous regimes, the Ethiopian state has become more heterogeneous in terms of comprising regional states established along ethnic lines. This has sometimes brought contestations among regional states and/or between regional states and the federal state on implementation processes of policies and programs such as the conservation of the Nech Sar National Park. Although regional states are not as such autonomous to take strong positions against some policies and programs that come from the federal government, internal heterogeneity of the state itself hints at possibilities that state intervention at times faces some challenges both from grassroots communities and from the state structure itself.

6.1.2 Development and Conservation in Ethiopia–Perspectives from the State

While discussing state strategies of implementing its developmental policies and practices, James Scott (1998) argued that authoritarian states design ambitious development plans and use coercive power in its implementation, although such 'high-modernist' practices fail because of the simplification of the simplification of practical knowledge, social relations and environmental conditions. When governments adopt a certain ideology, idea and model and channel it to the grassroots level, they often present it as a project designed to improve and transform the living conditions of society (Li 2007). Likewise, the successive Ethiopian governments used conservation and development as mutually reinforcing projects to improve the living conditions of local communities ensure the sustainable conservation of the biodiversity and promote tourism. The successive Ethiopian governments (the imperial, military and current regimes) used the concepts conservation and development as mutually constitutive phenomena in their rhetoric of establishing protected areas.

The past regimes emphasized the economic and environmental contributions of protected areas and presented the state as a key actor in achieving these goals. For instance, in its request for UNESCO's technical and expert support in the process of establishing national parks, the Ethiopian government under the imperial regime underlined its commitment to establish protected areas to boost the economy of the country through tourism, as well as contributing to biodiversity conservation (Huxley et al. 1963). Indeed, the establishment of the Nech Sar National Park was partly informed by the imperial regime's efforts to place the Guji agro-pastoral community under its effective administrative control. The military regime positioned national parks within its modernist discourses that juxtaposed development, nationalism and conservation as idioms of nation-building. It placed emphasis on forest, wildlife and soil conservation, albeit with utmost top-down approaches (Shibiru and Kifle 1998). The establishment of national parks during this period was further linked to the notion of conquering 'wild' areas as the case of the socialist military regime's often saying,'*Tefetiron be kutitir sir enawulalen*' (We shall put nature under control). This rhetoric implies the government's projects of controlling, domesticating and subduing the physical nature and its resources, although it broadly envisioning science and technology to implement its 'high modernist' economic schemes and social changes.

Like the post-enlightenment thinking in Europe that equated domesticating and subduing 'wilderness' with the victory of 'modernity' over 'primitivism', the establishment of national parks in Ethiopia was a conservation endeavor, an ideological manifestation and a strategy of political control. Keeping wildlife under the gaze of humans within national parks or game reserves by restricting their movements and behavior is tantamount to the conquest of the 'New World' from the seventeenth to the nineteenth centuries, whereby indigenous peoples were kept within restricted territories (Merchant 2003). I would like to reiterate that setting aside land in the form of national parks within the dominant nature-culture dualism imposes human's hegemonic control over the wildlife. Likewise, the 'modernization' discourse under the military regime itself was built upon the existing cultural supremacy and dichotomization between what were then considered by the mainstream society as 'civilized' and 'uncivilized' cultural practices.

Because it considered delineating humans from the wildlife as a sign of modernity, the military regime launched resettlement programs targeting the Guji and Koore communities in the 1980s. At a national level, the regime undertook an ambitious program of collecting peasants into collective villages, which was meant to 'develop' and 'modernize' the living style of peasants by facilitating easy delivery of social services (Taddesse 1995). Like a villagization program that was ostensibly planned under the rubric of 'modernization' and 'improving' the living standards of peasants, the government situated the resettlement program and in fact the eviction of the Guji and Koore from the Nech Sar National Park within the context of development. Moreover, its efforts to create better home for the wildlife that were thought to include the country among the progressive African nations of the time, such as Tanzania under Nyerere was also part of the 'modernist' project.

As a former official in Gamo Gofa province during the military regime recalls the 1982 eviction of the Guji and Koore communities from the national park, the socialist government was obsessed with its modernization discourse and ordered the eviction as part of controlling the environment and transforming it into modern ways of national park management. The informant described how government officials contextualized national parks within the modernization rhetoric, on the one hand, and attempted to justify the eviction as a process of 'improving' the lives of the people and of the wildlife on the other hand:

> Higher officials told us that people should not live with animals. They warned us that we had to control people's entry to the national park because it was considered as a sign of backwardness and embarrassment to the revolutionary and modernized state. The government said, 'the people should be removed in order not to distort the image of the country among foreigners who visited the country and also in order to conserve the national heritages such as the endemic wildlife'. Everything was understood according to the politics of the period. Today, the government talks

> about *beher behereseboch* [nations and nationalities]. In this context, everybody talks about rights of every group. People practice their culture and ways of life. However, during the Derg regime, those practices like nomadism were considered as primitive. Therefore, it was brought to the attention of the provincial administration that the people should be resettled somewhere and lead a better life where the government would provide them education, healthcare and other modern facilities. It was also to avoid human impacts on the wildlife (informant Ayele Danche, April 2012).

Moreover, pastoralist areas in the peripheral lowland parts of the country were seen as violent-prone and a threat to national security (Hagmann and Alemmaya 2008). Most of the pastoralist areas in Ethiopia straddle along national borders and given that pastoralist communities do not have the conception of compartmentalized boundary, there were actually apparent cases of arms smuggling and trans-border livestock raiding along Ethiopia's borders. Moreover, these border regions were where ethno-nationalist armed movements have been operating against the central government across different regimes. In this context, it can be noted that setting aside portions of a territory in the form of protected areas was part of the government's exercise of power in subduing dissidents, because it controls access to those areas. Moreover, the military state represented the territory as a violent prone area inhabited by 'warrior' nomadic pastoralists whose presence in the park threatened the security of the neighboring communities.

As I discussed earlier, political ecologists call for the greater scrutiny of conservation enclosures owing to the intricacy and inseparability of political and economic motives of actors from ecological objectives. Bryant and Bailey (1997: 62) noted that "in the case of the creation of a park, there is not only the delimitation of the borders of the new administrative entity, but also the appointment of a whole 'army' of park rangers and guards to ensure that excluded actors do not interfere with park management. These officials also keep potential 'troublemakers' in the area under close surveillance in a manner akin to a military operation." Likewise, the military regime took punitive measures against the Guji in Segen Valley (some 30 kilometer to the south of the national park) in 1975, accusing them of rebelling against the government (Asebe 2012d). The government used the same allegation against the Guji during the course of the 1982 eviction, with a Guji elder, Danbala Badacha, recalling the incident as follows:

> I have been living here since the last three governments [Haile Selassie, Mengistu Haile Mariam and Meles Zenawi]. All of them considered us as their enemy. They said 'Guji is shifta' [bandit/rebellion] but Guji is always peaceful. Actually, Guji does not bow down to its enemies. That is why they said so. Before they chased us

> from our home in 1982, the officials from Gamo Gofa province announced on a big public meeting that the shifta lived among the Guji. They also declared that unless the Guji were removed from the park, trade relations between the Gamo and Amara [Koore] would be impossible. They chased us from our home but that was not really for the sake of the animals (informant Danbala Badacha, March 2011).

Under the current regime, which ostensibly recognizes the rights of nations and nationalities to administer resources on their territories, the government juxtaposed both economic interest and conservation objectives when it comes to the Nech Sar National Park. These interests underpin the government's obsession with the narratives propounded by its predecessors. Besides the perspectives of the 'wild' and 'natural' south to which the government subscribes, its developmental state paradigm is also central in its policy formulations that rarely recognize the voices of local communities. However, as part of its initial notions of decentralization of state power to lower administrative structures, the EPRDF seems to have resorted to what Agrawal (2005: 15-16) describes as "governmentalized localities" and "the making of environmental subjects". By the term "governmentalized localities", Agrawal (2005: 15) refers to the process or strategy by which the state extends its control over the subjects and the environment by stretching state power and "political-economic relationships between centers, localities and the subjects." The making of environmental subjects is a process of inculcating dominant environmental values or discourses into the life worlds of the subjects through environmental education or coercive approaches.

Following the political reconfiguration along ethnic lines in 1991, the EPRDF government decentralized the management of the Nech Sar National Park to the SNNP regional state, whereby local administrative units were supposed to be given mandates of planning and managing the territory. As part of the general optimism of the early-1990s concerning the political 'blessings' of the regime, which were thought to promote democratization and development by harnessing ethnicity for the same purpose, the decentralization of the national park administration was also supposed to enhance local empowerment and mobilization towards conservation projects. However, decentralization alone does not guarantee community participation, empowerment and benefit. Rather, as Agrawal (2005: 7) suggests, "policies aiming at greater decentralization and participation are about new technologies of government. To be successful, they must redefine political relations, reconfigure institutional arrangements, and transform environmental subjectivities." Moreover, as Mohan and Hickey (2004) contend, any development program that undermines the cultural, economic, political and social rights of citizens cannot achieve any goal of social transformation, given that it faces resistance from the people whom it is supposed to serve. It has also been argued that citizens' participation should be more broadly conceptualized as a transformative political and citizenship right, through which "people extend their status and rights as members of particular political communities thereby increasing their control over socio-economic resources" (Mohan and Hickey 2004: 66).

In short, despite differences in their political and economic ideologies, there are significant commonalities between the three regimes in their representation of the territory and their approaches of addressing the questions of the local people to their customary rights. Fundamentally, the government has been considering the territory as a 'human-free' space by eroding the human history from the land. Informed by the political and cultural identity of the state, as well as development and conservation discourses, successive governments in Ethiopia have relegated the customary rights of the people in the periphery, including those in and adjacent to the Nech Sar National Park. Therefore, it is possible to argue that the notion of representing the Nech Sar as 'wild' and 'uninhabited' should be broadly understood as part of the 'North-South' dichotomy in Ethiopia, which has been instrumentally used as a political and socio-cultural schema of controlling the resources. In addition, the state used development and conservation rhetoric to delegitimize indigenous knowledge and ways of life of the people for similar purposes of ensuring its economic, political and environmental control.

6.1.3 Towards the 'Making of Environmental Subjects'?[30]

Over the years, state institutions and conservation organizations have worked to institutionalize and inculcate conservation discourses and approaches as mechanisms of 'making environmental subjects' through "involvement in new regimes of monitoring, enforcement, and regulation" (Agrawal 2005: 17). According to Agrawal (2005) and Fletcher (2010), the making of environmental subjects entails different strategies of creating people who care for the environment. The process includes neoliberal environmentality–the provision of incentives to resource users as an instrument of persuading them with conservation missions–and disciplinary environmentality–an act of inculcating environmental ethics in the life-worlds of resource users (Fletcher 2010). While agreeing to the two strategies in the making of the aforementioned environmental subjects, I push the arguments of Agrawal (2005) further that coercion is also part of the

[30] I borrowed the concept, "making environmental subjects" from Agrawal (2005) in which he referred to different technologies of environmental government that are intended to produce subjects that embrace values, discourses and practices of environmental conservation held by strong actors such as the state.

technologies used by states to create environmental subjects. As the case in the Nech Sar National Park demonstrates, the state's efforts to create subjects who 'accept' its conservation programs not only included techniques of incentives and ethics, but also coercion.

State officials have used multiple forms of environmentalities (neoliberal, disciplinary and sovereign) to suppress local resistance in the struggle over environmental resources. For instance, the following ethnographic data illuminates how the government used its visible and invisible hands to create environmental subjects. In March 2012, people from the park administration and some officials from Galana district of Oromia regional state visited a Guji village of Gode. This village was located around the southeastern part of the park, close to the hot spring (*Xabala*) that the Guji traditionally used for its healing purposes. By accusing the people of cattle trespass and encroaching into the park boundary through settlement, the government officials ordered the people to dismantle their houses in two days. On the third day, they came and destroyed sixty-four homes at gunpoint. In another incident, the government arrested eight Guji men and women whom the park administration allegedly accused of killing rhinoceros from Lake Chamo. Moreover, the park administration rewarded a Koore young man who reported the case to the park scouts with money. Therefore, by coercing people whom it found as reluctant in obeying its rule and rewarding those who were co-opted to the state's conservation rules, the government strived to create subjects who 'care' for the environment, due to either fear of coercion or in anticipation of reciprocity.

While the notion of commodifying nature and introduction of market-based approach of environmental resource management through ecotourism fits the neoliberal environmentality, the state's attempts to educate local people and its resort to coercive forces in enacting its authority go in line with what Fletcher (2010: 177) calls "disciplinary and sovereign environmentalities", respectively. Drawing on these conceptualizations, I will analyze how the Ethiopian government under different state agencies enforced different forms of environmental government or environmentality regarding the Nech Sar National Park vis-à-vis other contending actors, mainly the local communities.

Because development and conservation programs are often used by powerful actors such as states and multinational corporations as forms of control over the knowledge, the rights and resources of grassroots societies, political ecologists call for great scrutiny of these concepts and the way in which actors use them (Robbins 2004; Peet et al. 2011). I hold the view that although the problems are not inherent in development and conservation per se, but in the way, these notions are used and implemented, it has been often the case that some actors used these discourses to advance their own interests and deter others from perceived or actual fields of competition. This is why we need to cautiously address these concepts and their usage by different actors. According to Peet et al. (2011: 32), "the capacity to dominate or subjugate environment also entails stifling or controlling resistance from the side of local communities who rise against costs of environment related development projects or conservation activities." For example, in environmental management, the state exercises its power over the environment through its capacity to control actions of people or other organizations, thus revealing its 'sovereign' power (Peet et al. 2011).

Like competing political and economic interests between Oromia and SNNP regional states over the Nech Sar National Park, Robbins (2004) interrogates how different actors cast local or regional political questions and conflicts in ecological terms. Some scholars argue that social actors use narratives and discourses to legitimize their claim of entitlement to a territory in their pursuit to achieve their interests or deter other actors from getting access to the same environmental resources (Gossling 2004). Building my arguments on the notion that "any effort to environmental conservation in its many forms, for better or worse, is basically a form of environmental control" (Peet et al. 2011: 28), I will analyze how the state has used conservation and development narratives to control the territory and the people regarding the Nech Sar National Park.

Based upon its highly contested political and development rhetoric, I will focus on the current government's intervention in the Nech Sar National Park. However, I present the Ethiopian state as a divided political entity with different state structures and institutions constituting various views on the national park. As I have discussed earlier, ethnic federalism has created diversities of claims, interests and narratives among different groups and regions in the post-1991 Ethiopia by subscribing to ethnicity as a fundamental foundation in the socio-political and cultural reordering of the state. In most cases, ethnic groups began writing their history to redress past distortions and underrepresentation in the mainstream Ethiopian historiography. In the process, additions, omissions or corrections to geographical areas, historical incidents and people's history of origin and settlement became apparently evident. For example, a historical book published by Oromia Culture and Tourism bureau traced back Guji's settlement in the present day Nech Sar National Park area to the sixteenth century, although government officials in the SNNP regional state contest this part of the history of the Guji (Getachew 2007). In fact, Lewis (1966) takes this history further, arguing that the lowland areas to the west and east of Lake Abaya were part of the original homeland of the Oromo long before their expansion to different parts of the country in the sixteenth century.

By contrast, while the Koore elites and their ordinary people claim the entire Nech Sar area up to Bonke hill, the Gamo Gofa Zone has redefined its administrative boundary to encompass the entire park, as happened following the 1991 administrative reconfiguration. This contestation and claim of entitlement based on historical 'resources' goes in line with what Appadurai (1981: 202) argues as "discourse concerning the past between social groups is an aspect of politics, involving competition, opposition and debate." In this regard, while both Oromia regional state and the Guji community questioned the credibility of the notion of 'uninhabited' territory and raised questions about the customary rights of the people, those in the SNNP regional state also had their own historical 'resource' to counter the claim of their rivals. Such different interpretations of the past, including utilizing them as an historical resource for the purpose of claim over the territory, complicated the issues of management, utilization and customary rights of the people in and around the park. This shows the heterogeneity of the state, which could potentially challenge the smooth translation of any dominant state discourse into practice.

Because the state constitutes people with different interests who sometimes represent competing departments, views and interests it is naïve to imagine it as a monolithic entity with homogenous views among its structures and institutions. An insight into internal diversities and competing interests secures us from the risk of assuming an actor as a homogenous group with common interests, actions and characteristics (Bryant and Bailey 1997). There are clear differences within the state itself concerning the approaches to managing the national park, ranging from inter-departmental conflicts to inter-regional contestations over the national park. Nevertheless, different state departments and administrative structures at times converge in that conservation and development discourses or even resistance against these narratives have been systematically used to control the people.

Department wise, the Ethiopian Wildlife Conservation Authority (EWCA) under the federal state and the SNNP regional state Culture and Tourism bureau has encouraged conservation and tourism development in the national park. The EWCA has criticized investment bureaus for allocating land to investors inside protected areas and complained that they did not give adequate protection to wildlife and rangers. Rather, according to information from the EWCA, encroachment in the form of farming, grazing and extraction of resources are major challenges to the sustainable management of protected areas in Ethiopia.[31] As I discussed in chapter five, the EWCA has subscribed to the 'wilderness' notions for two major purposes: to attract tourists to the areas and justify the relo-

[31] http://www.ewca.gov.et/en/_sdpase

cation of local communities under the pretext that the coexistence of humans with the wildlife in and around national parks is incompatible. It would reiterate that knowledge concerning the environment is produced through combinations of narratives, discourses and frameworks that are constructed by a particular social actor to legitimize a particular form of knowledge and suppress counter narratives (Peet et al. 2011).

National parks in Ethiopia have been founded on the Yellowstone National Park model, which excluded human history from the territories; indeed, the management of most of the national parks remains largely exclusionist (see chapters 2 and 4). By adopting a mission of conservation and management of wildlife through scientific approaches for ecological, economic and social benefits, the EWCA has also set its vision within conservation and tourism framework and envisions "in 2020 to be one of the top five countries in wildlife tourism in Africa."[32] In order to achieve this vision, the authority has described income maximization in terms of tourism, wildlife hunting, a wildlife export license, as well as wildlife filming and collecting money from researchers.

During my fieldwork, I got the chance to talk to higher officials in the EWCA from whom I learned that conservation in Ethiopia is largely an economic project, while biodiversity preservation is considered as the means to this end. However, the government apparently switches its policies as it suits the interest of donors. For instance, while the front page of the EWCA's website depicted national parks in the country as 'wilderness', announced as "*Come and See the Real Wilderness in Ethiopia*", it has also published a self-contradictory statement on its webpage in which it recognized local communities' coexistence with the wildlife despite persistence of conflicts and harmony. While the former representation implies that these territories as 'uninhabited' and 'natural', the latter seemingly recognizes people's settlement in and around national parks. It went to the extent of defending human settlement in and around national parks as a question of survival for peasants and pastoralists who live side by side with the wildlife and experience day to day encounters, in contrast to Western tourists for whom wildlife tourism is a leisure time in Africa. The EWCA document further states:

> It is one thing to live in [a] European or American city, come to Ethiopia for holidays and watch dangerous animals from the security of a land cruiser. In addition, it is another thing to live with them side by side, have the crops and livestock, the only food for the family destroyed, and the grass, which provides the livelihood for

---

[32] http://www.ewca.gov.et/en/mission_vision_and_values

cattle, goats and sheep eaten by wildlife.[33]

From this dimension, one might wonder whether the government has made a paradigm shift in its approach of national parks management,–from exclusionist to a more inclusive approach that recognizes the customary rights of local communities. However, rhetoric is often different from action or practice. On the one hand, the rhetoric of inclusiveness the government advocates is a self-contradictory, because the government's resettlement plans and evictions local communities experienced under the same government are cases that contradict the seemingly pro-people perspectives of the government mentioned above. On the other hand, as the major discourse of conservation has moved from the classical protectionist approach to the so-called community-based conservation, which has gained the support of donors and influential conservation organizations such as the IUCN, the government seems to have adjusted to the 'realities' of the day. However, on the practical ground, the government still subscribes to the notions of 'fortress conservation' and excluded the local communities from the management, utilization and ownership of resources in the park, apart from strategically using the people as a local regulatory committee by recruiting scouts and committee members who inspect the activities of their own community.

While the above discussions illuminate the ideas, narratives and perspectives that the government adopted as technologies of channeling down its conservation programs, it is important to note what practical activities the park administration performs and how these activities have been undertaken. Such discussion hints at different forms of environmental governance that the state has adopted. The local level administrative unit of the Nech Sar National Park is the park administration, located at the outskirts of Arba Minch town to the east. This local administrative unit did not change much at different times, when the park experienced different administrative changes. The park administration was staffed with park rangers/scouts, biologists and administrative staff, whose number changed from time to time. During my fieldwork in 2012, the park administration had over forty scouts, two biologists, five administrative staff members, a tourism 'expert' and the warden.

The scouts were discharged to seven different outposts, three of which were located along the eastern side bordering the Guji and Koore communities. While the park biologists together with other administrative staff oversee wildlife habitat, location and behavior, as well as the status of pasture and water for the wildlife, the tourism 'expert' is responsible for promoting tourism through

---

[33] http://www.ewca.gov.et/en/_sdpase

media and giving orientations to tourists upon their arrival. The major role of the 'expert' involves reproducing the narratives of aesthetic values of the territory, 'emptiness' of the land from human habitation before the 1960s, as well as feeding the audience with the story of how the species have been endangered by human encroachment. Moreover, the officer capitalizes on the tourism potential of the national park, which would contribute to the national economy. Much of the park management activities were conducted by the scouts or park rangers, who were armed with automatic gun and patrolled the park the whole day. The scouts were responsible for inspecting the local people's day-to-day activities around the park boundary. In addition, with the entire staff members of the park administration, they participate in 'educating' local communities about environmental conservation ethics, benefits that can be recurred from tourism, as well as the punishments people would face if they transgress park rules.

This small government administrative bureaucracy has become a locus of power, sharing much in common with what Agrawal (2005: 15) describes as "governmentalized localities." While Agrawal's (2005) conceptualization of governmentalized localities refers to the establishment of local administrative units with closer ties to local communities and with stronger power vested in them, the Nech Sar National Park administration used park rangers who were recruited from local communities as its entry point, using them as instruments of despotic control in the making of environmental subjects. In fact, as I discussed in chapters four and five, the park administration under the past successive regimes exercised coercive power upon the local communities in its efforts to secure the territory from human influences. In this sense, the exercise of sovereign environmentality can be described in terms of the government's control over the people's customary rights of resource utilization and ownership through the territorialization of the park territory, relocation of the local communities, the enforcement of fines and the imprisonment of people whom it found 'guilty' of transgressing its rules.

Nevertheless, one major point to interrogate now is how to explain such state intervention beyond conservation discourses. Beyond the political-ecological discussion, it is also possible to situate the phenomena within the historical contexts of the birth of the modern Ethiopian state. The military conquest that brought the incorporation of the peripheral regions into the empire transformed the status of customary rights of the people, conferring all land to the crown (Donham 1986). From this angle of interpretation, all the actions of the government such as eviction of the people, restriction of their access to the resources, imposition of fines and other administrative measures can be understood as the exercise of the state's sovereign right over the subjects and territories under its jurisdiction. It is part of the state's continuous broadcasting of its authority to the periphery, despite the presumed completion of state formation in the late nineteenth century.

Back to the park administration's employment of coercive actions as strategies of extending its authority, the period from mid-1990s to 2005 was the time when the park administration attempted to quash, by deployment of hard power, the people's struggle for their customary rights. Nevertheless, its inability to suppress local resistance prompted the park administration and the government at large to resort to disciplinary techniques of environmental conservation. The park rangers, biologists, tourism 'expert' and the wardens of the park have been engaged in providing the local communities what they called 'awareness creation education'. In March 2012, I attended a workshop organized by the national park administration in collaboration with EWCA, within which local elders, members of youth associations, women and the entire park staff participated for two days. According to a presenter from EWCA who addressed issues of 'poaching', 'illegal' fire, cattle 'trespass' and farming 'encroachment' in different parks with particular reference to the Nech Sar National Park, national parks are protected not for the sake of prioritizing wildlife over the people but because the wildlife and other natural resources are fundamental national heritages in generating income. The participants of the workshop also mentioned that the local people should understand natural resource conservation as part of the global campaign against climate change, the effect of which has been locally felt through intermittent drought and flooding. In this regard, the park administration sought to produce environmental subjects by resorting to a different technology of environmental governance that Agrawal (2005) and Fletcher (2010) described as disciplinary environmentality. Neumann (2001) clearly argues that the state creates conditions that 'discipline' local communities by putting them under constant surveillance by state agents as well as by a few coopted members of the communities.

Nevertheless, as Guji and Koore informants commonly agree, the teachings of the park administration over the last two decades have been quite different from what they have experienced in their day-to-day interactions with the park. For these people, neither its ecotourism advantage nor its contribution to climate change mitigation is their priority. Deprivation from farmland, grazing areas and water wells is a more acute problem for the people than what the state envisions to receive in terms of environmental and economic returns. As will be detailed in the next two chapters, the local communities contested the state intervention through different strategies. This was why the park administration devised a different form of control by establishing local national park protecting committees.

The government and the local communities established local committees under local village administrations among the Guji and Koore communities. Committee members were not salaried but received incentives from the park administration in the form of daily allowances when they traveled to Arba Minch for meetings or reports. Although committee members were supposed to inspect within their respective villages, the park managed to manipulate the existing inter-ethnic mistrusts by covertly assigning local regulatory members to rival villages. For instance, when I was in the field in March 2012, a member of the Koore local regulatory committee reported to the park administration that a few Guji youth were involved in 'killing' a hippopotamus from Lake Chamo. Although the Guji informants dismissed the allegation and accused the Koore committee members that their allegation was motivated by existing inter-ethnic tensions, it ultimately brought the initial arrest of eight Guji individuals including three women and five young men. This strategy of control has been evident in many colonial systems, within which colonial authorities crafted local "regulatory community" to extend their influence to the local communities by tying the newly created regulatory body with resident subjects through social, economic and structural relationships (Agrawal 2005: 16). However, in the case of the Guji and Koore communities, due to strong social cohesion, values and sense of belongingness to one's community, members of the local regulatory committee were often reluctant to expose members of their own ethnic group. On the other hand, they expose or accuse members of any rival ethnic group as a mechanism of demonstrating their own loyalty, as well as concealing the practices of their own groups.

Put briefly, I argue that the park administration has exercised multiple forms of environmentality in its effort to create environmental subjects. Agrawal's (2005) conceptualization of the making of environmental subjects focused on the state's usage of disciplinary techniques and incentives to inculcate its conservation values in the life worlds of the people. I further I argue that the Ethiopian state has used sovereign, 'neoliberal' and disciplinary forms of environmental governance through its administrative agencies, aimed at creating people who 'care' for the environment through fear of coercion or in expectation of economic and social benefits, or as a result of 'awareness' concerning environmental affordances. While the different techniques of control the state exercised upon the local people were strategies of channeling down its conservation discourses to the ground, the representation of the territories as exotic, wilderness and uninhabited were discursive instruments in delegitimizing local claims of entitlement.

Moreover, objectification of the territory has been part of the 'othering' discourse of the state, which it inherited from the past. In this regard, by presenting itself as development agent and a guardian of the environment, the federal state downplayed any alternative views from the people. As I discussed earlier in this work, for the Guji Oromo for example, an appropriate development or conservation project is one that goes in line with their traditional values, ethics, customs, belief system and ways of life. Owing to the strong cultural and identity representations that it embodies, pastoralism remains central in their conceptualization of development. Therefore, any development or conservation scheme that threatens or alters this way of life is not acceptable by the majority of ordinary Guji men and women. By contrast, the government overtly expressed its intentions to change the lifestyle of the pastoralist communities into commercial agriculture and engaging them into daily labor work in the plantation activities, through which the people would be 'emancipated' from following the 'tails of the livestock' and begin to exercise modern life, according to the state perspectives. These competing visions of development are central in the contestations between the state and the local communities in areas of conservation for the sake of ecotourism.

6.1.4 The National Park as a Field of Contention between the Two Regional States

As I discussed earlier in this chapter, it is naïve to imagine the state as a monolithic institution with similar views on policies, programs and practices. As a major actor in a politicized environment with influential power, "the 'state' encompasses diverse agencies and interests, and further operates in a world of states." Therefore, "the state's role as an environmental manager needs to be understood as an outcome of intra-state, as well as inter-state conflict" (Bryant and Bailey 1997: 63). In this section, I will focus on conflicts over the management and ownership of the Nech Sar National Park that ensued between Oromia and SNNP regional states. The case of the two regional states is very insightful, partly because it shows the increased internal heterogeneity of the state after 1991 and partly as a phenomenon to probe into different strategies that both regions used to achieve their interests. Moreover, the disagreements between the two regional states were at times advantageous for the local communities, who were capable of converting the conflict to their advantage.

6.1.4.1 Oromia Regional State

To begin with Oromia regional state, it has persistently been claiming that the territory was an ancestral land of the Guji Oromo (Getachew 2007). This was partly why different government authorities in the region were reluctant to the resettlement plan of the Guji in the 1990s and 2004. The resistance from Oromia regional state can be seen as a political departure from the existing trend, because the politicians in this region have been criticized for their surrogate existence under the tutelage of the TPLF (Merera 2003). In his letter to the former deputy Prime Minister, the then regional president of Oromia underscored that the Guji lived in the area long before the park's establishment (see the letter in appendix). To unpack the 'mystery' behind Oromia's resistance to the resettlement plan of the Guji Oromo and to generally unveil the prevailing points of contestation, it is imperative to infer to a few of its referents of claims. From a long discussion with one of the higher officials in Borana Zone, I learned that there were two major reasons for the Oromia regional state, represented by the Oromo People's Democratic Organization (OPDO)[34] party, to resist or at least remain indifferent towards the conservation and development practices of the federal government in the Nech Sar National Park. The first reasons were economic and historical, while the second was political, involving issues of gaining legitimacy from the populace.

As Getachew (2007) describes, the fact that the Guji were prior inhabitants of the area, at least since the sixteenth century–long before the advent of the Ethiopian empire to the region, has given Oromia strong historical point of reference to claim for the inclusion of the territory into its border in line with the principles of ethnic federalism. Indeed, different historical sources attest that the Oromo lived around Lakes Abaya and Chamo long before the incorporation of the region into the Ethiopian Empire (McClellan 1988; Mohammed 1990). Others further argue that an area including the present day Nech Sar National Park was part of the cradle land of the Oromo people, from where they have expanded to other areas since the sixteenth century (Lewis 1966). Therefore, for a few Oromo political elites even within the OPDO[35] the incorporation of the territory into the SNNP regional state is commensurate to a compromise to the history of the Guji Oromo and a denial of their constitutional right. Moreover, there were economic interests within the OPDO political elites in Galana and Abaya districts to divert the tourist route from the SNNP region to their own by connecting a road to the national park from the east.

In addition to the historical and economic claims of entitlement to the national park, politicians at different administrative structures in Oromia regional

---

[34] OPDO is one of the four coalition parties that make the EPRDF, which has been on power since 1991. Although it is a party that ostensibly represents the Oromo people, OPDO's legitimacy remained questionable because of its surrogate existence within the EPRDF and for its failure to ensure democratic rights of the Oromo people (Merera 2003).

[35] The OPDO party has been often criticized for being surrogate within EPRDF and for its floppy position to defend the autonomy of the region (Merera 2003; Clapham 2009; Markakis 2011).

state strive to repair the legitimacy crisis of the OPDO party. Although it managed to recruit members through incentives and coercion, OPDO still suffers from lack of legitimacy and has limited popular support in Oromia. It remains critiqued by its critical supporters and detractors alike for being insensitive to address major democratic rights of the Oromo people and for its nominal position within the EPRDF. Thus, by 'defending' the constitutional rights of the Guji, thereby presenting itself as a champion of the interest of the Oromo people, OPDO attempted to gain legitimacy among the Guji and the Oromo in general. For instance, one of my OPDO informants made the following remarkable comments about the party as an insider:

> I worked for nearly fifteen years at different positions as an OPDO member. I know how people perceive about this party from outside and I know it from inside. Now, I tell you the inside reality of our party. I believe in its party programs and that, according to my view, is not the problem. The problem lies with the people who work on different positions from lower to higher administrative levels. Some join it to get their daily bread. Others still use the party as a shield for different interests. However, only a few work for the interest of the people. When we come to the Guji case, we clearly know that the park belongs to Oromia but the higher officials are obsessed by the politics of nations and nationalities rather than addressing the urgent problems of the people they represent. How do we respect and even talk about the rights of other nations and nationalities without our rights being respected? Our politicians are caught in dilemma between the issues of multi-nation federalism and self-government. Many of them prioritize their personal political interests. For instance, they do not want to enter into conflict with the federal government or other regional governments on small issues like Nech Sar. However, the land is not small for the people. It is only when they need the people like during elections that they seem positive to the questions of the people. When they get a loophole, they defend the right of the people and present that on big party meetings. They use it for their political agenda just to stay on power (informant: an official in Borana Zone, March 2012).

The above statements hint at the fact that political authorities in Oromia often pursue questions of Guji's ownership of the territory or retreat from the same question based on the nature of power balance between the regional state and the federal government, as well as in their pursuit to gain legitimacy from the populace. Again, this is not an assertion that political elites in the region held the same views on the political outcome of the resistance against the conservation program. The higher officials at the regional level systematically used OPDO's firm position against Guji's relocation as part of the party's quest to build legitimacy among the Oromo populace. Moreover, some insiders mentioned that Oromia used the Nech Sar National Park as a case in negotiating with the SNNP regional state on other ongoing conflicts over resources between ethnic groups from both regions. On the other hand, the position of local OPDO elites at district level highly depends on their short-term advantages envision from supporting the park ideal or resisting it. In fact, there were also genuine concerns among local government authorities to defend the rights of their people. For example, the Guji informants praise the former district administrator who indirectly worked towards the suspension of the relocation of the Guji in the late-1990s and 2004 for having strong sense of ethnic identity and consider him as their hero. This official did not ask for any individual benefit for supporting the questions of the local people, although it was common for many local government officials to seek reciprocity.

On the other hand, the current administration in Galana district uses the park issue as a political instrument in coercing the people in tax payments. For example, the district officials took part in the destruction of sixty-four Guji houses in Gode village in March 2012. A Guji elder from Arda Guddina village once metaphorically commented on the relationship between the people and the district officials as a relationship between a rat and a cat, as follows:

> We always live under conditions of insecurity. On the right side are the park people and on the left side are government officials. Both wait for some time and threaten us with our weakest part. They know our major problems are pasture and water. Government officials come from Tore and ask us to give them money. We do not know what they do with the money. They said it is for road, dam, tax, OPDO festivals and many more. We do not have money. We have only cattle. At times, they warned us that the government would displace us if we refuse to pay. Instead of destroying our entire life, we prefer to sell two or three among our cattle and give them what they asked. We slaughtered goats, prepared buna qala [traditional food prepared from roasted coffee beans mixed with butter and milk] and many things for the officials every time they came here. We offer them these all to escape their punishments. However, how do we continue to live like a rat and a cat? We always live under insecurity (informant: Shube Dalacha, May 2012).

It should be underlined that Oromia's resistance against the conservation program in the Nech Sar National Park was mainly informed by the political rewards that the actions were supposed to generate at local or regional levels, apart from some genuine commitments of a few political elites. On the other hand, some political elites in Galana district use the national park as an administrative weapon to coerce and control the Guji people and stifle local resistance.

To situate the views and contestations from Oromia regional state within the regional and national political dynamics, it is important to note how some authorities in the regional state have addressed the issue at different times. The Guji community in and around the Nech Sar National Park have tried to establish strong network with government authorities at different administrative levels in Oromia. The groundbreaking support that they received was first from the former district administrator in Galana Abaya, who guided local elders in their resistance against the resettlement program. This administrator was central in delaying the resettlement of the Guji in the 1990s and 2004. Second, the then regional president of Oromia, Abba Dula Gemeda, visited the area in 2008, before apportioning the eastern side of the park for the Guji with the consent of the president of the SNNP regional state. Within the local and national political constellations, the concession given to the Guji by the two regional presidents can be seen as a victory of both the Guji and Oromia in the contestation. However, Oromia's position regarding Guji's right to the land has not been consistent. Government officials in Oromia regional state either address or reject local people's questions for access to their customary resources depending on individual commitment of the officials as well as the power position of OPDO in the national political game.

However, there are contentious views regarding the position of Oromia, which some elites from SNNP regional state describe as "an obstacle to the government's development and conservation policies in the national park" (discussion with a former park Warden, April 2012). As an official in the SNNP regional state pointed out in her critique concerning Oromia's reluctance during the 2004 resettlement program, the park had no economic advantage for Oromia, although the region might have had hidden political motives:

> We could not understand why the people from Oromia were stacked to this bare land. It has remained a puzzle for us why officials who are seemingly educated give deaf ears to this national concern. Not to say for its economic reason, the land does not have oil, nor has it minerals. Nothing grows out of it except the grass, which the wildlife eats for survival. There is no historical fact at any time in history that would support Oromia's claim over the park. The park belongs to the south and it had been under the administrative province of Sidamo and Gamo Gofa even before the current federal arrangement. Then what should we conclude? It remains unanswered question for me, but Oromia might have had hidden agenda (former park warden, April 2012).

As an expert in conservation science and a technocrat who worked hard to translate conventional conservation theories into practice, the former warden of the national park emphasized the physical elements of the landscape without reference to its cultural, spiritual, historical and political significances. Accordingly, such ignorance of complex representations and purposes of the area for the local community often causes clashes of perspectives between conservation elites and local communities. Moreover, like many conservation and development practitioners, the technocrat has ignored the intricate political, economic and historical dimensions in the struggle over environmental resources.

6.1.4.2 The Southern Nations, Nationalities and Peoples Regional State (SNNP)

The SNNP regional state, which is the third state actor, has strong resemblance with the federal state in terms of its interests and the way in which it strived to justify its actions. Since the 1990s, tourism has become the major ambitious objective for the establishment of many national parks in Ethiopia, most of which are found in the southern region. According to Young (2012), the Nech Sar National Park was the second top in the country in terms of number of tourists visiting the park and income generated for the last five consecutive years (2007-2012). For example, from September 2010 to August 2011, about 22,709 domestic and foreign tourists visited the national park from whom it earned over one million birr (about 41,666 Euro).[36] As a result, despite persistence of conflict with neighboring communities, the economic value of the park has been very high compared to other parks in the region, such as the troubled Omo and Mago national parks.

Nevertheless, it is too simplistic to reduce the points of contestation on the part of the SNNP regional state purely to economic interests. The question that follows involves interrogating how the SNNP regional state addresses its claims over the national park. Like the federal state, which channeled down its developmental and conservationist discourses through the EWCA, the SNNP regional state also capitalized on both discourses and relegated local people's indigenous conservation approaches and livelihood styles as if they are at odds with biodiversity conservation. In addition, I have discussed earlier about identity provenances among the Gamo people in relation to their claim over the national park. As I noted in chapter three, the Gamo do not draw their claim of entitlement to the national park on any spiritual or cultural connectedness, but built it upon subjective identity constructions and administrative rights that they have gained over the territory particularly since 1991. Accordingly, for the Gamo Gofa Zone as an administrative unit, the Nech Sar represented aesthetic meanings and heritage, which gave a positive image to the people and the town of Arba Minch. Thus, the people's ascription of 'aesthetic' values to the land and considering the status of the territory as their knowledge of conservation were subtle claims of entitlement. Apart from their interest over the national park, the

---

[36] Annual Report of the Nech Sar National Park Authority, 2012 (see Appendix).

Gamo people had long trade relations with the Koore on the other side of the national park. However, the park's establishment and the presence of the Guji between the two groups have significantly hampered inter-societal relations.

The paradox is that despite the general representation of the south by the Abyssinians as 'wild' and 'uninhabited', the technocratic elites in the southern region subscribed to the same notion in their representation of the Nech Sar as 'empty' and 'natural' in defense against Guji's claim of entitlement. One might wonder if this representation is ontological or superficially driven by other motives such as political and economic desires. Political authorities and conservation technocrats in the region subscribe to these notions mainly to nullify Guji's claim for customary rights, as well as to conform to the broader narrative employed by the mainstream society in the representation of pastoralist ways of life as 'backward', because this narrative helps them to justify actions such as the relocation of the people.

While the views of many government elites from the SNNP regional state over the territory were similar, my conversation with a technocratic official who worked in different offices in the regional state was much instructive. This is because it clearly surfaced from the fact that the discourse of 'emptiness' of the land was part of the elites' management of meaning in the process of the contestation over the land. It was also a subtle instrument used by the officials to justify the project of creating 'human-free' space for the wildlife. My informant worked for over fifteen years in different positions in the national park administration, as well as the regional culture and tourism bureau. She was central during the resettlement program in 2004 and had been an important figure during the APF's takeover of the management of the park. She gained a master's degree in Forestry and Natural Resources Management from one of the universities in the Netherlands. During our interview, the official contrasted debates in the two competing approaches of conservation–the 'fortress' conservation and community-based approaches and attempted to explain challenges both approaches faced over the years in the case of the Nech Sar National Park.

Apart from such conceptual discussions with the technocrat, our conversation raised the views of the state concerning pastoralist peoples, which the official described as follows:

> As common elsewhere among pastoral communities, the worldviews and level of consciousness of these people [the agro-pastoralist Guji] are very much limited in scope. They think of what they get today and are not worried of what comes tomorrow. They bring thousands of their livestock and graze the park. After a week or so, they move to another place or stay at one place only for a season and go back again. At least the agrarian society is better in understanding its environment, soil

> fertility, land degradation and crop productivity. Because they depend on pasture and water, pastoral communities try to maximize what they get from a particular pastureland and water ground before moving to another place. Their livestock compete among themselves as well as with the wildlife. We learned from different historical sources that the Guji came only two or three decades ago. It was vacant land before that. What concerned us most was the future of biodiversity in the park. The government worked hard to conserve the resources and to contribute to development of the country. The people did not understand and share this vision. Our aim was double. Firstly, it was to conserve the biodiversity and through tourism to get foreign currency. Secondly, we aimed at resettling the people to improve their livelihood through improved agriculture (informant: a former head of Culture and Tourism Bureau, SNNPR, April 2012).

One would agree with my informant that the Guji did not understand and share the same 'vision' of conservation and development with the state. For the Guji people, particularly as the ethnographic data illuminates, conservation and development are inseparable from their cultural values and spiritual connectedness to the land. However, there is an elitist message in the informant's description of the human-environmental interactions, including the history of the human settlement in the area. Although the above description might have been informed by the informant's academic background, in which the classical notions of 'tragedy of the commons' seem to have influenced the official's perception about human-environmental relations, elite perspectives concerning the knowledge of resource management also visibly surfaced from the discussion. When the SNNP regional state took swift decision to relocate the Koore people to Abulo and Alfacho villages in 2004, government officials adamantly presented the program as a process of 'improving' the living conditions of the people, as well as boosting the national economy through tourism. Nevertheless, both conservation and resettlement programs proved to have failed. At present, almost ninety percent of the Koore have abandoned the resettlement site and returned to Tsalke and other adjacent villages. Since then, human predation on wildlife and encroachment to the park territory increased compared to before.

Despite differences within the regional state regarding questions of access to the national park, the post-1991 political reconfiguration has created a sense of common belongingness vis-à-vis the contending actors in Oromia. Moreover, its subscription to the discourse of conservation, tourism and economic development, which is advocated by the federal state, has been a major resource to the SNNP regional state in its claim for the national park and in granting it the support from the federal state. In short, the general position of many government officials and elites in the SNNP regional state is the claim that the national park belongs to the region due to administrative history since the imperial regime, identity marker and geographical location 'within' the region. Nevertheless, economic and political reasons were also the major factors for their claim of entitlement.

To summarize, the Nech Sar National Park has become an arena in which the state's competing notions of development and conservation encountered local critical views of human-environmental interactions. Despite internal variations within the state, the state practiced environmental conservation for the sake of its economic values through tourism and also used the practice for broadcasting its authority to the periphery. As will be discussed and analyzed in the next two chapters, local communities resorted to production of critical knowledge to delegitimize the state's 'accepted wisdom'. The discussion in this chapter ascertained that the successive regimes had similar goals and approaches with regard to the national park. While the imperial regime capitalized on the notion of 'wild' south, the military regime resorted to notions of 'modernization' that undermined the coexistence of people and wildlife. On the other hand, the EPRDF regime was confronted by its own political rhetoric of democracy and recognition to cultural practices on the one hand, as well as its ambitious conservation and development plans on the other. As a result, it used subtle ideas such as 'conserving' nature for future generation and 'improving' the living standards of local people to channel down its programs to the local communities. Accordingly, the EPRDF used 'developmental' rhetoric in extending its authority to the peripheries. However, all the three regimes shared commonalities in delegitimizing customary rights, denigrating the indigenous knowledge of the local people and limiting local people's engagement in the process of conservation and development practices.

Moreover, the discussions in this chapter show that the different kinds of depicting the erritory have been rooted in the socio-cultural, economic and political identity of the Ethiopian state, which was biased towards highland sedentary plough agriculture. Using these discursive representations as the instruments of delegitimizing customary rights of local inhabitants who claim to have had historical rights over the land, the state exercised different forms of environmental governance as technologies of control over the people and territories. However, neither its coercive nor disciplinary power was effective in creating environmental subjects. Aside from its rhetoric in 'educating' the local people about possible trickle-down economic benefits, the government has not institutionalized the community-based resource management that some consider as a neoliberal form of environmentality. This approach is thought to bring local communities to the conservation projects and missions in the expectation of economic and social benefits.

While the challenges it faced from local communities will be the subject of the next two chapters, this chapter has brought to light some level of complexity that ethnic federalism has brought to the socio-economic and political scenes regarding inter-ethnic and inter-regional relations. The increased internal heterogeneity within the state particularly after 1991 sometimes obstructed the federal government's exercise of power upon local communities. Moreover, the new political order has enhanced inter-ethnic competitions over common resources, with ethnicity having become an important signification in people's claim for participation in environmental resource management and entitlement to resource ownership. The failure of the federal government to settle inter-regional claims and contestations has also led to the withdrawal of the APF in 2007. However, the common denominator for the three state structures (the federal and the two regional states) is that narratives of conservation and development are either used or rejected for exercising control over the environment and the people. In the current context, the contestation over the Nech Sar National Park has been played on the socio-political field/arena established in the aftermath of the 1991 political reconfiguration, whether it is built on history of settlement, administrative jurisdiction or constructed identity claims. Under such circumstances, each state actor has been using strategies of management of meanings, discourses and narratives related to the territory that also inform their actions and practices. At the local level, government officials mainly consider ethnic membership of the local communities during the resettlement, eviction, compensation and punishment of people.

CHAPTER SEVEN: GRASSROOTS ACTORS, A BUSINESS/CONSERVATION ORGANIZATION AND HUMAN RIGHTS ADVOCACY GROUPS

7.1 Grassroots Actors

The category of grassroots actors or local communities broadly includes shifting cultivators, small-scale farmers, nomadic pastoralists, poor urban dwellers and fishers who resist, negotiate and at times come into conflict against the predation of dominant actors such as the state and business corporations in the contestation over environmental resources (Bryant and Bailey 1997). In the context of this study, I will mainly focus on the Guji, who predominantly practice an agro-pastoral livelihood style, as well as the Koore farmers. In addition, fishers, tour operators and forest resource users were also included into the study, albeit not as central actors. This differential emphasis is due to strong dependence of the first two communities (the Guji and Koore) on resources from the national park and owing to their power in challenging state intervention. As the park straddles the two regional states adjoining three ethnic groups and because of different livelihood engagements of these ethnic groups, the local actors are heterogeneous in many respects.

Under circumstances in which various actors with different power relations interact over a particular environmental resource, it is essential to understand social differentiations in many forms taking into consideration ethnicity, religion, language, gender and age to gain insights into how each category interacts with the environment and negotiates with local and global dynamics (Gezon 2006). However, I do not use economic activity as a defining marker of ethnic distinctiveness such as in the case of relating agro-pastoralism to the Guji and farming activities to the Koore. This is because such categorization not only masks internal diversities within the groups, but it might also unnecessarily place the blame on either of the groups in the politics of resource degradation.

In the study of grassroots actors, political ecologists often emphasize the impact 'powerful' actors perpetuate on the 'weak' local actors as a result of asymmetrical political and economic power relations between the two (Bryant and Bailey 1997; Scott 1985). However, by positioning grassroots actors on the weak side in the continuum of power relations, this perspective understates the agency of local actors in contesting and negotiating with dominant actors and their capability in formulating their ways of defining human-environmental interactions. Thus, by recognizing the heterogeneity of grassroots actors and acknowledging the knowledge and agency of members of these actors in defining human-environmental interaction, which is built on different notions of knowledge, I will discuss how these actors contest, negotiate or appropriate state discourses of conservation and development projects in the national park. Nonetheless, local communities' responses in terms of actions to intervention by the state and other non-state actors regarding the management and utilization of the national park will be the subject of chapter eight.

7.1.1 The Guji Oromo

The Guji Oromo agro-pastoralist community who inhabited the entire Nech Sar area before the establishment of the park were first evicted from their home in 1982 (see chapters 3 and 4). The establishment of the park that eventually led to the displacement of the people has dispossessed them their customary resource rights. In fact, the focus on physical displacement obscures other restrictions and dispossessions local communities face due to the enclosure of their customary lands through protected areas (Agrawal and Redford 2009). The state had its own justifications for setting aside the territory as a national park and for relocating the people from the area. Above all, it considered the traditional coexistence between humans and the wildlife as 'unholy' union because it was believed that the people would 'spoil' the natural beauty of the wilderness and threaten the wellbeing of the wildlife. By contrast, the Guji had their own conceptualizations of their surroundings and practiced different approaches of environmental conservation for centuries before the state intervention. In fact, among the Guji, conservation is not always intentionally planned practice in the context of modern biodiversity conservation; rather, it reflects an outcome of accumulated practices that are connected to their cultural, spiritual, economic and cosmological schemes. In this section, I will discuss and analyze how the Guji position their resistance against conservation approaches of the state from their epistemological and ontological understandings of their surroundings and their place in it. I will begin the discussion with the following questions as points of departure: How do Guji's understandings of their environment contradict with the state's notion of human-nature relations? What informs Guji's counter knowledge production in the struggle over the Nech Sar National Park?

To achieve a better understanding of the current interplay between the park and the Guji who inhabited parts of the national park territory, it is imperative to discuss three fundamental socio-cultural and historical subtleties of the Guji people. Firstly, the Guji had collective memory of evictions and restrictions the park administration perpetuated upon them in the 1970s and 1980s. Secondly, the social, cultural and economic importance of pastoralism among the Guji has been vital in shaping Guji's perceptions of their environment. Thirdly, an evident internal transformation within the Guji in terms of cultural and religious practices and livelihood engagements had also contributed its share to shaping the views of individuals towards the national park. These three socio-cultural features and historical memory of the ago-pastoral Guji community in the Nech Sar area convey competing notions of human-environmental interactions, but sometimes complement one another in view of the people's claim for customary rights to the land. One baseline piece of information that we have regarding the Guji-park relation is their persistent resistance against state intervention at the level of the discourse of conservation and development, as well as its practical implementations on the ground. However, the central point that should be interrogated relates to what informs the contestation and how the Guji position their claims of entitlement. I will discuss this in the following sub-headings.

7.1.1.1 State Mistrust and Memory of Past Eviction

The Ethiopian state has been considered among the people in the periphery as extractive and predatory state, with its coercive institutions such as the park authority and local government administrative units, who intermittently intervene into the ways of life of the people, compelling them to accept government political and economic policies. The Guji faced punitive attacks from the successive regimes for their non-submissive behaviors. For example, following the conquest in the late-nineteenth century all throughout the imperial period, the Guji in the highland areas were dispossessed of their land while the lowland pastoralist Guji were victims of sporadic state invasions (Hinnant 1977; Van de Loo 1991). In the process, as happened to other Oromo groups during those periods, the Guji experienced cultural marginalization, economic exploitation, administrative suppression and a stereotypical representation as violent and rebellion. During the military regime, the Guji faced one of the early brutal attacks from the government in 1975, in which the military massacred about 377 people in a day (Solomon 2002; Asebe 2012d). More related to the national park, I have discussed earlier how the military regime forcibly evicted the people from their home in 1982, not least involving the dispossession of their customary rights and restriction of access rights to the park territory all over the history of the park. As a result, the legitimacy of the state as developmental and guardian of the wellbeing of the people and the wild animals have been questioned by the majority of the Guji men, women and youth around the park.

Perceived and real imaginations of the state as a coercive institution have been vivid memories and day-to-day experiences among the Guji. For instance, my Guji field assistant once interjected in our discussion about the people-park interaction, saying, "The effect of this park on our people is very damaging. They displaced our parents and killed some people. If you go to Hiituu village, you would see a woman who lost one of her eyes by the government during the eviction [in 1982]. This is what we expect again if they come and chase us from our home. They kill people, destroy properties and do not care for the livestock" (Field Assistant, Edema Galgalo, April 2012). My field assistant was an eighteen year-old young boy, yet grew up with lived-in experiences and oral narratives about how his parents and the community at large were mistreated at the hands of the government for the sake of conservation and tourism. Most Guji men and women share similar views. Their memory of the past evictions and their current experiences of dispossessions of their customary resources enhance their resistance against state conservation discourses and practices. Moreover, despite changes in regimes in Addis Ababa, there was no significant change on the ground for the Guji in terms of their right to the land. Practically, they did not get healthcare centers, schools and roads under the successive regimes in the country. By contrast, all of the regimes were similar in terms of displacing, dispossessing and expropriating their resources, including livestock. Therefore, they question the legitimacy and credibility of the government's development and conservation rhetoric because neither the previous regimes nor the current government brought practical improvement to their living conditions. When it comes to the national park, given that the Guji did not benefit from tourism projects on the one hand, and rather the national park became the major factor for their displacement in the past on the other, the local people would less likely accept the state project unless the government uses coercive force, which will also be counterproductive.

Such collective memory of the evictions and the ongoing dispossessions of their customary resources have informed the Guji to carry cultural and historical idioms of representation together with a subscription to traditional forms of conservation and coexistence with the wild animals as critical forms of knowledge against state discourses. A sense of insecurity in the proposed resettlement areas, which was built upon past memories in which many people lost their lives and livestock, was among the major historical basis for Guji's defense against state intervention. In other words, the Guji considered the government's conservation and development projects as something that alienates them rather than a system that embraces their values, customs and needs. According to Guji informants, their long experiences with the state are sufficient evidence for them to consider it as an alien institution that seeks to grab their resources without reciprocating them for what it takes. This is why they often perceive state institutions, its agents and policies with great skepticism.

7.1.1.2 Situating Pastoralism in Defense for Customary Rights

Guji's relation to their environment, and particularly of those who remain engaged in agro-pastoral economic activity, is closely intertwined with the socio-cultural and economic significance of livestock herding. Without establishing the nexus between livestock herding and its socio-cultural and economic imperatives for the Guji, the people's contestation against the national park remains elusive. Moreover, place-based practices and rituals that largely contribute to maintaining the harmony between humans, non-humans and the divine being are also interconnected to herding practices. In chapter three of this work, I have discussed how the Guji position livestock herding in their worldviews, identity frameworks, creation myth, inter-societal relations, self-esteem, as well as cultural practices. In this section, I will reiterate some fundamental points of references that the Guji make in their contestation against the national park.

Among the Guji, livestock represents strong socio-cultural meanings in that "cattle herding and possession of large herd of cattle are associated with cultural pride, economic values (wealth), sense of Guji identity and provides social privilege in marriage arrangement and inter-societal relationships" (Asebe 2011b: 54). In terms of the psychological, emotional and identity representation of the self, the Guji consider possession of large number of healthy and productive animals as crucial because their interaction with their neighbors and their internal socio-cultural practices also strongly depend on the possession of large herd of cattle.

In the traditional context, the Guji made animal sacrifices during prayers, blessings, thanksgiving and rituals conducted to cleanse a person from a 'sin' and purify a certain environment from bad spirits. Moreover, they performed rituals to maintain spiritual connection between the people and *Waaqa* (God) and strengthen harmony between people and their surroundings–living and non-living things (Baxter 1991). For such purposes, riverbanks, mountaintops and sacred trees are important ritual sites among the Guji. Through these rituals and the sacrifices that accompany the practices, the Guji believe that they communicate with *Waaqa* who would reciprocate them with rain, fertility for the people and animals, peace and stability to the Gujiland and regeneration of pasture and water that are essential for herding communities. For instance, in the study area, the Guji used one of the hills currently located in the park, called Bonke hill as a ritual site. Year after year, the Guji conducted prayers and rituals that involved animal sacrifices for a few weeks, before they brought back their livestock to the wet season encampment in the plains of the Nech Sar (Ergansa). However, the Guji were restricted from access to the ritual site following the evictions in 1982 and 2004. In their struggle to gain access to their customary resources in the national park, the people capitalize on the economic, spiritual and cultural significance of the territory.

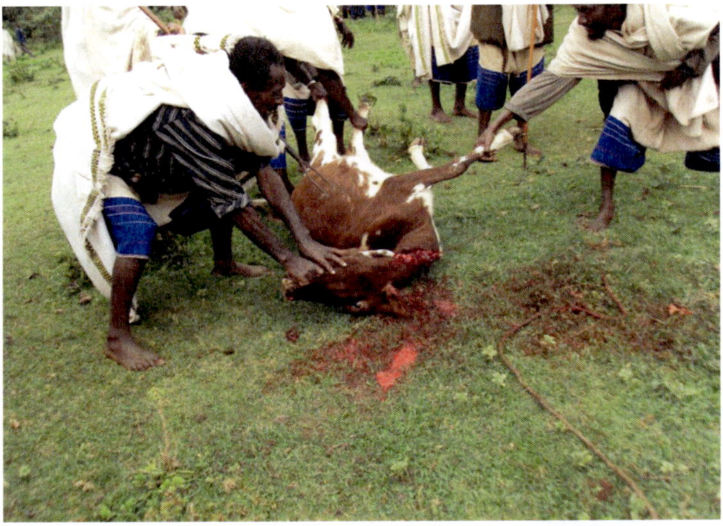

Figure 13: Guji Oromo slaughtering a Calve during a rite of transition in the *Gadaa* System (Photo: Asebe Regassa, March 2009)

Moreover, livestock possession is a marker of social prestige among the Guji. In the past, and indeed still today among the majority of the rural Guji community, bride prices are paid in kind, which mainly constitutes cattle. In fact, unlike in many areas in Oromia, they Guji pay bride price more in the form of reciprocity from both sides rather than as a payment for marriage. For example, regardless of the amount of 'money' that the bridegroom paid, the families of the girl accompany their departing daughter with at least ten to twenty cows. Marriage arrangements are often made between families of 'equal' social status, defined in terms of wealth in livestock and other social prestige that accompany wealth.

In April 2012, during my fieldwork in Arda Guddina village, I came across a story of a failed relationship between a schoolteacher and a Guji girl, which goes as follows:

> The schoolteacher, Gadisa (pseudonym) fell in love with a girl from Mi'o, one of the villages adjacent to the park. Although it was prohibited for Guji girls to practice love affairs before marriage, Sooretti (pseudonym) and Gadisa were in relation for couple of years. Eventually, they wanted to get married and thus Gadisa sent formal request to her parents via local elders. Nevertheless, Sooretti's parents de-

clined to accept Gadisa's request citing that the schoolteacher did not match their social status. Gadisa was from another part of Oromia with agrarian family background. Although some elders tried to negotiate so that Gadisa would pay the bride price in cash and jewelry, the parents finally canceled it because monetary payment would not substitute the symbolic meanings of livestock (reconstructed from discussions with Arda Guddina schoolteachers, May 2012).

The centrality of livestock in Guji's life as symbolic representations of their identity, livelihood sustenance and key elements in rituals defines the people's interaction with humans and non-human beings in their surroundings. The Guji have a well-established ethics of herding livestock. When they head their animals to grazing or water fields, herders restrain from chasing the cattle in order to avoid creating emotional chaos on the part of their animals and those in the wild. Unlike my own childhood experience when we used to enjoy watching and even instigating bulls to fight, Guji youth are educated by their community that such an act is unethical. According to the Guji, the fight among bulls would create instances of chaos that disturb the harmony of nature (spirits, wildlife, livestock and people).

During the evening of May 10, 2012, I joined my field assistant, Alemu, as well as his 'brother', Danbala (Alemu was 'adopted' to Danbala's family) in their search for a bull that remained behind the herd. After nearly two hours of searching in the bush, we found the animal in the park. Although Alemu and I wanted to propel the animal in order that we would arrive home before it was too late, Danbala insisted that we should walk slowly and quietly as it would otherwise invoke some sorts of anger on the part of spirits and the wildlife. He explained further on the point mentioning that the bull entered the territory of spirits at a wrong time and it would be inflicted with disease or even death if it dashes in the territory that is not its own. Danbala also refused to loudly call his brother, Dhugo, who went to a different direction for the same purpose of looking for the lost bull, given that shouting in the bush was prohibited. Rather, after we reached home, Danbala climbed on a tree and rustled. As it was a common sign of communication, Dhugo understood the message and came home.

Like most pastoralist communities in East Africa who believe that they were created as herders and all cattle belong to them (Igoe 2004), the Guji also have deep conviction and pride for being pastoralists. Guji's worldviews and their understanding of their surroundings are framed around herding and possession of large herds of livestock. As Baxter (1991) and Taddesse (2006) discussed in relation to the strong cultural and social attachments of the Guji to their livestock, my fieldwork experience also confirms the notion that Guji cosmology revolves around a tripartite nexus between *Waaqa*, the people and livestock and other non-humans. The relationship between the three is reinforced through prayers, blessings and thanksgiving rituals that are performed at specific sites. They believe that for the livestock to prosper, *Waaqa* created 'limitless' pasturelands and water. This is why the Guji claim to have a natural right over pasturelands and water sources as far as they could move their cattle. On the contrary, they traditionally believed that agricultural areas in the highland were exclusive domains reserved for other agrarian groups as agriculture was in the past a despised livelihood practice among the Guji.

Guji's respect for their cattle was part of the reasons why they traditionally prohibited consuming certain types of food such as wild animal meat. Apart from considering it as a sign of poverty, killing wild animals for food has strong provenance in Guji's high respect for their livestock and the belief that such an act would invoke punishment on their livestock in the form of disease, infertility or death of their animals. This is why the Guji strictly follow a tradition of cleansing a sin from a person who happened to have killed wild animals, because unless a they make purification and reconciliation rituals for the transgressor, the person or his family would face different kinds of punishment from *Waaqa* and the spirits in the surroundings. During the purification ritual–purification from sin of transgressing ancestral oath–they take a sheep or calf to a place where the wild animal was killed and slaughter the domestic animal, before sprinkling its blood to different directions and onto the face of the killer. In some cases, the person washes his hands with the blood of sacrificial animal. The people leave all the meat and skin of the sacrificial animal in the bush, with only the Watta–a caste group within the Guji–are allowed to take the meat.

Concisely, livestock herding is a cultural as well as economic practice for the Guji and any disruption in this mode of lifestyle is considered tantamount to intervention into their cultural identity as a group and a crisis to their livelihood. Besides their reference to history as inhabitants of the area long before the park's establishment, Guji's resistance against state restriction to their customary rights in the national park is geared towards the cultural and economic importance of the territory. A Guji elder once commented: "*Gujummaan keenya aada keenya. Aadaan keenya waa mara. Looni, jila, safuu, gaara, lageen fi mukkeen tunillee aadaa gabbisan*" (informant: Gagasa Adola, April 2012). Loosely translated, this means: "Our Gujiness is confirmed by our culture. Our culture constitutes everything. The livestock, sacred journey (ritual), customs, hills, valleys and trees [sacred trees] all contribute to continuity of our culture." In other words, according to Guji's conceptualization of the territory and their interaction with the animals inside the park, the economic and cultural significances of livestock herding are inseparable. One might think this does not show

Guji's connectedness to the land inasmuch as it focuses on the attachment of their cultural values to livestock. In this regard, I would explain the inseparability of pastoralism and their cultural values from the land in two ways. First, the Guji practice pastoralism within spatial, cultural and mythic frameworks conceptualizations of their surroundings. Thus detaching the people and their livestock from the space where cultural practices and mythic imaginations are based would disrupt the entire system. Second, the Guji believe that the government's resettlement plan would 'transform' Guji's lifestyle from agro-pastoralism to sedentary form of lifestyle, which in turn would disrupt the socio-cultural and identity provenances embedded in herding. This is why the Guji oppose the state's resettlement plans, conservation discourses and development rhetoric.

7.1.1.3 Religion and Education: Elements of Contestation and Negotiation

Since the mid- twentieth century, Gujiland experienced some external forces that brought some socio-cultural transformations within the Guji. Despite the multiplicity of internal and external factors for the socio-cultural transformation among the people, I will focus only on the role of Protestant Christianity and modern 'Western' education with particular reference to the change in people-park relations. In fact, increased inter-societal interaction (cultural diffusion), exposure of the youth to urban life, the engagement of the Guji in different livelihood options such as farming and fishing, as well as the introduction of the national park itself have all contributed their share to the change in some socio-cultural practices among the people and their perception of the environment. The introduction of protestant Christianity to the Guji land brought new worldviews to the people (Hinnant 1977; Van de Loo 1991). Moreover, although access to education among the Guji in general and particularly those in the Nech Sar area was very limited, children who went to school began practicing or imagining what they learned at schools, such as climate change, overgrazing and soil erosion, while it also gave them the opportunity to critically evaluate state interventions. These two external forces and processes enhanced internal heterogeneity within the Guji community, but also produced critical knowledge that challenged state intervention from different perspectives.

Regarding the effects of Christianity, it is possible to see from two different angles. On the one hand, it enhanced the utilitarian conception of human-environmental interactions based on the teachings in the Christian Bible. Guji Christians positioned their claim of entitlement to the resources arguing that God created these resources for human purposes, thus suggesting that they have natural rights to utilize it. While the Christians objected to traditional ritual practices, sacrifices and taboos condemning them as demonic practices, they were also against the government's exclusionist approaches. Guji Christians consider strict obedience of traditional customs as a sign of cultural conservativeness and in contradiction with the teachings in the bible. As common in Christians' justification of human mastery over nature based upon the Christian doctrine that vested authority on humans to control and subdue animals of the earth, birds of the sky and the fish of the Sea (Merchant 2003), Guji Christians began defining their relationship with their surroundings from religious perspectives.

However, there are intergenerational differences within the Guji Christians. The adults on their side understand their ancestral *aadaa* (culture) and *safuu* (norms) as something that is entangled to their Christian religious belief and practices. For example, during the drought that struck most pastoralist parts of Ethiopia in 2010 and 2011, the Guji *abba Qaalluu* (religious leader of *Qaalluu* institution) conducted series of public prayers to *Waaqa* for rain. Some informants stated that some Christians also participated in the prayer sessions. Gamade Halake, one of my Christian informants said that, "We, Christians usually pray to God in our churches but whenever there are large public meetings for common concern like the issue of rain, we attend the meeting with the non-Christians. However, we refrain from offering sacrifices. We never eat the food other people bring for the rituals. Those things are prohibited by our bible" Informant: Gamade Halake, April 2011). People such as Gamade do not see any problem in maintaining their ancestral *aadaa* and *safuu*, although not in a way that it contradicts with the new religious doctrine. On the other hand, some of the youth preferred the strict observance of the teachings in the Bible. Christianity in this context can be understood as an agent that detached the community from their spiritual connectedness to the land on the one hand, and as another form of critical knowledge on the other. Therefore, while it distances the people from their traditional values of nature by denigrating some rituals and belief, it also enhances people's claim for utilizing the resources including the wild animals building their argument on the Bible's prescription about human dominion over nature. Therefore, one might see the introduction of Protestantism as a socio-cultural transformation within the Guji community in bringing some changes in the people's understandings of their surroundings.

Followers of traditional belief interpreted the frequent occurrences of natural calamities such as drought, flooding and diseases as infliction invoked by people's deviance from ancestral norms, customs and beliefs. Adherents of traditional belief who are strict to Guji customs, values and rituals accused Christians for bringing 'sin' to the Guji land because "they kill wildlife for food, cut down sacred trees and denounce rituals as demonic practices. In Guji custom, these practices would evoke *Waaqa* for punishment" (informant: Aga Shota,

March 2011). These incidents illuminate the existence of cultural tensions between Christians and non-Christians and within Christians between the young and adult generation. The common feature among all different categories of the community is their views towards their surroundings. Whether from a Christian perspective or within the traditional belief, the people claim that they had undeniable right over the land, which *Waaqa* their ancestors and passed down to them. While members of Guji Christians claim their right over the land and its resources from a religious perspective that posits human dominion over nature, non-Christian emphasize the cultural and spiritual connected they have with the land.

Another point to interrogate in connection to these competing perspectives from members of Guji Christians and followers of traditional belief is whose perspectives to take in this context. Nevertheless, as I discussed in the first chapter of this work regarding the need to recognize multiple voices and interpretations, I argue that it is not possible to take a strictly distinct position as both arguments depend on the worldviews, beliefs and values of the subjects. Rather, it would be sound to suggest that any attempt to impose a certain discourse or policy without recognition of these multiple voices and interpretations would be a futile exercise in any development and conservation project like the Nech Sar National Park where actors with various ideas, perspectives and practices interact.

Likewise, the introduction of modern education, however limited it might be, has exposed a few of the young generation to different ideas regarding nature conservation. Although the establishment of school for the Guji children in and around the park was only in 2008, few children earlier went to schools in Arba Minch and Tore towns–the latter is located at a distance of over 400 kilometers from the park. Schools equip the children with different ways of understanding their surroundings. However, due to incompatibilities between home and school in children's domains of knowledge production, Guji children who go to school often face challenges to adjust themselves to these competing arenas of knowledge (Asebe and Kjolvolt 2013). While some managed to accommodate different views within particular contexts, the majority of schoolchildren I interviewed developed interest to the 'school culture' and thus considered some local customs and practices as backward. For instance, in one of my group discussion sessions with seven schoolchildren, we discussed on the importance of the national park, particularly regarding tourism.

It emerged that five of the participants agreed to the ideas about the issues of national parks for the sake of biodiversity conservation and tourism activities. In their arguments, they referred to what they learned in Civic and Ethical

Education and history books at school about conservation of national heritages. The youth seemingly subscribed to notions of tourism and its economic returns for the people and the country at large. On the other hand, the remaining two members of the group brought a different point of argument disentangling the two spaces of knowledge production, acquisition and exercise. This means that school and home are different contexts where knowledge is produced. In this line of argument, despite learning about the importance of national parks for economic development and biodiversity conservation at school, the schoolchildren were also worried of its impacts on the livelihood practices of their parents. They underscored that the knowledge they acquired from the school is useful only in a particular context and the same is true for culturally viable forms of knowledge. Accordingly, this is how the school created a critical mass to challenge state intervention based upon informed evidence.

In this regard, it is possible to analyze the impact of education on the children from two different angles. First, due to an absence of integration of local knowledge into modern school curriculum in the country, local communities consider the school as an institution that detaches children from their cultural roots. In this case, it facilitates internalization of 'developmental' and conservation discourses of the government by introducing sense of 'modernity' within the mind setup of the schoolchildren. In other words, the school might serve as the state's instrument of creating environmental subjects. For example, a grandmother from Dabaqa, one of the Guji villages in Abaya district, once commented on how the school 'spoiled' the behavior of her grandchildren as follows:

> In the past, children respected their fathers and mothers, their community and culture. Now, after this school thing came to our land, children do not listen to elders. They say 'it is our right'. In our culture, children learn the customs and values of their community at early age. Now, they began considering it as backward. Look at my grandchildren. They do not want to practice their ancestors' customs. They want to bring back what they learned at school. Nevertheless, that does not work with us. Our children respect the rules of the government and their teachers rather that of than our elders. I can simply say that the school spoiled our children and detached them from their roots. A girl who does not know how to prepare traditional food and beverage and a boy who is ignorant of herding livestock cannot be proper Guji (informant: Dalgee Robale, April 2008).

In a small community like the Guji around the Nech Sar National Park, the success or failure of few individuals becomes visible in the entire community. Likewise, the 'success' story of few Guji youth who secured jobs in the national park administration as scouts and of a young boy who managed to earn money as part-time tour guide has been commonly told among few young people who went to the school. In this regard, a short life history of my field assistant hints at how the Guji youth aspire to work in the national park. One of my field assistants, Edema[37], was studying tourism management in a private college in Arba Minch town. Besides his studies, Edema earns some money as 'informal' tour guide. Although my first contact with him was through his father, whom I visited in Arba Minch hospital, Edema was usually available in front of Arba Minch Tourist Hotel and Paradise Lodge. He used to wait for tourists who stay in these hotels. From there, he would guide them to the Guji villages on the other side of the national park. Unlike most youth of his age from his community, this young Guji had good command of English, Amharic and Gamo languages, in addition to his mother tongue language of Afaan Oromo. He has established a good network with few other youths from Arba Minch town who did not dare to take tourists to Guji territories because of the stereotypical representation of the Guji as still practicing the *midda* tradition, namely killing 'enemies' for honor. When Edema and his friends succeed in convincing tourists, Edema guides tourists or researchers to Ergansa/Arda Guddina village.

Although the park authority discouraged tourists from entering the Guji community, some tourists and researchers who were interested in experiencing human-park interactions opted to go through the 'informal' way. Edema and his friends share what they get from their guests. Through such economic ties, they have built strong social networks. Despite intimidations from the park authority that categorized Edema's activity as illegal, he continued with his part-time job for the last couple of years. In May 2012, the park authorities detained Edema accusing him of taking me to the Guji community. When he took me to the village, he shared with me his future career plans, namely that he planned to be a tour operator in the park. For Edema, "the park is a big national and international treasure. It is also an important resource for local people, because some already got job in the park as scouts and there is possibility for others in the future to get jobs either as tour operators or in other related jobs. I think it is still possible to divide the park into two for wild animals and livestock" (Edema Danbala, May 2012). Some Guji youth including Edema, consider the national park as one of their future employment institutions; indeed, they envision its conservation in line with their current and future economic interests.

Second, the local educated youth became more critical and aware of state manipulations in the struggle over the Nech Sar National Park. The schoolchil-

[37] This name is pseudonym upon request of the individual for his security. Edema was detained many times by the park authority accused of guiding tourists and researchers to Guji inhabited villages adjoining the national park.

dren have established informal networks with some government authorities in Oromia region and pass over information to the latter about what is going on around the national park regarding their community. They also took part in community meetings and helped the local elders when they traveled to towns for appealing against the national park authority in cases of detainment of members of their community or any other park encroachments. Although they did not have the knowledge and skills of legal procedures to help their community, the schoolchildren served as a bridge between their community and state officials in networking their elders with other educated Oromo who would support the people in the legal or administrative challenges they faced. For example, the pupils managed to get the official letter sent from the former regional president of Oromia to the former deputy prime minister of Ethiopia regarding the national park and the right of the Guji to be involved in the management and benefit of the national park (see the letter in Appendix). Government offices were not willing to give me the letter, although I easily obtained it from the school pupils who considered my research as part of their struggle to regain their ancestral land.

In short, the two elements of socio-cultural transformation among the Guji have brought different worldviews regarding the people's perception and interaction with their surroundings, mainly with the national park and its resources. While members of protestant Christianity began questioning some traditional customs and practices, the youth who went to school have started focusing on the economic benefits of the park rather than cultural and spiritual representation the elders attach to the territory. From this side, it opened a field of negotiation with the state discourses of environmental conservation, albeit from different perspectives. By devaluing the cultural and spiritual connectedness of the people to the land, Christianity seems to have brought the modernist understandings of nature-culture relations. However, both Christianity and education also produced critical mass who questioned the state's monopoly of rights over the natural resources.

Nevertheless, some commonalities bind these categories together. Whether it was for economic purposes of the territory as a grazing, farmland or as a future employment sector or for cultural and spiritual sacredness of the space, the claim for ownership of the territory has remained central. As a result, each category emphasizes upon the history of settlement prior to the park and subscribes to the current political model that grants ethnic groups with the right to own and utilize resources in its respective historical settlements. By subscribing to history, culture and spiritual practices connected to the territory, Guji elders dismissed the idea of state intervention as a threat to their life style that would in turn disrupt their identity, culture and economy. Likewise, the youth and a few Guji elders who were conversant of the current political constellation along ethnic lines were able to use the politics to their advantage by establishing informal networks with politicians in Oromia regional state at different administrative structures.

7.1.1.4 "They Came Yesterday and Told us What to Do and What Not to Do"–Views of the Guji on the State's Conservation Practices

In March 2012, just before the beginning of the rainy season, the park authority detained eight men from a Guji village in Maddo allegedly accusing them of setting fire to the national park. The village is located in the northeastern part of the national park, which the park authority claims to include within its boundary. The fire burned some bushy areas located between the main Nech Sar plains and the rugged hills to the east, where the Guji settled for their wet season encampment.

In another incident, the park authority and police force from the SNNP regional state in collaboration with Galana district administration (Oromia regional state) destroyed sixty-four Guji homes in Gode village to the southeast of the park. Gode was also a wet season Guji encampment site but due to its proximity to Sermale River and the farmlands in Gashe, Golbo and Qarsha, some people established permanent hamlets in this village. The park authority and Galana district administration accused the people for encroaching into the park territory and justified their action of destroying the houses to halt cattle trespasses that threatened the safety of the wild animals and sustainability of the park. However, local Guji community members and the village administration did not share this view. Local elders who objected the destruction and appealed the case to Borana zone were detained for a week by the district administration, accusing them of provoking people to rise against the actions of removing the 'illegally' built houses.

Both villages were located on the contested border between the park and Guji settlements, which were reserved for the people after the negotiation with the APF in 2007 that the two regional presidents approved in 2008. As I discussed earlier in this work, the park authority and different administrative departments in the SNNP regional state did not recognize the concessions given to the Guji by the APF despite the approval of the concessions by the two regional presidents. This is why the park authority claims all the Guji settlement areas, including the newly established Arda Guddina village, as included within its boundary.

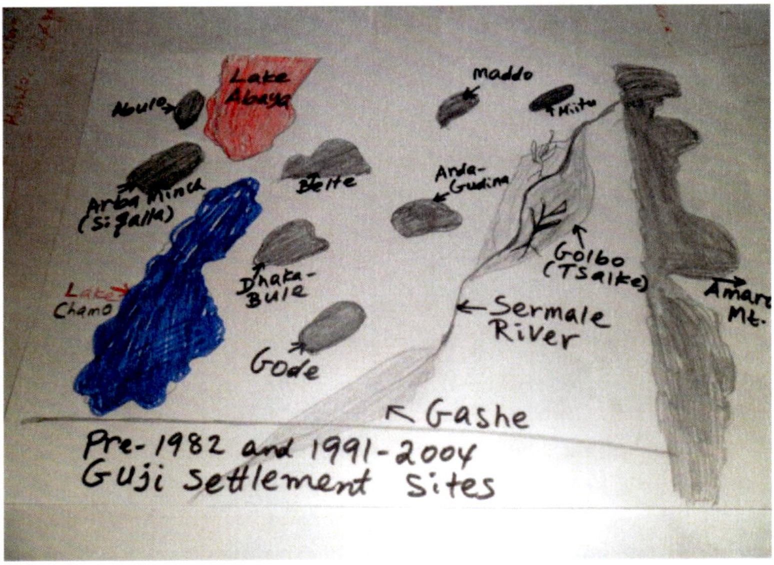

Figure 14: Guji Resource Mapping done through Participatory Mapping Method

Drawing on the above ethnographic information and other related incidents, I will briefly discuss how the Guji conceptualize the state conservation discourses. While Guji's actions against the state intervention will be dealt in chapter eight, in this section, I focus on the epistemological aspects of nature conservation by bringing into the discussion both state approaches and Guji's understandings of their surroundings. The above incidents hint at two competing perspectives and narratives on the part of the state and the Guji people regarding practices of burning pasturelands and questions of access to pasture and water fields that happened to be located inside the park territory. While Guji's conceptualization of their customary right to pasture and water was situated within their creation myth as herders, namely that they were given pasture and water from *Waaqa*, as well as their belief that they had a covenant with *Waaqa* to coexist with non-humans in their surroundings, the debate concerning the use of fire warrants further elaboration here.

The state and the Guji Oromo understand burning practices as forms of environmental degradation and mechanisms of conservation practices respectively. Therefore, the use of fire has remained a controversial dimension of conservation knowledge between the Guji and the park administration. Burning pastureland before a rainy season is considered among many pastoral communities in East Africa as an appropriate conservation approach, given that it enhances regeneration of fresh grass (Neumann 1998; Igoe 2004). Beyond its contribution to restoration of fresh grass following the rainy season, the Guji know that failure to burn pastureland for a year or two would destroy the entire pastureland, because the decomposition of dry grass would kill the grass from its roots and instead enhances the growth of horny bush tress (see picture below).

Figure 15: Horny trees grown in parts of the national park after the banning of burning (Photo: Asebe Regassa, April 2012)

In order to conceptualize how the Guji perceive environmental problems and how they opt to redress it, a glimpse of their reaction to state conservation approaches is imperative. The Guji understand changes in their environment at local and broader levels. However, they explain it quite differently from the state authorities. For Guji elders like Gagasa Edema, an 82 year-old resident of Maddo, Gujiland has undergone dramatic transformations in terms of physical resource coverage, socio-cultural dynamics and state intrusion into their affairs. When Gagasa's parents moved to Belte (now in the center of the park, see the participatory mapping above) sometime in the 1930s from Hiituu (some 20 kilometers to the northeast of the park), the entire Nech Sar plains were covered by long savannah type of grass as tall as two meters. It was a home for several wild animals, including elephant, rhinoceros, lion, buffalo, giraffe and zebra, among others. Gagasa's memory goes with Smith (1897 [2008]) who also described the area as one of Africa's spectacular territories with hundreds and thousands of beasts flocking together as if they were gathered in a kraal. In a retrospective comparison, Gagasa evidently observed environmental crisis in his surroundings. "Now everything is perishing out. Small rivers dried up, grassland was changed to horny bushes. Animals migrated to other countries and those remaining in the park are perishing. Animal disease has swept our animals including the animals in the park. We do not know what punishment it is but it came with these park people" (informant: Gagasa Edema, March 2011).

The Guji accused the park authority based upon its lack of conservation knowledge in many dimensions. Following the ban on fire in the 1990s, horny trees eventually invaded the grassland on the eastern side of the park. The Guji also attribute the death and reduction in number of animals in the national park to two major factors. Firstly, they associate animal death to parasitic mosquitoes and ticks. Although the park officials allegedly blame the people for the spread of contagious diseases from livestock to the game, in turn the Guji blame the park's resource management approach that made the environment conducive for the parasites to reproduce. The Guji argue that the failure of the park administration to burn the pastureland enhanced the reproduction of the parasites that enhanced the distribution of animal diseases. Secondly, the Guji contend that animals from the national park migrated away owing to the park authority's restriction of the animals from access to fresh grass by banning fire and exposed them to predators because of the displacement of the people. According to this latter argument, the displacement removed the people from the area, who were used as 'guards' to the prey animals. Migration of these animals is interpreted as an escape mechanism from 'a prison house' that kept together both the prey and predator through the territorialization of the space. This local conceptualization of national parks as a prison house is noted in Campbell (2005), who argues that the territorialization of protected areas entails enforcement of control on the movement and behavior of wildlife inasmuch as it imposes restrictions on local communities in their pursuit to access resources in the territorialized areas. In this regard, the Guji were capable of delegitimizing the state's conservation rhetoric by capitalizing on the state's weakest side.

As part of the general epistemological contradiction between the Guji and the state, a remark from a Guji elder from Gode, who had lost one of his houses by the government action in March 2012, illuminates both incidents and epistemological conflicts between the two actors:

> We lived here for many generations. I am now over eighty years old. My father was buried at Dhakaa Bule [inside the plains of Nech Sar]. My grandfather died at Hiituu [located to northeast of the park]. The majority of the people you see here in our villages have one or many burial places of their ancestors in Dhakaa Bule, Belte, Bonke, Maddo, Mi'oo, Ergansa and many others. They are all in the park now. In the past, our ancestors lived in harmony with the animals. They cared for the animals as they did for their cattle. This was because of covenant with Waaqa not to kill the animals. Of course, they killed some animals for cultural purposes

> such as midda. However, Guji never ate the meat from those animals. In those times, animals were abundant but there was no shortage of space for our cattle and other animals. These people [the government authorities] came yesterday and began telling us what we should do and what we should not do. They told us not to burn pasture. They banned our midda culture. Above all, they restricted us from access to our home. By blocking our access to spiritual and sacred sites, they detached us from our ancestors and Waaqa. Guji is now cut-off from the past. That happened during the military government. This government also cares much for the animals than the people. These park people consider us as children. They think as if we do not know anything about the animals. However, park problems came because of the park people. After they banned fire and restricted livestock from the park, the pastureland degraded. Mosquitoes and ticks infested the animals (informant: Chari Gamasa, June 2012).

The Guji understand and practice nature conservation as a holistic cultural, religious, economic and institutional practice that cannot be separated from one another. By referring to sacred spaces such as burial places and ritual sites, the informant linked historical referents of settlement and cultural and religious dimensions of the territories to indigenous knowledge of conservation. According to the elder, their ancestors maintained harmony between the humans and their surroundings through culturally embedded knowledge and spiritually connected relationships between the humans and non-humans. By capitalizing their counter narratives concerning areas of Guji's competence such as spiritual connectedness, sacred cosmologies and culturally embedded ecological wisdom, the Guji have challenged the government's allegations that they threaten the national park. They expressed these views in their conversations with park officials, government authorities and external observers (researchers, human rights advocacy groups and conservationists).

Guji's major counter narrative of the state's 'environmental degradation' thesis was built upon their ecological wisdom, which is also called 'sacred ecology' (Berkes 2008). As is common among all Oromo groups, the Guji symbolize 'green' with fertility, good luck, abundance and, above all, peace. Peace among the Guji is much beyond absence of conflict. The Guji people consider their homeland peaceful when their livestock and the people reproduce, pasture regenerate and water wells do not dry up. It is peaceful when predators do not prey on their cattle. It is peaceful when seasonal rain comes and goes on time; it is peaceful when there is no disease that threatens the people and their livestock. Gujiland is peaceful when they have harmony with their neighbors. While the latter aspect of peace is conventional, others are often subsumed into the general aspects of human security and have not been given due attention in the mainstream peace and conflict studies. Therefore, prohibitions against cutting sacred trees or generally restrictions against forest destruction and killing of wild animals are sanctioned by the cultural and spiritual norms and customary laws of the people. Guji's cultural and spiritual connectedness to their surroundings is believed to ensure peace between the people and their non-human surroundings.

However, it should be underlined that the Guji are not 'close to nature' in the strict sense of the representation of traditional societies as the 'noble savage', which subsumes the utilitarian aspect in people-nature relations (Berkes 2008). Within different socio-cultural and economic contexts, the Guji depended on their territory for different purposes. Grazing, farming and killing wild animals for cultural practices, food and business were common practices that indicate Guji's strong utilitarian tradition. However, the central point here is that despite their reliance on nature for their economic and cultural purposes, the Guji have maintained traditional values, ethics and practices of resource governance. The heterogeneity within the group that is increased by internal socio-cultural transformations and external forces is in fact another point of consideration in dealing with the weakening efficacy of traditional institutions, values, customs and practices in nature conservation among the Guji.

Despite changes in cultural practices and worldviews among the Guji regarding their conception of the environment, they still challenge the notions of state conservation practices. Accordingly, apart from contesting state intervention from historical and cultural perspectives, the Guji were capable of capitalizing on their own indigenous knowledge of conservation, which according to some elders, has maintained the human-non-human co-existence and particularly enhanced the conservation of biodiversity in the territory long before the advent of the state. The argument goes that due to the centrality of livestock in Guji's cosmological understandings of the world and their long experiences of herding, they have developed deep knowledge of resource management and utilization, herding skills, disease prevention and animal behaviors. In contrast to the mainstream academic discourses such as the "tragedy of the commons" (Hardin 1968), which downplays the concerns and knowledge of common resource users, the Guji have long history of governing communal resources such as pasture and water.

They had areas called *kaloo* reserved only for calves, goats and old cows. *Kaloo* is located close to homesteads and each household has a right of access for its animals, although responsibilities are also given to each household to monitor the pasture to prevent entry by other animals, apart from those mentioned above. Local elders have authority to sanction different rules upon individuals who transgress local rules of resource utilization and management. The people practiced the management and utilization of pasture and water in the major grazing areas through seasonal movements between dry season and wet season herding sites as well as through locally established customary laws. For instance, despite being understood as a common resource, the Nech Sar National Park plains itself was 'divided' into different portions according to its proximity to hamlet areas. The Guji did not draw boundaries but they knew which area belonged to which hamlet. According to Danbala Dhugo, one of my guides during a transect walk along the park border, the 'division' of the pastureland among different villages was made to avoid overgrazing that would damage the regeneration of the grass.

Nevertheless, most Guji informants also reflected on their own actions in contributing to environmental problems. While they are aware of the current resource depletion and its multidimensional causes, Guji elders pragmatically reflect on some actions of their members as part of the causes of the problems. Firstly, they mentioned the negligence to customary ethics and rules of resource management and utilization on the part of the youth. Secondly, elders who stick to traditional beliefs, customs and taboos interpret environmental degradation as a punishment from *Waaqa* for people's transgression of ancestral laws and customs. For instance, Dokale, one of my key informants from Hiituu, had six children and over ten grandchildren. According to him, only two of his children join him during his annual thanksgiving rituals, while all the remaining children and grandchildren including his in-laws have converted to Protestantism. Dokale complained about the behavior of his children and grandchildren in violating the ancestral laws. According to the informant, members of protestant families killed wild animals for food and demoted sacred spaces and trees. He observed that the following:

> Now it has become right for people to go for whatever religion they choose. That is not the problem. The problem is when people use religion as a shield and demote our sacred spaces. These children invoke infliction to our land by violating our ancestral laws. They kill animals and eat in the bush. Some of them bring home. We might think it is a cow or goat and eat. How do I go for rituals after eating such meat? That is impossible. This generation is lost. I do not know the future of Gujiland but I am afraid that much problem awaits us ahead (informant: Dokale Shanxara, April 2012).

At the beginning of this section, I have raised two interrelated questions regarding competing understandings of human-environmental interaction on the part of the Guji and the state, as well as the basis for Guji's counter knowledge production vis-à-vis the state's rhetoric of conserving nature and improving their living conditions. It has been ascertained that while the state's notions of human-environmental interaction regarding those in peripheral regions was partly rooted in the identity of the state and partly in its economic and political interests in controlling the space, Guji's conceptualization of their surroundings were informed by their cosmological, cultural, religious and economic schemes. More specifically, despite differences in political and economic policies at the national level, successive Ethiopian regimes subscribed to the nature-culture dualism in their representation of people-environmental relations in the periphery. On the other hand, despite internal dynamism within the Guji in terms of education, religion, livelihood and cultural diffusions, livestock herding continues to be strong marker of Guji's cultural, economic and identity representations. As a result, to maintain the health, fertility and reproduction of their livestock, the Guji practice rituals, prayers and sacrifices that includes non-human beings including sacred spaces. Therefore, for the Guji, the human, non-humans and *Waaqa* are inseparable in the way in which all form societies of nature. Thus, it is possible to conclude that Guji's understandings of their surroundings and their resistance against state conservation discourses were informed by their memory of their past relations with the state, the cultural and economic significances of pastoralism, a critical knowledge enhanced by education, religion and lived-in experience or knowledge called ecological wisdom. It is also important to underline that conservation is not practiced among the Guji as a planned activity for the sake of preserving biodiversity; rather, it is a cumulative outcome of religious, cultural and economic activities and governed by customary laws and eco-cosmologies.

7.1.2 The Koore

Similar to other actors discussed above, the Koore people have had claims of entitlement to the eastern lowland portion of the park. The Koore small-scale farmers who sustained their livelihood on rain-fed and irrigation agriculture in the Tsalke valley along the Sermale River considered the expansion of the park to this fertile low land as a threat to their livelihood. Despite competing narratives concerning when and how exactly the Koore moved downslope to Tsalke; some sources attest that the people began using the area for seasonal farming shortly before the mid twentieth century (Awoke 1985; Bediru 1986). Traditionally, there were remarkable ecological, cultural and economic boundaries between the highland Amaro district and the low-lying Nech Sar plains to the west, which the Guji inhabited as pastoralists. Koore's predominant engagement in peasant agriculture on the mountainous highlands of the Amaro district on the one hand, as well as Guji's pastoralist livelihood in the lowland Nech Sar plain areas on the other, created a well-defined ecological niche until the Koore moved to the Tsalke valley.

With the park's expansion to the east following the eviction of the Koore and Guji households from the area in 1982, the fertile Tsalke valley was included into the park boundary and this phenomenon brought direct contact between the Koore and the park. Although the Koore people did not settle beyond the Tsalke lowland, Amaro district exercised administrative authority over portions of the park during the imperial and military regimes. Thus, drawing on this history of political administration, Koore elites and ordinary people often claim that the park belongs to their district of Amaro. To better position Koore's contestation over the Nech Sar National Park, it is essential to underline that this narrow strip of land along the Sermale River has become a cause of disagreement between the Koore, Guji and the park. While the Guji used it for dry season pastureland and small-scale agriculture, the Koore used it entirely for agricultural purposes through irrigation, as well as to some extent by seasonal rain. The park administration also claimed that the area was a dry season retreat for wildlife, where animals would get water and fresh grass. Accordingly, it is within the context of Koore settlement history in the area, the economic significance of the area, contestation between the Koore, Guji and park and Koore's relation with the state that I will analyze Koore's views towards the national park.

7.1.2.1 "Tsalke Is Our Life": The Economic Dimensions of the Territory

In February 2012, when I entered the Koore villages of Darba Manana, Gumure, Tifate and Yero that adjoin the Tsalke valley, I was amazed by the abundance of fruits, vegetables and fresh maize in a local market in Darba. Given that February is a dry season in Ethiopia; it was uncommon to find on-farm fruits and vegetables, except from irrigated farms. Along my way to the villages, I observed degraded land on the highland parts of the district where the people did not practice any irrigation activities. By contrast, the majority of the households in these four villages have plots of land in Tsalke, from where they ensured their subsistence through irrigation agriculture. I was later informed that the fruits and vegetables in the market were brought from the contested Tsalke valley that adjoins the Nech Sar National Park. Local Koore informants recalled that the people in these villages did not have recent memory of food insecurity, apart from some three years food aid they received following the 2004 resettlement program.

Figures 16 &17: Farm Land in Amaro Highland; and Coffee Plants in Tsalke/Golbo Lowland

In addition to ensuring the subsistence demand of the people, many Koore residents in the adjacent villages see the Tsalke lowland as a source of their economic development and a safety valve during natural uncertainties. A story of Shibiru Ayele and his son Kasahun was illustrative in hinting at the significance of Tsalke for the people. Shibiru was a resident of Darba Manana village but he and his children had large plots of land in Tsalke. Shibiru traced back the period his parents moved to Tsalke to the 1930s. According to this elder, his parents and many other people in Amaro district initially moved to the lowland to escape from the invading Italian army (1935-1941). In addition, the Koore once again retreated to the lowland to escape the exploitative and despotic administrative system that was installed following the restoration of the imperial regime in 1941. Some informants mentioned that many people continued to cultivate the lowland areas to subsidize the heavy tribute levied upon them by the government. Cotton and maize were the major crops cultivated in those times, while they later began cultivating coffee. Shibiru and others in these villages continued going down to the Tsalke during dry seasons for irrigation, which they later expanded by planting fruit trees and other plants such as mango, avocado, banana, coffee and sugar cane for household consumption and business. Like all the villagers who moved to Tsalke, Shibiru did not abandon his land on the highland, because both areas were sources of livelihood security during times of uncertainties.

Population pressure and land degradation on the highland made it difficult for households with large family size like Shibiru–with extended family of over twenty family members–to confine to the highland. On the other hand, due to the swampy nature of the lowland with mosquito and tsetse fly infestation and prevalence of sporadic conflict with the Guji, permanent settlement in the lowland has been a major challenge over the years. Thus, the Koore were caught between competing uncertainties of life in both the highland and the lowland. For instance, there was fear and speculations of inter-ethnic conflicts in the lowland and state repression on the highland; deadly disease and destructive flooding in the Tsalke valley and unpredictability of the weather and yield on the highland; and state encroachment to Tsalke through the park and its 'incentives' of resettlement plans to Abulo and Alfacho villages. The Koore villagers continued their day-to-day routines within hope and worries/frustrations. It was within these dilemmas of life that the Koore people in these villages negotiated government conservation discourses regarding the national park.

Like other villagers in the area, Kasahun considers Tsalke as a life stream, situating it within the broader political, ecological and economic issues in the district across different historical periods:

> People first moved down to Tsalke to escape from feudal exploitation. There were problems of heavy taxation and administrative suppression in the highland areas. People also worked much and degraded the land to get high yield for their family and for the government. Gradually, the highland was degraded and gave low yield. It was not enough to pay the taxation and to feed the household. In 1976, some two years after the revolution, I quitted my school from Arba Minch and joined my family in Tsalke. That year, many people flocked to Tsalke because in the previous year there was no good harvest in the highland. Rain did not come on time. I remember people suffered a lot. In Tsalke, we planted maize, sweet potato and cabbage by irrigation. People planted those crops and vegetables that grew fast. After the harvest, we got sufficient food for the whole family until the next harvest in the highland. The mango, avocado, sugar cane, coffee, banana and ensete plants we planted in those times are now the source of our income. Every year, we get an average of over 100,000 birr [nearly 4000 Euro] from the fruits, sugar cane and coffee. We cultivate maize for household consumption but others are for the market. Tsalke is a gift for us from God. In the highland part of Amaro, people live on small plots of land that does not sustain their household demand. However, in Tsalke, except the problem from the Guji and the park, everything is very good. Now, there are serious problems and we do not know the future. The Guji claim that the land belongs to them. Two years ago, there was conflict with the Guji and that affected much of our market relations with them. The park also wants to include it into its territory as a whole for the animals. Another major problem is access to market. These are our major worries (informant: Kasahun Shibiru, February 2012).

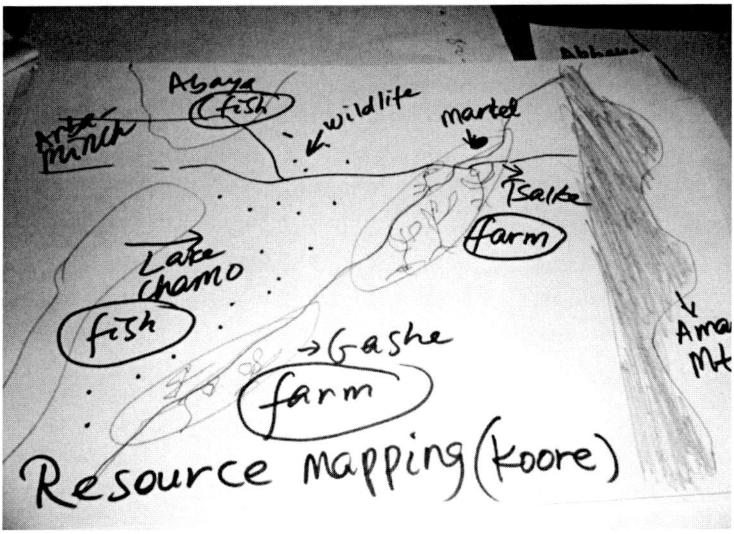

Figure 17: Resource Sites in/around the Park Important for the Koore (Source: Resource Mapping by Koore informants (May 2012))

For the Koore residents in the villages adjoining the park, the economic significance of the territory is central. From the above resource mapping, it became clear that areas such as Gashe and Tsalke along the Sermale River are important farmlands for the Koore, where they grow crops and fruits for subsistence and commercial purposes. Likewise, the two lakes (Chamo and Abaya) are also resource spaces for fishing activities mainly for business. The Koore used strategies of maintaining plots of land at the highland and the lowland for times of uncertainty. Confinement to the highland or the lowland was risky owing to uncertainties in government policies, inter-ethnic relations and weather conditions. The Koore capitalized on their dependence on this territory for their livelihood, with both the Guji and the park administration also competing for different purposes. Therefore, the Koore used the economic significance of the territory and their history of utilizing the area long before the establishment of the national park in their claim for entitlement to the eastern portion of the park, namely the Tsalke valley.

7.1.2.2 Situating History and Culture in Claim for Land Rights

In their claim for entitlement to the fertile agricultural Tsalke valley, the Koore positioned their arguments within historical and cultural references. In defining inter-group relationships or claims for resource ownership, different social groups or actors draw on historical incidents/references to justify their claims and to delegitimize any claims by their adversaries (Schlee 2002). Historical narratives are powerful legitimizing instruments that enhance people's agency to build their current positions on strong historical episodes and trajectories. As Appadurai (1981) argues, competing actors invest a lot in management of meanings, narratives and myth to counterbalance notions used by their rival actors. The Koore people's narratives on history of early settlement of their people in Yero, Darba, Tifate and Gumure villages were systematically used to extend the historical time span of their expansion to the lowland area, although the phenomenon was said to have been shortly prior to the mid twentieth century. According to a Koore elder by the name of Ayele Chamo from Tifate village, the people established settlements in the four aforementioned villages during the sixteenth century and expanded to Tsalke over the course of time. The elder used a metaphor of human body to explain how the two geographical areas or agro-ecological niches were highly interconnected. For elders like Ayele Chamo, Tsalke was part of their history of settlement that could not be separated from the main Amaro district:

> Our ancestors first settled in Yero, Darba Manana and others. They built Yero Medhanealem church [one of the ancient Ethiopian Orthodox Churches in the region]. Gradually they moved to Kelle and to Tsalke. Our Koore people settled in Tsalke just at similar time with other places in the district. Tsalke was a source of life for the Koore from earlier times and continues to be so. You know, a human body does not function well if any of its parts is missing. Darba Manana and other villages on the hill are like the head of human body. Tsalke and other lowland areas are like the heart, foot and other lower body parts. Can a person live without any of these? No, it is impossible. Our ancestors recognized that it was impossible to separate the different parts and that was why they included Tsalke into the whole body part to make it function well. Tsalke played an important role in saving the life of our people during drought and famine. Our people escaped from the fascist Italian occupation to Tsalke. Patriots like Abebe Mengesha and Beyene Tafesse retreated to Tsalke during the war. My father was killed in Tsalke by Italian forces and by the Banda [co-opted citizens]. For me, Tsalke is more than where we get our food. It is part of our history. It was where our ancestors went for hunting, it was where our patriots fought against the enemy, and it is where our generation still struggles against the Guji and the park. Is it not a historical place then? For me it is a historical place (informant: Ayele Chamo, February 2012).

Indeed, Tsalke served the Koore as a refuge in their retreat from foreign invaders and internal despotism and exploitation. Following the incorporation of Amaro into the Ethiopian empire in the late nineteenth century, local chiefs maintained their traditional authority but became authoritarian in levying tribute on their subjects. They oppressed them administratively and exploited the peasants in terms of economy. According to Awoke (1985), local chiefs levied unbearable taxation on the Koore peasants because of which some of them escaped to lowland areas where the state was unable to reach. Thus, Koore's downslope expansion was part of the broader political and economic systems of the period under the imperial regime. During the imperial period, as the above Koore informant mentioned, the lowland pastoral and agro-pastoral areas were almost out of the day-to-day administrative, social and economic burdens of the state. The past governments did not manage to put pastoral communities under effective state administration until the 1990s. Due to the cost of extending authority and control to peripheral regions, most pre-colonial, colonial and post-colonial states in Africa were fairly absent in these regions as a result of which peasants from the state-reach areas used the peripheries as exit zones (Herbst 2000). According to Herbst, African states across different historical periods did not have the technological, political and economic capabilities to extend power to the peripheries. In the context of Ethiopia, although the situation has been changed in the post-1991 period, the imperial and military regimes did not effectively control the periphery.

In the post-Italian occupation period (post-1941), the imperial regime restored and strengthened its administrative and exploitative state bureaucracy particularly in the conquered regions. These regions became the targets of state repression and exploitation in the post war period for two fundamental reasons. First, despite differences in the way in which Ethiopian people across the country reacted against the Italian occupation, these regions largely considered Italians as 'liberators' from the yoke of Amhara domination and exploitation. Within the short period of occupation, the Italians sought to win the support of the people in the subjugated regions by redistributing land to the peasants and by disposing Amhara feudal lords. Therefore, the imperial regime toughened its administrative system over these regions as a mechanism of reasserting imperial hegemony. Second, to consolidate its power both in terms of economic and political power, the regime had enacted brutal administrative and taxation policies that made the peasantry in those areas the major prey to the predatory state system (Markakis 1974; Donham 1986). Because of the state repressive system on the highland parts of the Amaro district, the Koore peasants were compelled to escape to the lowland areas by establishing social networks with the agro-pastoral Guji community in the area. Therefore, Tsalke was part of the exit options. Using *ilmoffacha* (adoption) tradition as another strategy of integration into Guji community, the Koore managed to establish settlements in the lowland areas.

Koore elders and political elites in the district have also raised some aspects of cultural connectedness of the people to the land. Despite meager information, elders traced back to the past when their ancestors were said to have been performing a ritual called *Tobe* ritual every year. According to their views, the Koore practiced the *Tobe* ritual to pray for rain and abundance. For followers of traditional religion, *Tobe* was a shrine where they worshiped the god of rain. Although the ritual is no more existent, Koore elders and elites seem to have subscribed to the cultural dimension of human-environmental relationship in line with the constitutional provision that grants ethnic groups the right to preserve their culture and exercise their religion. During my discussion with a chairman of Darba Manana village, he emphasized the economic significance of the territory but to substantiate the economic interest, the chairman referred to historical administrative borders of the district and some culturally meaningful places in Tsalke. According to the chairman:

> Nech Sar belongs to Amaro district. It belongs to the Koore people. In the past, even during the military government, our district administered up to Sanxaqa Dingay [Bonke Hill]. It was almost up to the forest close to Arba Minch town. The majority of the two lakes were under Amaro district. Our ancestors told us that our people used to go for rituals to Tsalke whenever there were droughts and other problems. They performed rituals and sacrifices to get rain. Later after the period of Italian occupation, people began cultivating the land through irrigation. However, today both the park and the Guji are catching it in hands with our people (Darba Manana village Chairman, February 2012).

Regarding the national park, the Koore peasants and political elites had somewhat common views at least when they expressed it on a 'public stage'. The people emphasized the advantages of the national park for the national economy, for the future generation and for the purpose of conservation of the wild animals while discussing the issue in a public or in a formal discussion. However, 'back-stage' information from the Koore residents depicts contradictory views about the park. For many men and women, the park is an institutional threat to their livelihood survival and they strived to challenge its existence in different ways. For the Koore, while historical reconstructions on their past administrative and settlement history were the major references for their claim of entitlement, they also sought to use the government's discourses of development, food security and conservation in their pursuit to maintain their traditional right over the land. For instance, while considering Tsalke as their life stream and source their economic development, the people covertly challenge the government's resettlement plan, given that it would dismantle their development foundations. In fact, they also openly expressed their views that the park is limited only to the grassland areas and they downplayed any claim of the park authority over the Tsalke lowland. In the next chapter, I will discuss how they contested, negotiated and at times appropriated government discourses of conservation, as well as dealing with their Guji rivals.

7.1.3 Other Local Actors/Groups

This category of local actors includes local social groups who had some degree of influence concerning the course of conservation practices and management approaches of the national park. For the sake of this work, I assess the role of self-organized fishermen, low-income town dwellers, those who depend on the extraction of forest resources from the park and tour operators in Arba Minch town.

7.1.3.1 Self-Organized Fisheries

Self-organized and unregistered fishermen were among the local actors with certain level of agency in shaping the conservation approach in the national park. The government considered this group as 'informal' fishermen to refer to their status of registration and organization in contrast to those registered and organized under 'formal' government administrative structures. According to some studies, 'informal' fishermen fulfilled the largest proportion of fish demand to Arba Minch town and beyond before the year 2004 (Dessalegn 2004; Bayisa 2011). During the 1990s, the fishermen expanded their scope of supply to major towns in southern Ethiopia up to the town of Hawassa while a few even managed to send it to the capital city, Addis Ababa. In those times, dozens of self-organized fishermen made their living by fishing business from the two lakes, which were partly included into the national park in the 1990s. Nevertheless, their limited financial capacity made the fishermen dependent on large businessmen, such as hotel owners, to secure loans for buying or repairing boats, hooks and nets. Beginning from the period of the preliminary implementation phase of the EU funded '*Southern Ethiopian National Parks Rehabilitation Project*' in the mid-1990s, the national park, in collaboration with Arba Minch town and Gamo Gofa zone administrations, banned 'illegal' fishing activities on the Abaya and Chamo Lakes. By banning the 'illegal' fishing activities, the government intended to encourage formally registered and organized fishermen, who would pay taxation to the government for their access to specific parts of the park. It was also necessary to regulate fishing activities because the self-organized fishermen were accused of catching premature fish, as well as using nets and hooks that indiscriminately catch the fish (Bayisa 2011).

However, according to members of the former outlawed fishermen who were later organized under registered associations, the government did not provide them with any alternative livelihood when it restricted the fishing activity in parts of the lakes. A chairman of Chamo Fishermen Association recalled the restriction and financial problems that he and his colleagues faced between 1998 and 2004 as follows:

> In 1997, I completed secondary school but I did not succeed to join the University. At that time, there was no private college in Arba Minch. Even if there were a college, I would not be able to pay tuition fees. We discussed with my former classmates and our friends from the town and started the business. Actually, we used to bring fish to hotels even when we were at school. The park already outlawed fishing from both lakes, which it said were part of its territories. We did not know where the borders began and where they ended. What we knew was that the Lakes were supposed to be free for all. Nobody was supposed to claim right over God-given resources like water and resources in it. Many times, they confiscated our fishing boats, nets and hooks. They detained us too. We moved further south but there were not many fish there. Some of our colleagues went to Lante area on Lake Abaya but there were already many fishermen. We became borko [literally: empty pocket] for many years. [I asked their sources of income during those difficult times and Hailu responded] Some of our colleagues went to different ways of life. They picked whatever came to their ways. Some of them even began looting tourists in the park. That was not a right thing but it was to save their lives for a day or two. However, some of us pushed on with fishing. We established good relations with scouts and got few hours on the Lakes. We also worked part-time in other activities in the town. Nevertheless, life was so difficult. Then African Parks came. I remember it did not lift the restriction. However, African Parks was better because it opened discussions with us and suggested for us to be formally registered as an Association. It supported us with money and expertise on how to begin the business. We formally established Chamo Fishery Association in 2006. The registration helped us to get access to parts of the Lakes, which were controlled by the park. However, the government at times asks us to pay much tax. They think that we collect fish as we wish. It is a difficult job. Anyways, life goes on like this (interview with Hailu Dacha, May 2012).

Two important areas of contestation and negotiation would surface out from closer analysis of the above quotation, which other former fishermen also shared. On the one hand, as it happened with the Guji and Koore communities, this social group also faced the 'tragedy of enclosure' built upon the classical notions of conservation through strict control of human influence on the 'natural' environment. In this regard, the fishermen challenged the concepts of territorialization or enclosure of the lakes, arguing that nobody was supposed to claim absolute right over these natural resources. On the other hand, by compelling them to be registered under different government offices/departments as associations, the government has put the fishermen under its control where they would be legible for taxation and other forms of administrative control. As Scott (1998) argues, one of the characteristics of modern states, and particularly in developing countries, is their deployment of coercive forces, placing their citizens/subjects under institutional and bureaucratic control to make them legible for taxation and administration. While the fishermen came to terms with some conditions of the government, particularly through registration as formal associations, they also contested the ban on fishing during the early years. At times, they prompted the government to facilitate them with some concessions such as micro-finance credit and formal access to some parts of the two lakes.

Drawing on the discussions with former 'informal' fishermen, I would like to further probe into how the fishermen who were denied access to the resources responded to the action. When I was in the field among the Guji and Koore people, my informants raised issues of security problems along their ways to the Arba Minch town until few years back. In most of the discussions on security issues, my informants underlined that there were gangs from Arba Minch town who were formerly engaged in fishing as their livelihood but they resorted to looting and killing tourists and any passer-by in the forest after the government banned their former economic practice. In order to get the link between the criminalization of fishing activity and the eruption of 'gang' crime in the national park, I used an extended case method. As I discussed in the methodology part, the extended case method extends the fields of investigation beyond a particular locale or beyond a micro level to establish a nexus between different local and extra-local social forces and processes (Burawoy 1991, 2009). According to this method, a social science researcher should follow a thread of information and establish links between different spots of data to formulate the link between the epicenter of the phenomenon and the edges of its radiating forces and processes.

I first began my academic investigation from the park authority and Gamo Gofa zone administration to get some clues. Based upon interviews and some documents, it became evident that there were security problems and attacks against tourists and park staffs by unknown gunmen in the years before 2004. However, none of the sources from both offices linked the crimes to the criminalization of fishing. Rather, they used the Guji as scapegoats for the attacks and included it in the agenda for the resettlement program in 2004.[38] Meanwhile, I had a chance to informally discuss historical trajectories in the national park with a park scout who has been working as scout since 1978. According to

---

[38] Minute of a Meeting held in Arba Minch (18/12/2003) to "Develop Implementation Plan for Resentment of People who illegally settled in the Nech Sar National Park".

this informant, security problems in the national park were major concerns to the park administration especially between 1991 and 2004. For people who closely observed different political and social developments that had direct or indirect effect on the park, the security threats in the park were inseparably linked to the management approaches of the park. My informant once interjected a clue to my question if the crimes might have been linked to the ban on 'informal' fishing stating that:

> Many people were killed in this park. During the period of transition, it became unsecured. Nobody knew the killers. Some five or six years later, all former fishermen became jobless. They all entered into the forest. They targeted tourists and passer-by people who come to the town. However, it seems also that there were some hands in it. It brought a big tension between Oromia and Gamo Gofa zone but it was solved. The last incident came out when two or three Guji men were killed some eight or now it was nine years ago (Senior Park Scout, April 2012).

Following the thread of pieces of information from here and there, I interviewed a former Galana Abaya district administrator who was supposedly involved in settling a large-scale tension between the Guji and Arba Minch town/Gamo Gofa zone in 2003. The story goes as follows. In 2003, two Guji men were killed and some five others were attacked and looted on their way back from Arba Minch town. The Guji men and women who escaped the attack speculated the identity of the killers. According to the district administer, the killers were later identified as members of former fishermen whose fishing practices were outlawed by the government. A few days after the incident, the Guji from the adjoining villages were fully armed and sieged Arba Minch town and vowed to attack the town in retaliation for the death of their fellow Gujimen. It was then that both Oromia and SNNP regional state administrations intervened and gave the responsibility to Arba Minch town and Gamo Gofa zone administrations to ensure the security of the people and the national park. The view from Oromia side attests that "the people in the southern region recognized that it was the former fishermen who turned into gang and criminal activities but they did not want to formally accept it, given that it was their fault in just leaving the youth on the streets. We also suspected that the Guji were targeted for their earlier refusal to the resettlement program. The issues of security were later used against the Guji to justify the government's plan to relocate the people" (Former District Administrator, March 2012).

From pieces of information that I managed to collect from different sources, it can be concluded that the criminalization of fishing activity prompted the youth in Arba Minch to resort to a different form of resistance and livelihood sustenance. Despite some officials' denial to recognize the link between 'criminalization' of 'informal' fishing practices and the security problems posed to the park, it later became evident that the government organized the former fishermen into formal associations and supported them with some start-up fund to avoid the risk of youth gangs. Nevertheless, even after registration, the persistence of strong business relations between fishermen and some government officials in Gamo Gofa zone had restrained the park from exercising its authority in line with its conservation approaches. For instance, although the park authority wanted to introduce regulations on types of hooks and nets to be used by the fishermen to avoid disruption in the fish habitat, government officials often ignored the park's proposals and complaints because they were connected to the fishermen for some business benefits (informant: Park Biologist, April 2012). In any case, the fishermen were able to challenge the park authority's imposition of its conservation rules but negotiated in terms of their practices of fishing, such as the case of formal registration and organization as associations. By resorting to criminal activities during the banning of informal fishing on the lakes in 1990s and later by creating networks with government authorities, the fishermen were able to contest their rights to the resources. Moreover, the government also negotiated its exclusionist approaches by granting the fishermen with fishing permission under certain conditions such as registration and regular payment of taxation.

7.1.3.2 Trapped between Daily Subsistence Needs and the Park–the case of Forest Users

The park has been a source of daily bread for many low-income Arba Minch town dwellers who were engaged in utilizing forest resources from the forest adjacent to the town. According to Dessalegn (2004) and Bayisa (2011), most poor women and youth in the town rely on the park to get income from selling firewood, charcoal and construction materials from the forest. These small-scale forest users provide almost the entire firewood demand of the Arba Minch town. However, an institution that had different interest over the same resource they utilize for their daily bread confronts these social actors. Fierce fines and restriction from park authorities were among the many challenges that these people have faced over the years. Under such circumstances, the people were able to use bribe and personal relations with park wardens, scouts and the police as strategies to enter the park 'undetected' or if captured, to escape punishment. According to Tadelech, a woman who has been making her living by selling firewood for the last ten years, it has now become common for some scouts to receive 20 birr (nearly 1 Euro) as a bribe for a bundle of firewood that could be sold for 50 to 60 birr. If she were taken to police stations, Tadelech would pay double the estimated market price of a bundle of firewood and would stay in jail for few days. As a result, according to the informant, paying the money to the scouts was safer and easier but it all depends on the personal behavior and ethnic background of the scouts. As Tadelech noted:

> The government wants to control everything. It restricted our children [fishermen] from the Lakes and it chases us from the forest. I lived for the last ten years by selling firewood. It was tough. Sometimes they detain us and other times they confiscate our woods. When I go home without the woods, the entire family suffers. We have no other sources of income. My child sometimes gets fish from Chamo but now he is out of the job (Tadelech Tibeso, June 2011).

African Parks Foundation organized these resource users and employed them to activities such as clearing invasive plants, building and maintaining drainage canals, soil conservation on rugged topographies and collecting dry wood for fuel consumption by the park staff (APF Monthly Report, August 2005). However, following the withdrawal of the APF by the end of 2007, part-time employees of the foundation were suspended and resorted to their former livelihood options–extracting forest resources from the park on 'illegal' basis.

Concomitant to the above statement, it is essential to note that the withdrawal of the APF had brought major structural change in the administration of the park. Based upon the intense competition between the two regional states particularly concerning the issue of the Guji, the administration of the park was rolled back to the Ethiopian Wildlife Conservation Authority (EWCA) in 2009. According to some of my informants from different offices in the SNNP regional state, the centralization of the park administration was considered as a denial of their constitutional right over the territory. Some park scouts even observed that Gamo Gofa zone and Arba Minch town administrations became too reluctant in taking administrative measures against the people whom the park found 'guilty' of 'illegally' extracting resources from the park territory due to their discontent with its centralization. The interdepartmental discrepancies in dealing with the park issue had opened opportunities for the resource users to continue their livelihood in utilizing the resources. As one park staff mentioned, there was no strong collaboration between the park administration and government departments in Gamo Gofa Zone and Arba Minch town. For instance, according to the informant, the police often release people whom the scouts capture with woods, stating that 'unless the Guji and Koore people stop encroaching the park and killing the wildlife from the eastern side, there was no value to restrict the people who used the forest for their daily bread' (personal communication with park biologist, May 2012). While the conflictual relationship between the people and the park authority emanated primarily from mismatches between the people's question for daily bread and the park administration's ambitions for conservation and tourism development, the interdepartmental tension between the park administration and some SNNP administrations is related to the politics of identity and questions of ownership. Such conflicts opened some windows of opportunities for these low-income town dwellers to maneuver the system into their advantages.

7.1.3.3 Local Tour Operators

As business-oriented local actors, local tour operators have also contributed to the construction of different images about the territory and in lobbying the park administration to follow strict protectionist approach. There are seven different tour operators in Arba Minch town with specific focuses on national parks in southern Ethiopia, namely Nech Sar, Omo, Mago and Maze national parks. According to a chairman of "See us Tour Operator", Nech Sar National Park is the most visited park among domestic and foreign tourists. Despite problems of security and accessibility, the scenery and wildlife in the park are very attractive for tourists. However, according to the chairman, the presence of the Guji on the eastern side of the park has become a big threat to the park, in terms of posing security problems to their organization and tourists, as well as degrading resources in the park territory through hunting and cattle trespass. As a result, tour operators in the town have often lobbied the park and Gamo Gofa zone administration to relocate the Guji to a far distance to secure the park (interview with 'See us Tour Operator' Chairman, May 2012). As it could be understood, tour operators romanticized/glamorized the park to market the territory to foreign and domestic tourists. Behailu, a deputy chairman of 'Rift Valley Tour and Travel Association' described the Nech Sar and other national parks in the south as follows:

> Our region is rich with natural areas for tourist attraction. When a tourist sits in Arba Minch town, for example in Paradise Lodge or Bekele Molla hotel and gazes down the valley, the person would love to go to see the beautiful landscape in Nech Sar. Nech Sar has natural beauty that other parks do not have. By the way, it is not only the topography. The wild animals are spectacular. Omo and Mago national parks are far south. Tourists even visit these parks to view the culture. The Mursi, Surma and many other people are tourist attractions. Nevertheless, in the Nech Sar, tourists do not want to see the people and their livestock. Their culture is not like those in far south. Our tour and travel association benefits from making use of the natural beauty of our region. The people, their culture and ways of life, and the wildlife and topography are what we work on. We contribute to our country's development through tourism. We pay taxation, contribute to development of the country and play our role in building national image (Behailu Ayele, May 2012).

Although they do not have direct administrative relationship with the park authority and local communities, tour operators contribute a lot in shaping the perception of tourists about a particular national park by romanticizing nature and at times exoticizing cultural practices of people who live in or around the national parks. For instance, Laura White, an American tourist whom I interviewed at Tourist Hotel in Arba Minch was briefed by local tour operators about the Omo and Nech Sar National Parks. Based upon the information she received from the tour guides, Mrs. White expressed her impressions about the two national parks as follows:

> Last week, we were in Northern Ethiopia. We visited Lalibela, Gondar, Axum and Bahir Dar. These were great historical sites. We were amazed. We arrived here yesterday. Our local tour guides told us about Nech Sar and Omo. I believe these are wonderful natural sites. I saw the two Lakes and parts of the Nech Sar from that Lodge up there, Paradise Lodge. I want to see zebras and other animals. We will have boat trip on Lake Chamo. We look forward to seeing the beautiful culture of the people in Omo valley. We have read a little bit about them but have never visited the place. I think Ethiopia has beautiful places, history and of course good people (informant: Laura White, May 2012, Arba Minch).

When I was in the field in March 2011, few foreign tourists were taken to the Nech Sar plains bordering the Guji people. According to the sources, the tourists saw cattle trespass into the park territory and whereby the livestock were grazing with the wild animals. This became a shock to the tourists because it contradicted their prior orientations about the area as 'pristine nature'. Upon their return to the park administration's head quarter in Arba Minch town, the tourists complained to the park officials about the cattle trespass not mainly due to its conservation impact, but rather because conflicted what they were told before. The report and complain reached the office of the EWCA in Addis Ababa and instigated a revival of strong restriction against the Guji cattle trespass. The park authority used such complaints to justify its actions such as the 2012 destruction of Guji houses and its plan to launch large-scale relocation of the people. In this regard, tour operators played dual role in influencing the management approach of the park. Firstly, they lobbied the authorities to ensure the security of the tourists by relocating the people. Secondly, they reproduced an exoticized image of the territories to attract tourists. However, when the image constructed by tour operators was ascertained in sharp contrast with the reality on the ground, 'conservationists' and nature tourists began advocating for the fierce protectionist approach to conform with the imagined 'naturalness' of the park. It could be learned from the description by the above American tourist that she situated her descriptions of the tourist sites in northern Ethiopia as historical and her expectations of those in the south as spectacular natural sceneries. This was part of the north-south dichotomy reproduced in the mainstream conceptions of these cultural and geographical areas as vertically opposing spaces. Tour operators' representation of the territory as aesthetic and pristine nature reveals how the discourse of wilderness conservation is instrumental in various actors' claim for the legitimacy and credibility of their practices. For example, the tour operators sought to justify that resettlement of local communities should be considered as a legitimate and rational action to save the beauty of the territory in this case.

7.2 African Parks Foundation and International Human Rights Advocacy Organizations

In chapter five, I have discussed how the Dutch company called African Parks Foundation (APF) took over the management of three national parks in southern Ethiopia (Omo, Mago and Nech Sar National Parks) in 2004 and 2005. Currently, it operates in six African countries (Chad, Congo, DRC, Malawi, Rwanda and Zambia). The APF clearly specified its conservation missions that it works for multidimensional purposes of conservation–preservation of biological diversity, generation of ecosystem services and economic benefits through tourism and other interrelated purposes.[39] APF is a business and conservation company that attempts to foster conservation through business and vice versa. In contrast to its protectionist approaches it strived to implement in 2004 when it took over the Nech Sar National Park, it seems that the APF shifted to a neoliberal approach of environmental governance whereby it opted to the so-called community-based conservation and inculcation of conservation values through education as it stated on its website updated in 2013:

> It is increasingly being realized that the continuity of conservation endeavors can be hugely influenced by the attitudes and actions of the local community. Only when conservation initiatives can directly contribute something towards answering the needs and prerogatives of local people will they act in the interest of conservation. Throughout our portfolio of parks, community outreach has always been intrinsic to our conservation endeavors –education centers instill an environmental consciousness through explaining the importance of conservation efforts, community funds help provide social infrastructure and small business enterprises, community camp sites provide direct economic opportunity within our parks. The greatest benefit of all is the provision of employment opportunities and training. In some places African Parks is the largest local employer of staff, and each job usu-

[39] http://www.african-parks.org/Conservation+Approach.html

ally has a huge knock-on effect supporting extended families.[40]

The APF's involvement and the Ethiopian government's 'commitment' to resettle the local communities, as well as the local and external reactions it created, were important aspects of investigation in the contestation and negotiation among competing actors in this section. Such interaction entailed the exercise/deployment of different forms of power and inter-actors' network in which local communities were also involved in the contestation.

When the APF signed an agreement with the Ethiopian government in 2004 to take over the management of the park, the Western perspectives of 'natural' and 'unspoiled wilderness' of national parks in Africa was implicitly inscribed in the agreement. To meet its preconceived imaginations of the territory, the APF included a clause in the agreement that put the responsibility on the Ethiopian government to relocate indigenous inhabitants so that the territory would 'regain its natural beauty'. While the APF strived to play-off the game by 'refraining' from involvement into the relocation process of the people, the dominant notion that considers the state with a moral responsibility to safeguard its citizens was behind the idea of giving the assignment of relocation of the people to the Ethiopian government. However, under 'developmental' state paradigm, human rights violations by the name of development are often overlooked. Like other privatization programs in Ethiopia, the local people did not have a say in the process of the agreement with the APF; rather, it was an agreement discussed and signed between the company and the Ethiopian authorities.

Regarding the future of over 10,000 Guji and Koore individuals and over 7000 livestock, the position of the APF was initially ambivalent due to its international obligations on the one hand, as well as its business and conservation motives on the other (Dowie 2009). In other words, while its international obligations compelled the company not to be directly involved in violations of human rights in the form of displacement, the company was alerted by the massive strain the population and livestock would pose to the wild animals and other biodiversity in the park and the implication of such activities on its tourism interests. As a result, it systematically pulled out from the trap by adding a clause in the agreement that gave responsibility of resettling the local people to the Ethiopian government.[41] In all the three national parks (Omo, Mago and Nech Sar), the APF agreed to administer, the Ethiopian authorities hastened to evict the local communities to handover 'empty' land to the company. This was how the

---

[40] http://www.african-parks.org/Conservation+Approach.html
[41] http://www.matthijsblonk.nl/paginas/AfricanParksEthiopieEng.htm

resettlement programs and restriction of access to customary land of the Guji and the Koore were enacted in the case of the Nech Sar National Park. Moreover, also in the Omo and Mago National Parks, traditional resource use, access rights and resource ownership of eight adjoining indigenous communities were criminalized (Turton 2011).

According to BBC News on 26 September 2005 titled "*People Problems for Ethiopian Game Park*", the continuous encroachment of the park by local communities who depended on resources inside the territory for their livelihood has been the major challenge to the APF's conservation and business discourses. However, according to the report, the APF designed alternative solutions to employ women and men to cut down invasive horny plants and made "a fund available that will have $4,000 a month to help the communities".[42] In the same report, the BBC News mentioned another early challenge to the government's privatization policy, namely that "During the May [2005] elections the opposition attacked the government for "selling off the park"". This means that while local communities and human rights advocacy groups criticized the government for its resettlement plans, opposition political parties presented the transfer of the national park to the APF was equivalent to selling the country's heritage.

Despite carrying out fierce administrative measures against local resource users by confiscating their boats, hooks, sickles, axes, firewood and prosecuting them for their alleged 'crimes' in encroaching the park territory, the APF claims to have organized urban communities whose livelihoods were jeopardized as a result of the restriction. According to its monthly report in August 2005, forty-eight individuals were employed in the park on part-time basis to collect invasive wood, while another forty-eight individuals were organized by the company to make their living by selling the wood collected from the park (APF Monthly Report, August 2005). Moreover, the company used innovative strategy by employing former resource users in rehabilitation of degraded lands in the plains of the Nech Sar (APF Monthly Report, April 2005). This type of job opportunity was in addition to the people that the APF employed as park staff. While the Koore mainly recall the APF's financial contribution, provided as compensation for their resettlement, the Guji praise the company for listening to their problems, recognizing their customary rights and to some extent opening job opportunities for a few individuals from the group.

However, the resettlement plan did not go smoothly because Guji's resistance against the resettlement program on the one hand, as well as international pressure against the APF's involvement in the resettlement deal on the other, prompted the company to come to terms with Guji's questions of custom-

[42] http://news.bbc.co.uk/2/hi/africa/4272388.stm

ary right in September 2007. According to local sources, the APF attempted to reverse the old exclusionist approach at least by engaging the Guji and other local communities in part-time and full time employment opportunities and more importantly bringing them in dialogue and negotiation on issues of conservation and access to some grazing land and water grounds. In September 2007, the company reached an agreement with Guji elders and 'granted' the people portions of the park on the eastern side for settlement, access to pasture, water and medicinal salt-bore around *Xabala* (hot spring). The negotiation between the company and the Guji people angered some authorities in the SNNP regional state and some even described the negotiation as the APF's move to sell the park to Oromia regional state.

While the APF was involved in the management of the park as part of the commoditization of nature in Africa, it was confronted by international human rights advocacy organizations such as Survival International and Refugees International that partly contributed to its termination of the agreement in December 2007. The company's involvement in management of the three highly contested national parks in southern Ethiopia and particularly its indirect support of the resettlement plans instigated a series of criticisms and condemnations from human rights organizations and the local communities. In December 2004, Refugees International published a critical article titled "*Ethiopia: The Human Cost of Tourist Dollars*".[43] This article was published a few months after the APF signed the agreement and took the responsibility of managing the Nech Sar National Park. According to Thompson and Winer, over 10,000 Guji and Koore communities were relocated from their homes. The article criticizes the company for its covert involvement in displacement of the people. It recommended that, "Donors to the Africa Parks Foundation, which include the U.S. Department of State, insist that the government and the Foundation work with local people and the former residents of Nechsar National Park to find conservation-based solutions that provide a future for both wildlife and people" (Thompson and Winer 2004: 2).

Despite local resistance and pressure from human rights organizations, the Ethiopian government initially pursued with its resettlement and eviction programs that instigated another round of condemnation from advocacy groups. In April 2005, following the forceful eviction of the Guji Oromo, Refugees International's bulletin came up with another round of critical article on the APF and the Ethiopian government (Thompson 2005). The article explicitly indicated that the resettlement process involved semi-voluntary and involuntary reloca-

[43] See Thompson and Nicholas Winer (2004) "Refugees International Bulletin" http://www.refworld.org/docid/47a6eeafd.html

tion of the local people whereby the government used arbitrary force to assert control over the territories in its haste to empty the land for the company. Drawing upon empirical findings from fieldwork among local Guji and Koore communities, Refugees International suggested that both the government of Ethiopia and the APF should reconsider the relocation of the people, stating that:

> The government immediately halts forced expulsion of people from Nechsar Park, compensate those who have been expelled, provide timely humanitarian and economic assistance to people resettled or still living in the Park, initiate new consultations with local residents, and permit limited grazing and agriculture within the Park by local people to continue for the time being. The African Parks Foundation and the Ethiopian government acknowledge that the removal and resettlement of people living within or adjacent to Nechsar National Park was poorly conceived and managed. Planners for the park seems to have regarded local people as a barrier to be overcome and shunted aside in favor of wildlife, but a planning approach which stresses the total ecosystem, including people, is probably more appropriate for Ethiopia and likely to generate more local support (Thompson 2004:2).

In October 2007, Stephen Corry, director of Survival International, wrote an advocacy letter to Peter Fearnhead, Chief Executive Officer of the APF, with direct reference to human rights violations in the Omo national park. In the letter, Corry referred to international laws to which Ethiopia and the Netherlands (APF's home country) were signatories and asked the company to refrain from violating international human rights of the people in and around the Omo national park. In its advocacy works, Survival International made explicit that the APF would face legal challenges and legitimacy problems if it continues to breach the fundamental human rights of the indigenous peoples:

> One international agreement ratified by Ethiopia (and by the Netherlands) is the UN Convention on Civil and Political Rights, Article 2(2) of which states that "in no case may a people be deprived of its own means of subsistence". Does APF seriously contend that the crops cultivated on the West bank of the Omo do not form an essential means of subsistence for the Mursi? UN General Comment No 23 confirms that the cultural rights enshrined in Article 27 of the Convention include the right to pursue "a particular way of life associated with the use of land resources, especially in the case of indigenous peoples. That right may include such traditional activities as fishing or hunting and the right to live in reserves protected by law." […] I understand that the Bank (World Bank) may have already rejected an APF approach about the Nech Sar National Park, at least in part because of concerns about APF's likely impact on local communities (Survival International Letter to APF, October 31, 2007).

Although Survival International's advocacy mainly emphasized the Mursi in Omo valley, it had put company in a trap of legal and legitimacy questions by addressing the letter to influential individuals, organizations and institutions such as Rodolfo Stavenhagen, UN Special Rapporteur on Indigenous Peoples; Royal Netherlands Embassy (Ethiopia); Regional Environmental Office for East Africa; the US Department of State and ILO. As a result, APF fell under fierce criticism from international human rights organizations for its subtle engagement in the resettlement deal and from Ethiopian authorities for its negotiation with the Guji community. After few years of struggle to manage competing interests from different actors, the APF terminated the contract in December 2007.[44] In its termination statement, the APF mentioned lack of collaboration on the part of Ethiopian authorities, the persistence of complex inter-ethnic conflicts, claims of ownership and a flaring up of criticisms from human rights organizations as major reasons for its early termination of the contract in Ethiopia.

To summarize, it has emerged from the discussions and analysis in this chapter that competing actors produce narratives, discourses, ideas and actions about human-environmental interactions and use it as strategy of legitimizing their actions or delegitimizing the actions of their rivals. Contestations over environmental resources necessitate actors' production of knowledge to challenge or delegitimize counter knowledge production by other contending social actors. Although local tour operators and the African Parks Foundation subscribed to the aesthetic values and the so-called 'biodiversity degradation' thesis in their justifications for modern conservation approaches, the fundamental motive behind these actors' call for external intervention were economic reasons. On the other hand, local communities had cultural, economic and historical reasons for claims of entitlement to the land and for resisting the state's conservation discourses and practices.

Drawing on empirical evidence from the field and building on the theoretical underpinnings in political ecology perspective discussed earlier, I conclude that grassroots actors are not passive recipients of external intervention or conservation discourses and programs. Rather, they articulate their contestation against dominant state discourses based upon their ways of understanding, interpretation and definition. As Robbins (2004) noted, the recognition of local critical voices entails a different form of understanding how dominant notions of scientific knowledge are challenged from the perspectives of local actors who capitalize on their indigenous knowledge in their interaction with their environments.

Likewise, most of the local actors discussed in this chapter capitalized on

---

[44] http://www.conservationrefugees.org/APF_withdrawal_statement.html

their own definitions and conceptualizations of the territory and counteracted the state's notions of nature-culture dualism. However, it should be underlined that the local communities did not prescribe to similar notions human-environmental interactions. Despite their internal heterogeneities, some local actors also produced collective narratives and discourses regarding conservation knowledge, history, customary rights and cultural and spiritual connectedness to the land. This is particularly evident among the Guji Oromo. As the next chapter will illuminate, the local actors not only challenged the discursive narratives of the state and other non-state actors but also practically resisted its implementation and at times negotiate or appropriate conservation and development practices of the state.

CHAPTER EIGHT: STATE INTERVENTION AND LOCAL RESPONSES

Introduction

In the preceding chapters, I discussed how competing notions of nature conservation came into the scene following Ethiopia's adoption of Western ideals of national park in the 1960s. The establishment of the Nech Sar National Park as a territory considered 'wild' and 'uninhabited' through an exclusionist approach had confronted with local communities' counter narratives about their history of settlement, knowledge of conservation, customary rights, as well as cultural and spiritual connectedness to the land. Added to these counter narratives were also questions related to the right to participate in the planning, management and utilization of the national park and issues pertaining to the right to control and decide their destiny–rights of self-government, which encompasses the right to administer their resources. In chapters six and seven, I discussed how various actors' produced narratives and counter-narratives and their understandings of human-environmental interactions. Political ecologists argue that state development or conservation discourses encounter critical knowledge from below (from grassroots actors) that contextualize, contest, negotiate or appropriate dominant discourses based upon their own local values, knowledge and interests (Peet et al. 2011). Apart from encountering the state at the level of ontological and epistemological representations regarding human-environmental interactions, grassroots actors also resort to different forms of resistance to curb intervention from the state or to get access to state development and conservation projects (Scott 1985).

In this chapter, I will analyze different forms of state intervention and local responses within the context of contestations and negotiations over the Nech Sar National Park. The chapter goes beyond describing contestations and negotiations between the state and the local communities over the land and resources by positioning the discursive struggle between the two categories of actors within the context of state encroachment/engulfment of the periphery, as well as the responses from the latter to such phenomenon. The chapter attempts to answer the following key questions: What were the essences of government intervention in the Nech Sar National Park? How do grassroots actors respond to external interventions? How can the interplay between the intervention and responses be understood?

8.1 Forms of State Intervention

Seen from a critical perspective, state intervention in the form of national parks and other development projects is considered as "an effort to integrate and monetize the people, lands, and resources of the periphery so that they become […] rentable–auditable contributors to the gross national product and to foreign exchange" (Scott 2009: 4). Likewise, as Li (1999) argues regarding state intervention in the life worlds of indigenous peoples in Indonesia, resettlement programs for the establishment of national parks on indigenous territories were justified by the state authorities as the 'will' on their part to improve the living standards of the people. Nevertheless, state intervention particularly in developing countries has been criticized for "the objective has been less to make them [peripheral people] productive than to ensure that their economic activity was legible, taxable, assessable, and confiscatable or, failing that, to replace it with forms of production that were" (Scott 2009: 5). An important point to interrogate at this level is the question why state intervention in developing countries gears towards exploitation and subjugation of their subjects. Before moving to the forms of state intervention, it suffices to mention that the nature of state-society relations in postcolonial Africa was a replica of the elitist, hierarchical, predatory, exploitative and hegemonic system during the colonial era. Thus, it is fair to argue that the nature of the state in Africa entails the deployment of coercive power in its quest to exercise control over its subjects and resources to maintain political and economic power, which in turn is essential in the nation-building projects.

There are growing trends in appropriation of lands traditionally owned by pastoral or agro-pastoral communities in Africa through two interrelated approaches. The first is land acquired "by agro-industrial companies or foreign states promising to use it for highly efficient commercial agriculture", and the second pattern is land acquisition by conservation groups who advocate to protect wildlife both for biodiversity conservation and as a pathway to economic development through lucrative tourism (Galaty 2013: 143). I also agree with Behnke and Kerven (2013) that hydropower dam constructions in many parts of Africa deprive pastoralists of their land rights by restricting access to the major river basins that have been used by pastoralists. Land appropriation and subsequent relocation of pastoralist communities entailed cultural and economic 'transformation'. This is because most states and aid organizations considered the sedentarization of these communities as an alternative development program, which in fact disrupts the people's cultural connectedness to the land and herding practices. Saugestad (2005) has made remarkable notes on the state relocation programs of indigenous people in Botswana under the notion of 'improving their lives' within the modernist state agenda. It was stated that, "The relocation from the Game Reserve fits into a vision of Botswana as a modern, prosperous and homogenous state, where hunter-gatherers have no place. It is consistent with a development model that sincerely wants to 'improve the lives of its citizens' as quoted in the introduction, but which also claims to know what is best for them" (Saugestad 2005: 7). Postcolonial states in Africa used modernization and development discourses as strong state projects in their justifications of relocating peasants, pastoralists and other communities engaged in different life styles. While the notion or rhetoric of 'improving' the lives of local communities has been presented as a state vision of development, the modernist bias against lifestyles that are different from mainstream community in the state machinery has made marginal communities victims of state-induced projects and programs.

In Ethiopia, the current Ethiopian Peoples' Revolutionary Democratic Front (EPRDF) government has managed to fully engulf and incorporate the pastoralist lowland peripheries and has begun 'taming' and commoditizing their lands and resources since the 1990s. While the government opted to 'subduing', 'conserving' and 'marketing' some territories and resources such as national parks, intervention also took a form of dam construction, plantation agriculture and sedentary farming in other parts in the peripheries (Markakis 2011; Turton 2011). However, in both cases, whether conserving nature in its 'natural' setting or making resources legible and confiscatable, state intervention has revealed a hegemonic process of subjugating the periphery.

In the case of the Nech Sar National Park, state intervention took different forms. It ranges from discursive narratives of denigrating the local people's customary rights to practical resettlement programs that dismantled the people from their economic, cultural, spiritual and social spaces. Owing to internal heterogeneity within local communities regarding claims of entitlement to the territory, as well as differences in terms of state intervention, analyzing the intervention is much complex than one would expect. However, the core aspects of state intervention were characterized by two interconnected schemes. First, successive Ethiopian governments had amorously envisioned to commoditize the resources in the Nech Sar National Park (the landscape scenery and wildlife) in the form of tourism. Like colonial and postcolonial states in other African countries where the myth of 'wild' Africa was translated to practice through national parks (Neumann 1998; Brockington et al. 2008), successive governments of Ethiopia also intended to 'conquer' and 'subdue' the territories in the South that have been described in the mainstream historiography as 'wild'. Secondly, the current government inherited the hierarchical ordering of different ways of beings, which can also be traced back to the ideological bases of building the Ethiopian state. Specifically, the political, economic and socio-cultural identity that considered sedentary agriculture as superior to pastoralism was inherited from its predecessors. Moreover, under the developmental state, citizens are reduced to subjects, which mainly entail the representation of citizens in the periphery as illegitimate beings to decide upon their own development agenda.

As I discussed in chapter five, the developmental state paradigm of the EPRDF government undermines multiple ways of conceptualizing development discourses by putting ultimate power on the government to design what it considers 'right' to its subjects. As a result, the government advocates transformation of the lifestyle of the people in the peripheries by changing them into sedentary farming according to its own visions of development. The EPRDF government has launched massive villagization programs in the peripheral regions such as Somali, Gambella, Benishangul Gumuz and SNNP regional states as part of extending development projects to these peripheries. Thus, as Ferguson (1994) argues in respect of development projects in Lesotho, for example, development as part of extenders of state power to the society, particularly in areas where state institutions rarely function.

Herbst (2000) also contends that postcolonial states in Africa invest much on broadcasting state power to the peripheries through political infrastructure or development projects that enable the states to exert control over their subjects. Therefore, the Ethiopian government's intervention programs through conservation projects and resettlement of people into sedentary form of lifestyle also part of the broader attempts of state to engulf the periphery. In fact, that does not entirely nullify the government's development and conservation policies but they are part of the broader state projects of controlling the periphery. Indeed, political ecologists caution us to question intervention by the state and non-state actors under the guise of nature conservation because such process entails exercise of dominating power that stifles the voices of grassroots actors (Robbins 2004). Exercising domination on the part of the state is necessitated due to contradictions between state visions of development and locals' spatial-temporal visions of themselves. In contexts where local communities' understandings of their own life styles and their interaction with their environment clearly contradict with the state' perception of its role to transform society, the interaction often leads to the exercise of various forms of power on both sides. Nevertheless, the last thing that often happens in circumstances where both parties fail to negotiate their positions is domination of one side because it produces subjugation through realization of its vision at the expense of the compromised vision.

In the following few paragraphs, I will present some ethnographic data on the state's intervention in the form of enclosure, deprivation, relocation and enactment of fines and prison of members of the local communities around the Nech Sar National Park. Such state intervention entails control over multidimensional spaces of the people. By doing so, I will analyze how the Guji and Koore people responded to these forms of intervention. Beginning from the 1960s, enclosure remained the common feature in the management of the Nech Sar National Park. Enclosure, in this context refers to the conversion of former common-access lands and resources into private or state-owned territories/properties and thereby enclosed through fences or administrative territorialization (Campbell 2005; Scott 2009). Although the Nech Sar National Park was not fenced, it can be considered as enclosure, given that it has been territorialized as a national park where imaginary 'boundaries' were drawn to delineate human settlements from the park territory.

With reference to enclosure of rangelands in Borana and Guji Zones of Oromia regional state in Ethiopia, Tache (2013) describes different patterns of enclosure. While the first pattern is related to customary enclosure managed and utilized by the local community under customary rules, the second category is community-based enclosure, utilized by the local community but co-managed by the government or NGOs and the community. The third type of enclosure is private enclosure, managed, owned and used by individuals (investors, wealthy pastoralists or small-scale agro-pastoralist households). The customary enclosure was mainly practiced among the Guji and Borana Oromo and other pastoral communities in southern Ethiopia to ensure reservation of pasture for emergency seasons for the needy classes of their herds (calves, milked cows and sick animals). On the other hand, the private enclosure essentially restricts different members of local communities from access to pasture land, water, sacred spaces and other resources as in the case of the national park.

With reference to the Nech Sar National Park, although it is not yet gazetted and physically fenced, the park authority established several outposts in all directions encircling the park boundaries. The park administration armed the rangers with light automatic gun to patrol the entire park outskirts the whole day. As the Guji and Koore inhabitants mentioned during my fieldwork, the relationship between the people and the park authority has been tense since the 1970s owing to the exclusionist conservation approaches that the park administration follows. The institutionalization of authoritarian system of resource governance in the national park was part of the broader political system in the country, which has been marked by strong state predation over its subjects. Soon after it assumed state power in 1991, the EPRDF government resumed engulfment of the periphery that its predecessors failed to fully accomplish. Unlike its predecessor regimes that opted to strong centralization policy as a form of nation-building, the EPRDF government penetrated the periphery through its political reconfiguration that it later used as an instrument of accessing the periphery through administrative and economic systems (Clapham 2006a; Markakis 2011). Until the late-1990s, the areas around and including the Nech Sar National Park were part of territories not yet 'pacified' by the regime. I argue that the plan to resettle the Guji and Koore people, in 1990s and 2004 was part of the new regime's pacification and incorporation of the peripheries into the center to make them more legible for administration and commoditization of their resources.

The government used conservation and development discourses interchangeably to justify its actions. According to the views from government officials, the government took relocation measures to create a better environment for the wildlife, enhance tourism, as well as improving g the living conditions of the relocated people by providing better social services in collectivized villages. At this point, it should be noted that the modernist and developmental state paradigms in Africa have similarly entailed state domination of development projects without any room for popular participation because postcolonial states in the continent were caught up with issues of development and nation-building within the context of elitist political culture inherited from their colonial predecessors (Saugestad 2005). Therefore, the state presents itself and is portrayed by international organizations alike as a legitimate institution capable of carrying out projects that would 'improve' the life styles of its subjects as well as guard the sustainability of the environment.

As I discussed in chapter five, the resettlement plan failed in the 1990s due to inter-regional disagreements between Oromia and SNNP regional states, as well as local people's resistance against the government plans. From the broader national political dynamics, the government's failure to exert its full authority over the area partly shows how the new political reconfiguration contributed to local communities' claim for autonomy and right of resource ownership. After a decade on power, the EPRDF government consolidated its political and economic power and established strong administrative bureaucracy to channel down its policies to the peripheries (Vaughan 2003; Markakis 2011). Unlike in the 1990s, EPRDF used both soft power and hard power during the 2004 resettlement program and removed both the Koore and Guji from the park core areas. By soft and hard power, I concur with what Ash (2009) describes as the exercise of power in the form of dialogue, negotiation, persuasion and participation of the local communities, as well as the exercise of coercive force to implement state intervention/projects, respectively. While the government's attempts to persuade the local people through few selected elites was part of the deployment of soft power, its coercive actions in the form of forceful eviction can be considered as the government's use of hard power in the process of extending its authority to the periphery.

In the context of this study, I conceptualize the government resettlement programs and other forms of restriction against the Guji and Koore people as systematic deprivation of the people from their economic, social, cultural and spiritual spaces as a form of hegemonic environmental control, which in turn is part of the government's assertion of domination over the formerly 'ungoverned' territories. It is evident that the logic of conservation within protectionist approach has sought to mediate human-environmental interaction and classifies 'appropriate' and 'inappropriate' activities and "categorizes landscape on evaluations of what it would look like if kept free from human interference" (Campbell 2005: 289). In the governance of protected areas, new forms of territorialization and regulation are put in place regulating the movements of people, animals and classifying 'legitimate' and 'illegitimate' users of resources in these areas. The institutionalization of bureaucratic systems of punishment, checkposts, registration and surveillance are mechanisms of control of the material, economic, cultural and political spaces linked to protected areas. Thus, contestations over protected areas are part of the complex interplay between different actors far beyond normative resource competition because in defining degrees of access to formerly 'common' spaces, the establishment of protected areas becomes a struggle over place, with political, economic and cultural implications at different scales (Few 2001).

Likewise, the struggle over the Nech Sar National Park between the local communities, the state and non-state actors should not be understood as a mere competition over resource utilization. It was rather a discursive political, cultural and economic struggle between the center (the state) and the periphery (local communities) on questions of rights over the land, which hints at the exercise of power–aspects of domination and resistance. By relocating the Koore and evicting the Guji in 2004, the government had put the people under its control by depriving them from access to their economic resources. For example, in the case of the Koore, the relocation reduced the people to the status of aid dependence for food and other social facilities in the new resettlement sites. Despite the participation of a few local Koore elders in the discussions for the resettlement, the Koore ordinary men and women came to terms with the relocation mainly because co-opted local elites maneuvered them. Whether it was coercive or voluntary arrangement, relocation deprived the people from the Tsalke valley that they considered as their life-stream and a safety corridor both against natural and human induced disasters. Moreover, the relocation to Abulo and Alfacho villages detached the Koore from their well-established social networks in Amaro highland and in Arba Minch town. According to Gule Mamacha, one of my Koore informants who returned from the resettlement site, the relocation of the people had paralyzed their economic capacity, social network and wellbeing. He summarized his views as follows:

> The government promised us to provide all facilities. But, after we arrived in Abulo, everything was different. There was no water for irrigation. The land was dry. Rain did not come for two years. In Tsalke, we used to produce the whole year. In Abulo, our children starved and our women suffered a lot traveling long distances to fetch water. Life was completely difficult there. We were economically torn out. We had to wait what came from the government every two or three months [food aid]. That was the worst in my life. It disconnected us from our kin members in Darba, Tifate and Yero. There was no road to come to these areas. It took us two days on foot to reach for funeral of our relatives or just to visit them. After two years, many people returned to the highland and some went back to Tsalke but the land was occupied by the Guji. That caused the conflict two years ago (Informant: Gule Mamacha, March 2012).

On the part of the Guji, the eviction that removed them from the core areas of the Nech Sar plains primarily restricted Guji's access to wet season pastureland and water routes in the park territory (Taddesse 2004; Dowie 2009). Blocking the wet season grazing land implies confining the people and their livestock to the mosquito and tsetse fly infested Sermale valley where neither the people nor the livestock would easily survive. Moreover, the blockage deterred the Guji from access to the a hot spring water locally called *Xabala* that the people used for healing purposes against malaria, cold, influenza and stomachache. The hot spring area had also a salt bore called *Haya* used for livestock (Getachew 2007). The Guji traditionally used the *Haya* as a healing mineral salt for their livestock and for its purpose in increasing milk production. Although these resources were essential for the Guji people and their livestock, any transgression of the park rule in the form of cattle trespass, hunting, farming encroachment, honey production and burning pastureland would result in severe punishment from the park authority.

However, due to international criticism and internal resistance against the government's coercive actions during the 2004 resettlement process, the government resorted to a subtle strategy of expelling the people from the park area. As one local government official remarked, the strategy was to indirectly force the people abandon the area by themselves. According to this official, "the park authority used hidden strategies to force the people out of the park by restricting them from basic social services such as healthcare, road, water, access to market and by instigating inter-ethnic conflicts" (informant: local government official in Galana district, March 2012). Thus, the government intervention strategy was changed to covert approach of deprivation and isolation to put the local people in a condition where they would realize the hardships of living adjacent to the park. For example, in a letter sent to the SNNP office of the regional president, the Gamo Gofa zone administration requested the regional government to intervene in Guji's construction of permanent settlement and establishment of social facilities such as health center, school, flourmills and water services[45] (see appendix).

The background phenomenon to this intervention was as follows: After they were expelled from the Nech Sar core plain areas in 2004, the Guji agro-pastoral community began establishing some basic social services because access to these services from Arba Minch town was almost blocked by the state except along the risky boat transportation on Lake Abaya. Few individuals established flourmills and began building business shops in Golbo/Tsalke. After they reached on an agreement with the African Parks Foundation in September 2007, some ten families built residential houses made of tin roofs in the newly established village adjacent to the park boundary. It also followed that a primary school and health center were built in Arda Guddina and Golbo villages in 2008 and 2010, respectively, partially funded by Oromia regional state and partially contributed by the local community. These all signals of 'development' and transformation among the Guji alerted both the park authority and different administrative departments in the SNNP regional state. As a result, the authorities began what I call strategies of deprivation to compel the Guji to leave the 'park territory' by their own. This strategy included isolating the people from nearby markets and healthcare centers through control over the routes to these social services. For example, the government conservation and development projects In Lower Omo have alienated the indigenous people from their customary lands, eventually leading to the material and spiritual deprivation of the people (Turton 2002, 2011; Markakis 2011). It has also been evident in other parts of the world that the deprivation or alienation of indigenous peoples from their material and spiritual bases has been used by technocratic elites and development organizations alike to compel these people to live in conformity with ideals of modernity (Adkin 1998).

In the study of encounters between state and local communities within the context of development interventions, local people have often been represented

---

[45] Letter from Gamo Gofa Zone Administration to the regional president office under Ref. No. [Ga Go Z A D/3509/23], dated 2/10/2000 E.C.

on the contours of resistance–overt and covert (Scott 1985, 1990, 2009). However, as Long (2001) and Li (1999, 2007) contend, interactions between actors with competing perspectives and interests compel both sides to rearticulate their discourses. In the process, competing actors contest, negotiate or come to terms with some perspectives and practices based upon the nature of power positions of the actors involved. The underlying argument in Long's (2001) and Li's (1999, 2007) inter-actor relations is that both grassroots actors and state/non-state actors position themselves on dominating or resisting sides depending on the power position of their rivals. Despite power being unevenly distributed among various actors, it is evident that all actors (grassroots communities, states, business corporations and international organizations) have different sorts of power (Sharp et al. 2000).

Depending on the nature of local resistance, external pressure and internal administrative dynamics, the park authority has been using different strategies to control the space and expel the people from the area. Among the government strategies, I will discuss the blockage of market routes, the imposition of restriction on establishment of social services and issues of security. In the past, the people crossed via the park territory when they traveled to Arba Minch town where they sold their livestock products, in turn buying different things from the market. Another local market was located in Darba Manana, in Amaro district. However, access to this market was highly contingent on the nature of inter-ethnic relations between the Guji and the Koore that has mostly been characterized by conflict. Moreover, the Darba Manana market was used by the Guji mainly for the purchase of food items such as ensete *(ensete ventricosum* or commonly known as the Ethiopian false banana) and maize, as well as selling livestock products to the Koore, albeit not for livestock marketing. Another small local market is located in the contested territory in Golbo/Tsalke where mostly Guji and Koore women trade some local products.

Figure 18: Guji and Koore Women in Golbo/Tsalke Market (Photo: Asebe Regassa, May 2012)

Because of the blockage of the road via the Nech Sar National Park, the Guji and Koore opted to use risky traditional boat transportation on Lake Abaya, which was manually operated and vulnerable to waves and attack by crocodile. In March 2012, nine passengers (seven Koore and two Guji) died after a boat sank down into the lake. The boat carried 14 people and over one thousand kilograms of coffee from Amaro to Arba Minch town by the time it got the accident. As there was no road that would connect Koore villages adjoining the national park to the administrative district in Kelle on the eastern side of the Amaro Mountains, the people used Arba Minch as their major destination for market and in getting other social services. Although the people took the risk of traveling on the lake, the boat transportation did not solve the problem because it was only used for light goods. As a result, both groups were unable to take their surplus production to the market–as it happened in 2009 when they got good harvest. On the contrary, in the last three consecutive years (2010-2012), the local people were hit by severe famine because of the 2010/2011 drought that had hit eastern Africa in general. In both contexts–during periods of high yield and low production–lack of access to the market has been the major problem for the people.

Figure 19: Guji individuals traveling on local boat (Jatara)
(Photo: Asebe Regassa, May 2012)

As agro-pastoralists, the Guji needed accessible market for their livestock. Nevertheless, they had to travel long ways to Gedeo zone while taking their livestock to market that had also economic and physical challenges. From a political ecology perspective, the questions of road are beyond the issues of the government's lack of technical or financial capabilities. Rather, such restrictions and blockage show the government's systematic strategy of controlling the physical space–the route to the market places–to compel the people to comply with the government's conservation visions. By blocking the route to livestock market in Arba Minch and eventually forcing the people to divert to the long and tedious route to Gedeo zone in the east, the government exercised its dominating power over the subjects. Apart from the time and energy that they lost to and from the market, it also had significant economic repercussions, as the livestock were susceptible to weight loss on the way.

The blockage of the route had also a dimension of inter-regional contestation over the territory. For example, the park authority and Gamo Gofa Zone administration were said to have refused to permit the entry of construction materials for school and health center in the Guji villages in 2008 and 2010. The incidents happened when Galana district administration of Oromia regional state planned to establish a primary school and a health center for the Guji community. Although the district administration took the case to the regional governments to gain permission to transport the construction materials through the park boundary, none of the regional governments solved the case. As a result, the construction materials were transported via Amaro Kelle to Darba and then by donkeys to the villages. In some cases, the people also used the traditional boats called *Wogolo/Jatara* (personal communication with a government official in Galana district, March 2012).

After interviewing the head of park scouts, it substantiated my conclusion that the park authority's strategy of depriving the people from social services was to force them abandon the area by their own. The official who was working as a park Warden during my interview strongly argued that:

> The park had potential to become a major tourist destination. But, the Guji pastoralists are among the major threats to the game in this park. The government planned to resettle them but they refused. We know the people suffer a lot because they do not have clinics for their cattle and people. They use the risky transportation to come to the town. Last month nine people lost their life in the lake. But, the major question we have is that why are these people stuck to that place? When the government planned to resettle them, other people shouted as if it was violation of their rights. Now, they complain they do not have services. If they want to live better life, they should abandon following the tails of their cattle and start modern life. There is vast land in their region. We have the information that there is empty land in Oromia, around Tore and in Abaya district. They could settle there. Now, there are two options for them. They either live in poverty in the current illegal settlement or choose to go by themselves. This is national park. It is national heritage (informant: Head of scouts, April 2012).

Despite hardships they faced because of the enclosure, the Guji and Koore communities were comfortable to live in the area and continued different approaches of ensuring their customary rights. According to the views from many of the local communities, they were not poor but do not have equal power with the state to change the course of state intervention.

Security problem was another strategy that the state used to make life adjacent to the park difficult for the people. During our discussions, local communities often complained that the government has been creating security problems in the area in many dimensions. From 1991-2003, the national park was considered as a home of bandits where town gangs were engaged in robbery, lootings and attacks against local people who traveled to and from Arba Minch town through the park territory. According to local informants from Guji and Koore communities, the government did not give adequate protection to the people during those times. In the post 2007 period, following the withdrawal of the

APF, conflicts between the Guji and Koore have become major security threats for the local people. Although the central cause of the conflict between the two has been questions of resource ownership, government's reluctance in settling the disputes seems to be part of its strategy of discouraging the people to live in the contested territory. However, as the following analysis of their strategies to counter the state intervention reveals, the people were able to translate their agency into resisting power and at times succeeded in curbing government conservation practices, eventually compelling it to negotiate and compromise on some of its ambitious programs.

In short, state intervention through the national park has been premised on the ideas that conservation and development projects improve the living conditions of the local communities and ensure the sustainability of biodiversity. However, because it encountered different understandings of human-environmental interactions from the part of the local communities, the government resorted to exercise dominating power in the process of channeling down its conservation and development projects to the grassroots level. Nevertheless, it is naïve to consider the establishment of the national park and persistent state intervention from purely environmental perspective. Instead, the empirical data in this study reveal that successive governments in Ethiopia had political and economic interests and brought the conservation rhetoric as means to these goals.

8.2 Forms of Local Responses to State Intervention

Despite political ecologists placing much emphasis upon the political and ecological oppression grassroots communities face from state and non-state actors, recent works in the field have begun to consider the agency of local communities in resisting and eventually influencing the course of external interventions (Bryant and Bailey 1997; Neumann 1998; Peet et al. 2011). It has been argued that regardless of their subservient status under different forms of dominating power, grassroots actors possess power that they deploy during times of resistance, which Sharp et al. (2000) call it as resisting power. Notwithstanding the precarious impacts of environmental degradation and enclosure owing to disruption of indigenous conservation regimes, "grassroots actors have more often than not been at the losing end of environmental struggles with their lot in a politicized environment one largely characterized by marginality and vulnerability" (Bryant and Bailey 1997: 151).

Beginning from the colonial conquest all through postcolonial eras, grassroots actors in developing countries have been engaged in different forms of resistance against colonial rulers and postcolonial leaders who usurped their customary rights of resource ownership, utilization and management. However, unlike peasant rebellion that takes a form of direct confrontation against state and/or non-state actors, grassroots actors often opt to different strategies of covert resistance, which is commonly termed as 'everyday forms of resistance' (Scott 1985). As Bryant and Bailey (1997: 151) argued that: "Much of this [everyday] resistance took the form of covert activities since public opposition usually incurred a highly repressive response from colonial states–a response, moreover, routinely resorted to by many postcolonial Third World states as well." Nevertheless, in contrast to Scott's (1985) conceptualization of everyday resistance as unplanned and trivial response that rarely changes intervention from the dominant actors, I argue that this form of resistance has a potency to undo state projects or at least in compelling the state to come to terms with local demands. I take the position that grassroots actors do not automatically resort to resist state intervention. It is naïve to consider local actors as always placed on the resisting end and the state on the dominating side. Rather, resistance is an outcome of a process of interactions between competing perspectives and practices, whereby aspects of dominating power and resisting power come into interaction whenever different actors fail to negotiate or come to terms with ideas of the other,. At times, the state takes the position of dominating power but in some cases it turns to resisting power because based upon their agency in maneuvering resources at their disposal, grassroots actors might possess also dominating power.

Although I agree with Bryant and Bailey (1997) on their position that grassroots actors opt to covert forms of resistance due to the repressive and coercive responses from states, I depart from them regarding their conceptualization of some changes in livelihood practices among local communities in the face of resistance against intervention as *adaptation strategies*. According to these authors, practices such as economic diversification, adoption of new crop varieties, planting trees, selling out livestock, migration from the enclosed areas and storage of crops from good seasons were considered as strategies of adaptation (Bryant and Bailey 1997). Instead, I argue that these practices should also be understood as *innovative approaches of resistance* that grassroots actors strategize in their contestation or negotiation with other actors. As the empirical data in this section demonstrates, the Guji Oromo planted fruit trees and established school, flourmill and market within the contested territory as a pragmatic strategy of resisting further expansion of the national park, for example. Reducing the complex strategies and knowledge of the people to mere adaptation mechanism is too simplistic because for the Guji these strategies helped them in resisting the state intervention besides securing their livelihood.

In this section, I analyze multiple ways of local responses to state intervention in the struggle over environmental resources. Owing to their persistent struggle in negotiating and contesting ideas and practices of environmental conservation espoused by the state, I will focus only on the Guji and Koore people. However, while the central actors in this analysis are the Guji, Koore and the Ethiopian state, I will also discern internal heterogeneity within each actor. Moreover, the study does not categorize each actor as an adversary to the other. As a result, some inter-actor networks will be important points of investigation whereby some social actors utilize social capital, ethnicity and political frameworks as resources in their exercise of power (domination or resistance).

8.2.1 The Guji and their Responses to Conservation Enclosure

In the last forty years since the establishment of the Nech Sar National Park, the Guji experienced serious repercussions of the tragedy of enclosure. The enclosure resulted in eviction of the people from the land and restricted them from access to cultural, spiritual and economic resources and more recently it led to inter-ethnic conflict because of competition over resources along the eastern part of the park boundary. Enclosure and territorialization of indigenous people's lands are precarious due to these people's strong reliance concerning the resources in the commons for their livelihood sustenance, spiritual wellbeing and cultural practices (Bryant and Bailey 1997; Campbell 2005). The Guji had memories of coercive state force under the military regime. In 1975, the government ambushed and killed about 377 Guji residents of the Segen valley–some 25 kilometers to the south of the Nech Sar National Park–allegedly accusing the people for rebelling against the regime (Solomon 2002; Asebe 2012d). In 1982, the government forcefully expelled them from the national park to which they returned during the period of transition in the early-1990s (Dowie 2009). According to Galgalo Edema, a Guji elder who lost close relatives during the 1975 massacre and who also lost dozens of livestock during the 1982 eviction:

> All the three governments were similar in considering us as an enemy. Haile Selassie ignored us as if we were not important. Mengistu [military regime] attacked us as if we were the enemy. Meles government [EPRDF] burned our houses and removed us from our home. If you ask me which one is better, my answer is that all are similar. In fact, they had also some differences. At least we got those who speak our language now [referring to Oromia regional state under ethnic federalism]. They do not solve our problems but at least understand when we cry. The government came with paper and pencils just like you [referring to the ethnographer]. Later they came with guns. During that time, we did not have these mod-

> ern guns. When Meles came to power, they came again with paper and began counting our cattle and fruit trees. Our people did not go to school. We did not know what they wrote down and how much they counted. After they counted everything, our cattle, goats, fruit trees, size of our farm and our people, they told us to look for alternative places for settlement. Since then, we do not trust those with gun and paper [government authorities]. Because we had the scar from the past, we did not want to confront the government. The government is always powerful. It has the education, the money and the equipment [technology]. We appealed to the government itself. I was among those who went to Addis Ababa to put our application in front of the government. We pledged to Oromia regional government that we had nowhere to go because Ergansa is our ancestral land. It was where our ancestors lived and passed away. It is where we live, herd our cattle and farm some crops. Our life is based here. But, they did not listen to our issues. Only four years ago, Abba Dula Gamada [regional president of Oromia] came and gave us this small land for our cattle and settlement (informant: Galgalo Edema, March 2011).

Like many Guji elders in the study area, Galgalo revealed three fundamental issues in the above conversation. Firstly, the elder described the Ethiopian state as an alien entity that engulfed the Gujiland from outside. In his imagination and experiences, he dichotomously positioned the state in conflictual relationship with the society. Secondly, Guji's relationship with the state in the national park area was informed by their past memories of the encounters. Lack of trust and legitimacy on the part of the state eventually reinforced Guji's responses to state intervention. Thirdly, the elder highlighted historical, cultural and economic significances of the in his stories in that the contestation encompassed complex issues beyond mere struggle over economic resources.

The Guji in and around the park unanimously agree that they have never been beneficiaries from the park. Some officials in Galana and Abaya districts share the views of the local people and argue that the exclusion of the Guji from the planning, management and benefit of the park in contrast to other ethnic groups such as the Koore and Gamo was unfair. At this point, it is essential to note that apart from some internal differences, there was a consensus among the Guji that access to the resources in the park was supposed to be their customary right. However, the state's coercive measures in the past had compelled them not to overtly confront it. This is why they used different strategies of resistance to reclaim their rights to the land. In the following sections, I will discuss Guji's responses to state interventions in three categories, namely situating resistance within the political framework, everyday resistance and innovative strategies.

8.2.1.1 Situating Resistance within the Political Framework

From my conversation with Guji elders during the fieldwork, I learned that they were cognizant of two important political situations about the EPRDF government. First, they were aware of the fact that the government controlled financial and technological power of coercion and co-opted local elites for exercising its coercive force over the people. Second, they were aware of the political reconfiguration articulated based upon ethnicity. In other words, they were knowledgeable of possibilities of playing the 'ethnicity card' in their claim for their customary rights over the land. As a result, the Guji initially chose to use the formal administrative procedure to challenge the government for their customary rights. From 1991-2008, the Guji in the area remained out of any formal administrative structure only stranded between the park to the west and Amaro district to the east. Although the Guji considered absence of the state as a relief from its coercive institutions, they were also devoid of social services that the state was supposed to provide to citizens, although such expectations are rarely met in such contexts. The only state institution they experienced in their surrounding was the park authority that often threatened them with eviction as part of its exclusionist resource management approach.

Despite such periods of isolation, Guji elders were aware of inter-regional 'friction' regarding the park and used it to their advantage. In other words, by requesting formal administrative integration into Oromia, which SNNP regional state initially rejected, the Guji used the political space within the context of ethnic federalism that eventually led to the formal recognition of their settlement in portions of the park territory in 2008. Beginning from the mid-1990s when the resettlement plan was on board, the Guji have collaborated with few Oromo intellectuals and local government authorities in Oromia in seeking advice on the bureaucratic procedures. For example, the local Guji inhabitants consider a former higher official in Galana district as their hero for his persistent support and guidance including writing letters of appeal on their behalf to the Oromia regional president and office of the Prime Minister during the 2004 resettlement program. In this regard, the people demonstrated their agency by playing the 'ethnicity card' within the new political context, thus requesting the government to be formally administered under the Borana zone, Oromia region, and gaining the right over the territory as historical inhabitants of the land.

As a former head of Culture and Tourism Bureau of the SNNP regional state noted, based upon the appeal from the Guji agro-pastoralists, the government established a team of experts to estimate their properties (fruit trees, sugar cane, coffee and other resources) to decide whether resettlement would be feasible or not. The team calculated the properties to worth equivalent of twenty seven million birr per year (nearly three million euros according to the exchange rate during that period). The government official explained how the partnership between local Guji elders and Oromia regional state authorities made the proposed compensation payment difficult and eventually led to the suspension of the resettlement plan in addition to external pressures on the government. The former government official argued that:

> According to government regulations to compensate land-based resources, the amount calculated for one year should be multiplied by ten years and finally the payment would be made accordingly. In this case, our region was asked to pay 270 million birr for ten years estimation. That was huge money but we agreed to pay. The Guji initially agreed to take the compensation. These people were innocent but they had strong relationship with officials from the region [Oromia]. Later, they refused to accept even the 270 million birr compensation. They further took the case to the prime minister's office and as a result, the resettlement plan was suspended. Finally, they got portions of land in 2008 (informant: former head of culture and tourism bureau, April 2012).

By establishing networks with local government officials within the new political rhetoric of ethnic federalism, the Guji elders asserted their claim for customary right over the land and to some extent convinced the Oromia regional state. It was said that some local government authorities played significant role in raising the amount of compensation to make the proposal of the resettlement difficult for the government to decide. Eventually, the two regional presidents arrived on the scene in April 2008 and formally recognized the concession previously given to the Guji by the African Parks Foundation. Accordingly, all different Guji hamlets around the park were organized under the Arda Guddina village, joining formal government administrative structure under Galana district, Borana Zone in 2008, for the first time since 1991. The Guji considered this as a major victory, although the park authority and some government departments in the SNNP regional state did not recognize the decisions of the two regional presidents.

Despite some improvements regarding Guji's access to pasture and water after the concession, the park authority continued posing insecurity to the people. It has been discouraging the people from building permanent settlements and establishing infrastructure in the contested territories. Restrictions on the Guji by the park authority and some government departments in the SNNP regional state were perceived as strategic move to threaten the people and compel them to abandon the region. As a result, the Guji resumed to a different form of struggle/resistance against the park authority and other related actors. This was a subtle strategy to delegitimize park laws through everyday strategies of resistance.

8.2.1.2 'Everyday Resistance'

As grassroots actors in different parts of the world recognized brutal retaliatory actions from colonial and postcolonial states against their overt resistances to assert their customary right to environmental resources, these actors resumed covert forms of resistance, often on an individual basis, to undo political and economic practices by the state or non-state actors. Despite some differences in operation between peasant rebellion and everyday resistance whereas the former is often collective and direct in confronting prevailing economic and political norms, both strategies are used by grassroots actors to reassert their rights to environmental resources. As the Guji case illustrates, the role of everyday resistance in incapacitating the effective implementation of state projects should not be underestimated. As is common among pastoral and agrarian societies in East Africa where colonial and post-colonial states have disrupted indigenous resource management regimes (Neumann 1998), hunting specific animals, cattle trespass and setting fire to the park represent the major elements of everyday resistance among the Guji. Apart from directly responding to the state intervention, the Guji took these actions as resistance against the entire state encroachment. This has been informed by historical narratives and experiences the Guji accumulated since their early contact with the Ethiopian state in the late nineteenth century.

Memories of deprivation, marginalization, domination and exclusion were at best the common features the people in the peripheral regions experienced under the successive regimes. As a result, the people in these regions consider the state as something to be avoided, escaped and resisted. In the previous chapters, I have discussed that the Guji unanimously had the feeling of being excluded from any benefit from the national park under the successive regimes. Except during the period when the African Parks Foundation recruited few Guji men as scouts and negotiated with the people on their rights to get access to pasture and water in some portions of the park, the park authority had exerted its maximum effort to exclude, threaten and evict the Guji from the land. Despite some differences within the Guji, particularly few members of the youth who anticipate future job opportunities within the park, the Guji people across age, gender and religion held the view that the park threatens their livelihood and alienated them from their customary right. They also associated the park authority with the coercive state system that snatches their right over the land but does not give them something else in return.

Nevertheless, it should be clear that the people were skeptical to express their views on the park to an outsider due to their grievances with government 'experts' who, according to the Guji, surveyed their opinions and eventually concluded that the presence of the Guji would be a major threat to the wild animals in the national park. Therefore, access to their 'backstage' information or what Scott (1990) calls 'the hidden transcripts' has been difficult but through time they became comfortable to share with me not only their views but also actions.

The Guji had the knowledge about the initial objectives for the park's establishment, namely for the conservation of an endemic Swayne's hartebeest. Moreover, they clearly understand the economic motives behind government conservation programs. In other words, they are conscious of the economic value of animals such as Swayne's hartebeest, Zebra, Lesser Kudu and Antelopes for the government in generating income through tourism. The Guji people aim at deterring both the conservation and economic practices of the government in the national park through everyday resistance. According to information from park scouts and some Guji informants, the Guji resumed killing the wild animals mainly in response to the government's policies of enclosure and displacement. During my fieldwork in 2011 and 2012, I received many oral reports from the park scouts about Guji youth being arrested in suspicion of killing some game in the park territory. It is important to note multiple purposes and motives of hunting among the Guji. First, it shows Guji's confrontation with government institutions and represents the people's efforts to delegitimize government conservation practices and laws. Second, the people kill the animals for food and business as substitutes for the loss of their livestock because of the enclosure. This is also an indirect strategy of reclaiming for environmental entitlements.

The striking exception in Guji's hunting as a form of resistance against the state was the zebra. They do not kill zebras for reasons they metaphorically explained as the animal having collaborated with them during the eviction in 2004. I observed during my fieldwork that herds of zebra were grazing together with Guji cattle and stayed besides their kraals at nights. Government officials in the park administration explained Guji's restraint from killing the zebra from cultural perspectives, because traditionally the Guji considered zebra among culturally prohibited animals for food. Although cultural restriction was true, the Guji explain the prohibition of killing zebra from a historical incident of partnership between the people and the animal. For the Guji, the zebras were their 'friends' during the period of hardship and had since continued their partnership. The story goes as follows:

> In 2004, the Guji were evicted from the Nech Sar plains in the process to transfer the management of the park to African Parks Foundation. Following the eviction,

> many herds of zebra also moved with the people to the eastern outskirt of the park. Since then the animal left areas such as Belte, Dhaka Bule and Bonke hills, and migrated to Arda Guddina, Gode, Golbo, Gashe and Maddo areas. The informants explained that while other animals 'chased' the Guji, the zebra demonstrated their unreserved solidarity in facing the eviction with the people. Through such solidarity, the animals were believed to have cooperated with the Guji and challenged the government's tourism schemes. That means, the migration of the zebra threatened tourism in the national park. The zebras voluntarily accepted the hardships that Guji cattle faced because of enclosure of the pastureland. This was a sign of solidarity. Because of such solidarity, the Guji do not consider zebras as foes (reconstructed from group discussion among Guji elders, May 2012).

According to the Guji, the zebras challenged the state conservation practice by refusing to be commoditized–to be watched by tourists. Therefore, like other human actors who resisted government intervention and with whom the Guji wanted to collaborate in their struggle, the zebras exceptionally collaborated with the people and their livestock in facing the tragedy of enclosure. Except the zebras, other animals particularly the Swayne's hartebeests were targeted as revenge to state policies of eviction. For example, when the APF terminated its contract by the end of 2007, the number of Swayne's hartebeest was said to have been over forty, although the animals were dramatically reduced to seven in 2012. Despite multiple factors and involvement of many actors in the predation over the animal, backstage information from the Guji indicates that the animal is the easy target in order for the people to destroy the 'identity' of the park. By killing the animal that was the major conservation objectives of the national park beginning from its establishment, the Guji attempted to attack the identity of the park and destroy the fundamental bases of the national park, namely biodiversity conservation and ecotourism development.

Political ecologists and scholars on local resistance have addressed cattle trespass to protected areas as a common strategy of everyday resistance used by grassroots actors to reclaim their customary resource rights (Scott 1985; Bryant and Bailey 1997; Neumann 1998). In the case of Meru National Park in Tanzania for example, the Meru people used cattle trespass as covert resistance against colonial and postcolonial conservation policies (Neumann 1998). Apart from enabling the people who lived in and around national parks to sustain the lives of their livestock, cattle trespass was one of the resistance strategies against protectionist notions of conservation. In the Nech Sar National Park, cattle trespass has been reported as a major threat to the park causing overgrazing and trembling of the grass (Freeman 2006). After the 2004 eviction, the Guji settled at the eastern side of the park around Gode, Maddo, Arda Guddina,

Golbo, Gashe, Hiituu and Mi'o, although these areas were woodlands with limited access to pasture and water. During dry seasons, the Guji moved their animals to the western escarpment of Amaro Mountains. The challenge comes when the Guji bring back their cattle to these settlement sites. During these times, they had to use different strategies to gain access to parts of the Nech Sar plains. The Guji used nighttime grazing and access through social and ethnic networks with the park scouts as major strategies of challenging the government's exclusionist conservation practices and securing their rights to the resources. In this regard, the resistance focuses on delegitimizing the state institution and its laws. To gain detail insights into the complexity of this form of resistance, it is important to discuss nighttime cattle trespass and the strategy of social network in detail.

An encounter in an evening of 22 April 2012 gave me an opportunity to probe into issues of nighttime cattle trespass to the park boundary. On that particular evening while my field assistant, Garoma, and I were returning home from a visit to Maddo village, we came across a herd of cattle heading to the direction of the park around 7pm. It was unusual for me to see herders heading cattle to grazing field at that time while it was supposed to be a time for collecting cattle to the kraal. We saw three young Guji herders each armed with AK47 guns. As a schoolteacher, Garoma lived in the Guji community since 2008 and thus he knew about the nighttime cattle trespass, which he shared with me. In addition to this encounter, in-depth interview with the young herders enabled me grasp this strategy of resistance. An eighteen years old Guji boy made an interesting remark on the contestation saying, "we take at night what they [park scouts] restrict us during the day" (informant: Galgalo Boka, April 2012).

Figure 20: A Guji boy with his AK47 gun (Photo by Asebe Regassa, April 2012)

Another very salient strategy the Guji used to gain access to water and pasture in the park was using ethnic or social networks with scouts who were also from the same group. There was strong social network and sense of ethnic belongingness between the Guji and few scouts. The relationship was built on mutual reciprocity and social harmony. Above all, some local scouts had families, relatives and clan members who lived adjacent to the park. Secondly, due to the limited salary that they received from the government, which was less than 1000 birr (nearly 50 Euro), the scouts were compelled to maintain their social ties with their community to keep some cattle with the latter to subsidize their living. In other words, besides the reciprocity between those families who kept the cattle for the scouts and the scouts who allowed the people to enter into the park boundary, the scouts had direct benefit from cattle trespass as their own personal cattle were also among the whole livestock that would gain access to the pasture and water sources. The local community was also a good shelter for the scouts from whom they would get milk, coffee and other foodstuff without payment during their patrol around the villages.

Such relationships and arrangements are often carefully negotiated and arranged because if the scouts are aware that their relationship with the local communities would lead to dismissal from their job or imprisonment. The scouts were in a dire dilemma between competing socio-cultural, economic and political settings. On the one hand, disloyalty to the state would result in punishment in the form of imprisonment or dismissal from their job. On the other hand, breaking local norms and values in exposing members of the local community who transgress park laws would result in disassociation from the society. Within the tense inter-ethnic relations, exposing members of one's own ethnic group is also considered as betrayal and co-optation into the 'enemy' group. Thus, the scouts play tactics of contextually switching loyalty.

Everyday resistance entails transgression of rules and regulations set by rival actors. In the process of reclaiming their customary practices of resource management and in their quest to enhance regeneration of fresh grass for their livestock, the Guji practiced burning grassland adjacent to the park territory. In some cases, the Guji were accused of setting fire to the park as a direct response to actions of the park authority but they were said to have done these actions in the absence of state agents. Issues of fire have been reported as one of the major causes of biodiversity degradation in the park, with people's motives of setting fire often described as their lack of knowledge of resource management (Freeman 2006). However, the Guji had traditional knowledge of resource management for which they commonly practiced burning pastureland. The analysis of Guji's 'backstage' informations indicates that Guji's setting fire to the park was partly a covert confrontation against the park authority and partly a strategy of ensuring source of their economic life–the livestock herding–by burning grassland to enhance regeneration of fresh grass. Although this covert resistance did not always guarantee the perpetrators of the action unnoticed, it served them in increasing the opportunity for their livestock to get fresh grass once the rainy season started and also increased the pressure on the park administration that eventually prompted it to negotiate its conservation practices.

8.2.1.3 Innovative Strategies of Resistance

When grassroots actors change their livelihood practices or adopt new technologies and ways of life because of enclosure of their customary lands and resources, some political ecologists refer it as adaptation strategies. As I discussed earlier, some scholars consider changes in economic practices and introduction of different strategies by local communities under circumstances of environmental enclosure as strategies of adaptation (Bryant and Bailey 1997). Although some of these activities might be described as traditional adaptation strategies, the case of the Guji is more complex than simple reduction to adaptation. I rather consider Guji's diversification of their economies, change in their ways of life, establishment of schools and other 'modern' infrastructure as *innovative strategies of resistance*, although the overall process might be considered as the people's adjustment of their life style in conformity to the state's development visions.

To elucidate my argument, I reiterate some of the arguments used by the state to justify resettlement programs because the Guji pragmatically responded to these notions and practices through their own ways. First, the government considered Guji's engagement in pastoral activities as a 'backward' form of economy that was supposed to be transformed into sedentary agriculture. In this regard, the Ethiopian government has launched massive sedentarization program of pastoralist communities in the periphery under the guise of enhancing easy distribution of social services, which would be otherwise difficult with scattered and mobile agro-pastoral communities. Secondly, the park authority and other government officials used Guji's seasonal movement to justify the presumption that the people did not have customary right over the land because in Ethiopia, agrarian forms of land utilization continued to be the basis of resource ownership rights (Donham 1986; Turton 2011). The denigration of indigenous peoples' ways of life, governance institutions, livelihood practices and cultural values has been common in many parts of the world where nation states portrayed the values, knowledge and practices of indigenous people as obstructions to modernization and development (Stewart-Harawira 2005). The Guji articulated their resistance within the government's development discourses and arguments of property right by appropriating the state's perspectives, albeit positioning it within their own ways of countering the notions of the state.

Moreover, the Guji were well aware of the impact of limited education opportunities to their community on their struggle over the territory. According to Guji elders and youth, their deprivation from modern education affected their struggle over the land in two ways. From the perspectives of the youth, the park authority used Guji's lack of education as an excuse to exclude them from job opportunities in the park. On the part of the elders, they recalled their encounter with the government as an encounter between competing notions of knowledge. In the preparation for the resettlement programs in the 1990s and 2004, the government authorities came with pen and pencil, counted their livestock, fruit trees, as well as the number of people living within and around the park boundary. The government sent technocratic 'experts' who understood human-environmental relations only from physical interactions in the form of grazing, water, farming, hunting and extraction of forest resources, on the one hand and who perceived conservation as a practice of delineating human and non-humans. By contrast, the Guji, who had a different conceptualization of their place in their surroundings, were unable to read what the government 'experts' calculated, measured and counted but they were told to put their thumb print on the papers. Later during the resettlement plan, the park authority used the data as a documented 'evidence' for threats that could be posed by the local community in the form of overgrazing, hunting and farming encroachment.

After presenting some ethnographic data on Guji's practices such as a shift in economic activities and the establishment of social services, I will analyze it at two broader levels, namely as an aspect of resistance and as part of the broader adaptation to the government's conservation and development visions. From resistance perspective, Guji's engagement in practices like planting fruit trees, establishment of social services like school, healthcare centers, houses and flourmills could be seen as techniques of challenging the state's resettlement programs. The Guji shifted from pure pastoralism to agro-pastoral livelihood around the mid twentieth century, although they increased the scope of their agriculture through irrigation and sometimes rainfall with the increased pressure from the park authority concerning their transhumant seasonal movements. For such practices, the Guji extensively used the eastern escarpment of the park (Golbo, Gashe, Korde and Hiituu) for the cultivation of subsistence and commercial crops. They commonly cultivate fruit trees (mango, avocado, banana and coffee), root crops and sugar cane in this lowland area through irrigation and rain fed agriculture.

Apart from ensuring food security of the people, engagement in agriculture challenged the government's resettlement programs in two ways. First, the people began 'settled' agriculture with innovative skills of diverting streams for irrigation, which the government also advocates to introduce to pastoral communities. The Guji are skeptical of the resettlement as would displace them from their 'sedentary' agriculture. Second, from a development perspective and, most importantly, from the government's views of counting and measuring properties for compensation, the large fruit trees made it infeasible to demolish all these properties for the sake of the wildlife. For example, the two regional presidents were amazed when they arrived on the scene in April 2008 by the dense fruit trees cultivated in Golbo/Tsalke valley. This was partly why the regional presidents came to terms with Guji's settlement in parts of the park. In their defense against any resettlement plans proposed by the government, the Guji capitalized on their agricultural practices and appropriated the government's rhetoric of development. In doing so, they questioned the government's resettlement program that ostensibly aimed to 'improve' the living standard of the pastoralist people arguing that they were already under lifestyle 'transformation' at their current location and there was no be sound reason for the government to dismantle their agricultural investments.

Moreover, in addition to securing livelihood, the Guji are engaged in agriculture (farming and planting permanent fruit trees) as a strategy to claim customary right over the land, which would otherwise be impossible in purely pastoralist context. A Guji elder, Gagasa Edema, whom I interviewed while he was working in his farmland (*ficha*) in Golbo considered the fruit trees as insurances for his children's right over the land in the future. He contrasted notions of land ownership among the Guji on the one hand and agrarian society and the state on the other hand:

> Guji did not know dividing land into portions like this. Land is what Waaqa gave us to live on it. In our culture, malkaa [access route to a river or a lake] and marra [pasture] are common for all Guji clans. It is only around homesteads that people had enclosed kaloo [enclosed rangeland] because that is reserved for vulnerable livestock. The Amara [Koore] and other people of the hill had their private ficha [farmland] and we borrowed it from them. When the government people came here, that was long time ago during Haile Selassie; they said only the Amara lived on this land. We told them we were the pioneers in Ergansa [Nech Sar plains]. However, they told us the land was empty. They only saw farmland and houses and then considered the Amara were legal inhabitants. For us that space was our ancestral home. It was where our ancestors were buried. It was where we made our jila [sacred rituals]. The government people could not understand it. Finally, they evicted us from our home. Even last time, some four years ago, Abba Dula Gamada [regional president of Oromia] and Shiferaw Shigute [regional president of SNNPR] saved us from eviction because of our farm. For them, cattle could move to other places. However, what about these mango, coffee, avocado and sugar cane? They agreed that we stay on our land. We do not know if another government comes in the future and wants to displace us. However, at least these trees will be evidences for our children in their struggle for their land right (informant: Gagasa Edema, May 2012).

Regarding using permanent settlement and resources as strategies of claiming for land ownership, local Guji elites in collaboration with political authorities from Oromia regional state have established small administrative villages in Golbo and Arda Guddina in 2008. In both areas, few houses were built from tin roofs with many more under construction. It could be fairly contended that the construction of flourmills, clinic, school and administrative offices was among the major strategy of entrenching Guji's settlement in the area. A park official remarked regarding how it is going to be difficult to resettle the Guji because:

> They [the Guji and political officials from Oromia] had strong cooperation. First, they built the school and market. Later they started clinic and some houses. Now they formally erected the flag of Oromia. Who can move these people from the land? It will be very difficult. If it were only dealing with the people and their livestock, it would have been easy. Now, the houses, flourmill machine, school, clinic

> and administrative office are all there. They also planted massive fruit trees. Sometimes, we feel guilty to decide on resettlement. I think they got advice from their educated people in Oromia on what to do in order to stay on the land (informant: Shalaqa Admasu, April 2012).

On the other hand, these practices could also be framed within broader processes of local communities' adjustment of their life style to live in 'harmony' with the state. In this regard, Guji's adoption of farming culture, semi-sedentary way of life, introduction of school and government administrative bureaucracy would inevitably create subjects whose tastes of life, worldviews and understandings of their environments would strongly resemble that of the state technocrats who privileged agriculture over animal herding, settled agriculture over transhumant pastoralism and scientific knowledge over indigenous forms of knowledge. More specifically, while the introduction of the school was assumed to enable the Guji in educating their children who would challenge state authorities in the struggle over the national park, modern schools also have a tendency to create a generation that internalizes state developmental and conservation discourses and denigrates its own community's ways of life. The same is true with the establishment of government administrative villages in the area. These administrative offices of the government at the village level were used as institutions through which the government channels down its policies and programs for implementation.

Although it could be seen as a victory for the Guji in winning the support of the regional presidents from Oromia and SNNP regional states in the establishment of the administrative villages, it was also another way of the government's entrenchment of its authority to the local community through its administrative bureaucracy. Since then, the village administration was given responsibilities to oversee activities of the local communities regarding their interaction with the national park. Therefore, the complexity of the situation makes it difficult to reduce the people's shift in economic practices and introduction of some new ways of life merely to adaptation. Rather, it can be understood both as innovative approaches of resistance as well as adaptation strategy. Therefore, our understanding of these approaches depends on ethnographic information on why and how the people do these practices.

Another strategy of resistance the Guji employed in their struggle to regain their customary rights to the land was capitalizing on their self-image. The Guji were traditionally represented among their neighbors as warriors and were feared particularly for their practice of *midda* tradition, namely killing game animals and enemy men for honor (Taddesse 1994; 2006). Killing for honor was practiced by many communities in Ethiopia and was not unique to the Guji.

Nonetheless, the case of the Guji has been emphasized by their neighbors because the Guji used the story of *midda* practice as a complement for their self-pride and self-defense and often frustrate their 'enemies' by capitalizing on the story in their conversations with outsiders. Although the practice of killing for honor was abolished because of protestant Christianity, the Guji still use their self-image to defend their territories from intruders (Asebe 2012d).

When I was planning for the fieldwork among the Guji villages in 2011 and 2012, I was warned by outsiders that I might risk my life for entering the Guji villages as the people were thought to be unfriendly. In fact, such warnings were not new for me as I encountered similar impressions from members of Guji's neighbors in the past as well. On the contrary, from my experiences of fieldworks in different Guji territories, I enjoyed Guji hospitality and honesty once they established friendship. I was welcomed as their special *imala* (guest). Among the Guji, *imala/imaltuu* (guest) is a most respected and welcomed person and receives special treatment. The type of food prepared, the place the *imala/imaltuu* sleeps and overall companionship he/she receives from the community is quite different from what any relative or clan member is given. It was the outside image and stereotypical representation of the Guji that made them to be feared. The striking aspect of Guji's self-image is their capitalization on the 'negative' image others subscribed to them as a strategy to defend their space from outsiders.

In this regard, security problems created during the transition period in the early-1990s and following the disbandment of 'informal' fisheries in the late-1990s had created a general impression among the park authority as well as the people in the Arba Minch town that the park has been one of the most insecure national parks in the country. Although the Guji did not have direct interaction with tourists, they used the existing sense of fear among the outsiders to discourage people from entering the national park. In other words, the Guji invested in the reconstruction of self-image, which was believed to repel tourism practices in their homeland. In this regard, the 'invention of tradition' among many Oromo groups in the post-1991 period which also constituted the killing of big game animals for honor was among the major reasons behind non-Oromo communities to develop the fear that the tradition of killing enemy men for the *midda* tradition might still be practiced. By 'invented tradition', I refer to traditions that reemerge in response to novel conditions but drawing their reverence on the past norms, practices and values and adjusted to changing socio-cultural, political and economic contexts (Hobsbawm and Terence 1983 [2012]). In the context of the revitalization of Oromo traditions, I use the concept 'invention of traditions' not as a complete construction of practices without any reference to the past. Rather, as Hobsbawm and Terence (1983 [2012]) remind us, the concept also implies traditions that are reconstructed from the past with traceable knowledge about their functions and practices and that are formally adjusted to changing circumstances.

Lastly, Guji's experiences of deception by the government technocrats for their lack of modern education initiated the people to appropriate education as a resisting power. As I discussed earlier, the Guji held consensus concerning the fact that they were excluded from possible benefits from the national park. However, opinions were diverse within the Guji community whether they had to claim for benefit from the national park or altogether denounce its existence. This difference stems from internal heterogeneities in terms of age, education and religion. For a few youth who went to secondary school, the job opportunity that six Guji individuals got as park scouts is part of their future career plan. In a community where education was only introduced less than a decade ago and where there are no other employments sectors for the youth in their locality, the scouts are their career models.

In one of my group discussion sessions with primary school students in Tore, the students underlined prevalence of discrimination in terms of employment opportunities against the Guji on pretexts that their people are not educated. Therefore, according to the students, education is the best way to tackle unequal representation in job opportunities in the park and other government offices. Apart from job opportunities, the Guji youth were keen to challenge the park authority for their right over their ancestors' land once they were educated. For the majority of the Guji youth in the area, its exclusionist approaches rather than its intervention per se represents the major drawback of the state conservation programs. According to the youth who took part in the group discussions, issues of fair employment opportunities, redistribution of the benefit, consultation of the people on their basic demands, establishment of social facilities and recognition of Guji's customary rights in the present areas of settlement, would be addressed only if the Guji reach on similar platform with other actors through education.

Elders and local government representatives on their part felt that the government deceived them during resettlement plans in the past because their people were unable to communicate with government officials in both writing and reading. These issues touch upon many fundamental points such as power relations between the state and society in which the former used the people's lack of education as a loophole for domination. In the context of the current political reconfiguration, it hints at the local people's demand for participation, equality, autonomy, decentralization and equity in resource distribution and issues of customary rights.

As a result, the Guji requested Oromia regional state for establishment of a primary school in the newly established Arda Guddina village, which came into effect in 2008 with financial contribution from the region and with active contribution from the local people. In 2012, during my fieldwork, three hundred children attended the primary school and other three satellite schools. In a local context of inter-ethnic relations, Guji elders and local elites were very cognizant of the difference in level of education between them and their rival communities as well. A remark by a chairman of Arda Guddina village on a meeting with primary school students was instructive here:

> I am pleased to be here today. I am happy to see our children have started to read and write. This was what we have been longing for long. We were chased out of our homeland several times because we did not know the language of the government. The government people always come and tell us to sign on a document that we do not understand. Look at our rivals; they went far in education. Their children went as far as Addis Ababa [the federal government]. From there, they discuss on issues that affect us. Do we have a single person in higher government? No! It is good that our Oromo brothers and sisters like our son who came to visit us here [referring to me, the ethnographer] are everywhere. They saved us from displacement in the past. Now, it is your turn to struggle for your own destiny. This land is the life of our generation and of the future. Unless you are able to know what government people plan, you, your family and the future generation will not live here. We are in continuous struggle with the park and our rivals. However, they are powerful because they are educated. I am grateful to the teachers who are paying much sacrifice. They do not get access to town. They live here without getting water, market and health center. The next teachers should be among you, the children of the Guji. We are also ready to mobilize the people to contribute for the construction of additional classrooms and other facilities. The family should not keep their children at home (Arda Guddina village chairman, May 2012).

Figure 21: Guji children in classroom attending lesson, Arda Guddina Primary School (Photo: Asebe Regassa, May 2012)

This primary school was established at the border between the Guji and the park territory but within the areas that the latter claims to include into its boundaries. The establishment of the school in this particular place can be interpreted from different angles. First, by 'appropriating' the state's discourse of modernization through modern education, some of the Guji informants considered the process as a gradual step of emancipation from the asymmetrical power relations between the state and their people. They were aware that lack of education affected them in their struggle against the government intervention due to language barriers and their inability to present their cases at the state court. Moreover, by establishing the school, which is a symbol of struggle for knowledge within the national park's de facto boundary, the Guji partly exercised their authority over the park administration. In terms of administrative structure, the establishment of the school brought to the scene institutional competition between the park authority and Galana district administration that administers the primary school. Although discourses of 'modernization' through education were appropriated from the government, the Guji were very cognizant of the power it would nurture them in their struggle against their rival actors–the state and neighboring ethnic groups.

Put briefly, one might conclude from the analysis that the Guji responded to dominating power from the state pragmatically in a subtle yet successful approach. Indeed, as I will discuss in the last section, they compelled the government to come to terms with their demands in some critical cases, such as the government's permission of Guji settlement in Arda Guddina in 2008. Nevertheless, the national park administration's continuous encroachment to Guji customary land and its reluctance in adopting community-based resource management have threatened the people's existence in the area on the one hand, as well as prompting them to consider the park as a property of an enemy to be destroyed on the other.

8.2.2 The Koore and their Responses to Conservation

Koore's encounter of the Ethiopian state with reference to their settlement in Tsalke was a contact between an ever-expanding political center and a retreating periphery. State encroachment to their homeland was first under the imperial regime that prompted the Koore in the highland to use the 'exit' option because of which they moved downslope to the Tsalke Valley. In traditional state-society relations in Africa, the society in the peripheries of the state administration commonly used strategies of exit or escape because the state has been continuously perceived as an alien to be abandoned rather than an entity to be embraced. The second encounter was under the military and EPRDF regimes in the lowland. While the imperial regime pushed the Koore to the Tsalke through its coercive administration and exploitative economic system, the next two regimes followed the Koore in the lowland (Tsalke Valley) through the national park. These regimes used local elites whom they socialized into their political ideologies in translating the central state's discourses of development and conservation into practice.

Following its submission to the invading army of the central state in the late nineteenth century, the Amaro king retained his traditional authority but was said to have become too authoritarian (Awoke 1985; Bediru 1986). The traditional king and his authorities oppressed the Koore in Amaro highland through imposition of heavy tribute, corvée labor and other forms of administrative and economic burdens of the feudal system. As a result, many Koore families from adjacent highland villages moved downslope to the Tsalke either to escape from the state oppression and exploitation or to subsidize their income and pay taxations to the state. Under the EPRDF regime as well, Koore local elites were highly co-opted to the so-called developmental state discourses of the government. As a result, they gave deaf ears to their subjects' different ways of interpreting development. Despite the rhetoric of decentralization in empowering local administrative units to administer their resources in line with local values and norms, the new political reconfiguration has at times resulted in a system similar to what Mamdani (1996) calls 'decentralized despotism'. Like British colonial rulers who created traditional authorities in conformity with its assumptions of pre-colonial African social structure, the regimes in Ethiopia also coined local elites who would serve the state in its exercise of dominating power over the citizens. In the process, the central state only devolved its coercive administrative system and extractive economic schemes to the peripheries through newly designed local administrations.

The Koore people have mixed reactions to the government intervention through nature conservation in the national park. Differences in attitude to the national park were based upon education, livelihood and the shifting inter-ethnic relationships with the Guji. The national park is simultaneously viewed as a threat and a safety corridor among the Koore. For Koore peasants who heavily rely on irrigation practices in the Tsalke lowland areas, the park is seen as a threat to their livelihood and customary rights. Memories of the eviction and resettlement programs in 1982, 1990s and 2004 still linger in the people's day-to-day interaction with the park. In 1982, the military regime burned their houses, crops and fruit trees. It literally invaded their home and criminalized their customary rights over the land. From these vantage points, the presence of the park close to their fertile agricultural land that the park claims to include into its borders is considered as living close to an enemy neighbor. Tenure insecurity has made the Koore peasants to minimize their investment on fruit trees although the area remains the major source of their economy. For example, during the 2004 resettlement program, the Koore abandoned their mangoes, avocados, sugar canes, sweet potatoes and other crops and fruit trees. From 2004 to 2007, the park administration nominally included all the agricultural land along the Sermale River into its boundary until the APF gave the land to the Guji in 2007. Moreover, Koore's sense of insecurity was posed both by the park and the Guji as the latter particularly increased their engagement in the contestation over the Tsalke since the 2004 resettlement of the Koore.

On the other hand, amidst rival relationship between the Koore and the Guji in which Tsalke is central in their contestation, most Koore residents in Tsalke and its adjoining villages perceive the national park as a shield against Guji's threat. The Guji currently live between the Koore settlements and the park. The Koore believe that any actions from the park administration would first affect the Guji before it reaches their people. Moreover, any action of the park authority against the Guji that would most likely relocate the Guji to other places is speculated as an opportunity for Koore rivals because it would reduce Guji's competition over the Tsalke irrigation land. The Koore see the Guji not only as rivals over the Tsalke Valley but also as sources of security problems in their trade relations with the Gamo and other groups on the Arba Minch side. During times of inter-ethnic conflict between the Koore and the Guji, the former would be disconnected from Arba Minch because the only path to the town was crossing the Guji territory–a narrow strip of land between the park and Amaro district. For example, following the 2010 conflict between the two groups, the Koore were entirely isolated from Arba Minch, while the same was true for merchants from Arba Minch who wanted to trade with the Koore. This is why some scholars and Guji elites associate the aggressive actions from the SNNP regional state to resettle the Guji as a fixation to ethnic dissociation to clear the Guji from the buffer zone between Amaro and Gamo Gofa (Dowie 2009).

Apart from deterring Guji's potential competition over the agricultural land, the park is also seen as an employment opportunity for a few young Koore individuals who at least completed secondary school. Under circumstances of high unemployment rate in the district, where over fifty percent of college graduates are unemployed, some of the youth cast their eyes on the national park as their future employer as tour guides or scouts (interview with deputy administrator, March 2011). Therefore, it is within these two mixed representations of the park that the Koore came to terms with conservation discourses and practices of the government by employing different sorts of strategies. Under such conditions of uncertainty and insecurity, the Koore have mainly devised two different techniques of 'survival' since the 2004 resettlement. While the first strategy was collaborating with the park authority and appropriating the government conservation discourses both in practice and rhetoric, the second was playing the 'blaming' game against their Guji rivals to win the impression and support of the park authority.

8.2.2.1 Coming to Terms with State Conservation Practices

Like the Guji, the Koore also had recent memories of conservation-induced displacement that removed them from the land. Since the 1980s, the government used different strategies to implement the conservation project. In addition to its coercive approach in implementing its policies, the government used approaches of co-opting local elites to persuade their respective subjects. As a former administrator of Amaro Special District confirmed during my interview in 2012, local elites including those in the district administration were prompted to accept the conservation discourses and implement the plan of resettling the people because it was an order from the higher administrative structures. Local elites' co-optation to the government system was mainly due to asymmetrical power relations and a lack of competence. In its supposedly policy of 'empowering' ethnic groups in the peripheries, the EPRDF filled administrative structures and party profile with people from a low academic background, including primary school teachers, secondary school dropouts, unemployed secondary school graduates and low scale businessmen (Markakis 2011). For people in such category of life, political loyalty to the party was the only means of survival in the government system, which was also a guarantee for their future promotion and economic security. This was why, according to the former district administrator, "people at the district level were unable to challenge the resettlement plan because it came from the higher level. Resisting it would have been taken as rebelling against the government. At district level, we were there only to implement what came from the higher structure. This is how things are done in our country" (Former Administrator of Amaro district, March 2012).

On the other hand, as part of the EU project for '*Southern Ethiopian National Parks Rehabilitation*', during the 1990s the government promised the Koore to fulfill their material and social demands, including financial support, establishment of school, clinic, market and construction of road and water supply (Turton 2002). The ambitious program included establishment of an irrigation project in the resettlement site. Similar promises were given to the people during the 2004 resettlement plan when the APF took over the management of the park. In any case, be it through coercion, co-optation or persuasion, local elites mainly constituting government cadre members played a strong role in resettling the Koore peasants. However, this does not reduce the agency of ordinary Koore peasants in resisting the action during the process. Given that actors' decisions and actions are largely influenced by the power positions of other actors (Long 2001), the Koore peasants became victims of the asymmetrical power relations and thus came to terms with the resettlement plans from 2004-2007.

After they faced hardships in Abulo and Alfacho resettlement centers and with fresh memory of the past evictions, the Koore began devising different strategies to live in 'harmony' with the park administration. Government officials at the district level and the Koore peasants who returned back from Abulo and Alfacho resettlement sites unanimously confirmed difficulties the people faced pertaining to lack of water, absence of fertile land for agriculture and the subsequent food insecurity. The Koore lost their land in the Tsalke lowland during the resettlement, which the Guji partly controlled later. As a result, the returnees faced another uncertainty of life. Regarding the state, the Koore were aware of its monopoly of coercive force. As the past historical trajectories of the park-people relations demonstrated, the government exercised hard power against the people. In the case of the Guji, their resistance was mainly effective in enabling them maintain part of their traditional grazing land and access to Lake Abaya due to their strong agency in winning the support from Oromia regional state. On the contrary, apart from the nominal district administration, which was also co-opted to the government system, the Koore did not have the support from their regional government. Therefore, the Koore used different strategies to comply with the government's conservation regulations but in a way that would also enable them to gain access to part of the farmland in the contested valley.

In 2007, shortly after the APF reached on an agreement with the Guji over access to pasture and water, the Koore established a local committee called "*Committee to Protect the Safety of the Wildlife and Other Resources in Nech Sar National Park*". According to a chairman of Darba Manana village, the local people took the initiative of establishing the committee to oversee and monitor 'illegal' encroachment into the national park and renew the relationship with the park administration. The committee was given a mandate to report to the village, the district administration and the park authority concerning issues related to 'illegal' activities such as hunting, cattle trespass, farming, practices of burning grassland, and fishing practices within the park boundary. The village chairman further described the role the committee played in renewing relationship between the people and the park because "the committee worked hard in exposing illegal activities mainly performed by the Guji and some members of the Koore and Gumayde ethnic groups. As a result, the park officials developed sense of trust on the Koore and became a bit reluctant to expand the park's boundary to the Tsalke Valley" (Darba Manana Chairman, February 2012).

In their 'front stage' or 'public transcript' (Scott 1990) presentation about the national park, the Koore conveyed strong conservationist messages to their audiences. For instance, during three separate focused group discussions among Koore peasants in Darba Manana, Tifate and Tsalke, the participants initially focused on their strong commitment to conservation of the wildlife in the national park mentioning that the national park is a national heritage and source of foreign income for the country through tourism. They explained much about the side effects of hunting, saying that it compels the animals to migrate to other countries and in diverting tourists away from their region. Apart from exposing hunters and cattle trespassers mainly from rival ethnic groups, the local committee played significant role in 'educating' the people about the government's version of conservation.

One incident worth further elaboration is the role the committee played in winning the impression of the park authority particularly by exposing members of rival groups. In 2012 when I was in the field, five Guji individuals were arrested allegedly accused of killing a hippopotamus. Despite the accusations behind the plot that ranges from personal conflict to inter-ethnic hostility between the Guji and Koore, it emerged that one of the Koore committee members reported to the park authority that Guji individuals killed the animal. Subsequently, the park authority dispatched its scouts and the police force from Arba Minch town that finally arrested the people suspected of the killing. According to my sources from the park authority, the person who reported the incident was given two thousand birr (about one hundred Euros). While I will shortly return to the 'blaming game', it suffices to mention is Koore's agency in maneuvering the power game. By proving their loyalty to the park through exposing 'illegal' activities, the Koore people were able to show their alliance and partnership with the park authority and won its support. The latter on its part heavily relied on the Koore local residents and committee members in its fight against the other strong local actor–the Guji.

The Koore's collaboration with the park authority does not rule out the prevalence of violations of park laws by its members and it does not indicate their covert or 'hidden transcripts'. As I discussed earlier, the Koore perceive the presence of the park at their proximity as everyday forms of insecurity and uncertainty. Besides, the people complained on absence of trickledown effect from the park in terms of benefits from tourism. During a key-informant's interview, I managed to grasp some 'back stage' views and information from the Koore peasants regarding the park. For example, my informant Shadi Macho, an elder in Tifate village, disclosed issues of hunting and fishing within the park territory and aspects of hiding hunters within the community. Referring to hunting, Shadi mentioned that:

> If people come across an animal that can be used for meat, it is difficult for them to ignore the meat. If you were given food by the time you are hungry, would you refuse? No! That is the same. Our people do not go for hunting but when they see the animals, they kill for food. Now, it also became a business. It is money. The young people go for hunting or fishing to sell the meat and fish to make money. But, elders insist that we should not enter into conflict with the park. The park has also problems of its own. It has been there since Haile Selassie period. We did not see anything positive from it. They [people in the park administration] are there only to threaten us. I think our children are also sometimes right when they argue that they want to get what the park denied them (informant: Shadi Macho, March 2012).

It was also ascertained that Koore scouts and even the community often hided 'poachers' when scouts or police officers from other ethnic groups entered the community for searching perpetrators. In contrast to the public image that they presented as conservation friendly and in harmony with the park authority, the 'back stage' opinions and practices were quite different. The remarkable difference between the Guji and the Koore is that while the former covertly work to eliminate the park by attacking the basic objectives of its establishment–the endemic Swayne's Hartebeest, for example–the Koore were largely involved in resource extraction for economic reasons. However, in any case, the Koore used strategies of appropriating government rhetoric to convey their 'partnership' with the park authority mainly to ensure their land right in the Tsalke Valley.

8.2.2.2 Playing the 'Blaming Game' against the Guji

Another strategy that the Koore used in confronting the park's encroachment to their customary land involved casting the blame on their rival Guji agro-pastoral community. In March 2011, I made my first interview with an official from Amaro district Agriculture office in the department of wildlife and natural resources. The following excerpt from the interview illustrates how the Koore elite presented the Guji as a culprit in defending the right of the Koore to the land:

> The issue of Nech Sar always strikes my mind. The park is a national heritage and even it is international. We have animals that are found only in Ethiopia. I personally like the park. It is home for innocent animals that do not have any home other than the park. People can live somewhere else but it is difficult for the animals to move somewhere and begin new life there. The major danger to the park came from nomads. These Guji nomads roam in the desert with cattle. You know? Agrarian societies are wise. They care for tomorrow but nomads want to maximize what they get today. Regarding hunting, burning the park and grazing cattle in the park, all these crimes are being done by the Guji nomads. After we told our Koore people not to kill wild animals, they stopped. But if you go to the park, it is almost empty now. The animals, including the endemic Swayne's hartebeest, were gone. Unless the nomads are relocated to another place, the future of the biodiversity in the park will remain in danger (interview with Kassahun Beyene, March 2011).

Local Koore residents in the villages adjoining the national park also shared similar views in their discussion about human encroachment to the park territory. One important case was related to hunting for business. In the field, I got insider information about 'secret' market place where wild animal meat was sold. Women were said to be the central actors in the business but it was initially difficult to get the detail from where they brought the meat. This information first emerged during a focused group discussion with secondary school students in Darba Manana but the students refused to go further to explain about the issue. Based upon the clue that I got from the students, I probed further in one of my discussions with Koore women who were thought to have the knowledge about the business. The women concealed the involvement of their men in the killing but stated that the meat came from Guji areas. In other words, it was initially presented as if the Guji killed the animals and sold it to the Koore in which women from both sides take part in the trade activities. However, after a long period in the field, I came to know that both groups, and particularly the youth, were engaged in killing animals for such business and household consumption. In any discussion about violations of park laws and any threats upon the game, the Koore unanimously blame the Guji for such transgressions and practices. As I explained above, the Koore local committee and scouts from the same group linked many of the transgressions of park rules to the Guji. By casting the blame on the Guji, the Koore managed to change the attention of the park management to the Guji. Moreover, as it was true among the Guji, the Koore also used park scouts from their community as a shield in getting access to the park resources. For instance, I learned from the discussions that scouts often conceal people who transgress park rules but expose others even without having concrete evidences based on their ethnic membership.

I conclude that local responses against state intervention take various forms including covert and overt resistance, negotiation and appropriation. In the case of the Koore people, the people's initiative in establishing local committee to oversee human influences on the national park can be interpreted as a pragmatic strategy of appropriation of state conservation discourses to enhance access to the farmland in Tsalke Valley. Likewise, Koore local community members and government officials in the district strategically used the blaming strategy whereby they used the Guji as scapegoats for all the problems in the national park. Although it is not supposed to nullify Guji's engagement in different forms of resistance through which they attack the national park, the strategy used by the Koore is mainly a tactic to divert the attention away from their members.

8.3 Ethnic Politics, Conservation Enclosure and Local Resistance Nexus

The historical antecedents in the establishment of the modern Ethiopian state necessitated a continuous broadcasting of power from the center to the peripheries for the last century or more. However, despite relentless efforts on the part of successive Ethiopian regimes to pacify the periphery in their quest for converting the resources in these regions into commoditized goods, the pastoralist and agro-pastoralist lowland regions of the country remained out of the effective control of the state until the 1990s (Markakis 2011). Nevertheless, in addition to the socio-cultural and economic bias of the state against the pastoralist lowlanders, these communities faced punitive attacks from the center whenever the latter had some security concerns in border regions (Hagmann and Alemmaya 2008). The periphery has been marginalized in terms of getting social services and economic benefits from the state. State presence in pastoral areas in eastern, southern and southwestern Ethiopia under the previous regimes was mainly in its coercive and punitive approaches of pacifying inter-group or cross-border conflicts and extraction of their resources, particularly the livestock. While discussing legitimacy problems of colonial and postcolonial states in Africa, Herbst (2000) argues that these states failed to reach the periphery in delivering social, economic and security services, apart from sporadic interventions in the form of coercing the subjects to comply with the interests of the center. For the Koore and Guji people, whose contact with the state was through enclosure of their customary resources, eviction and resettlement programs, successive Ethiopian regimes shared common features as encroachers and extractive institutions.

Despite the bogus ambitions of nation-building projects under successive regimes in Ethiopia, which constituted at times attempts to homogenize the entire multicultural ethnic groups into the Amhara identity, sense of ethnic belongingness continued to overshadow national identity among diverse ethnic groups (Markakis 1994). For ordinary men and women in remote parts of Guji and Koore territories where state institutions rarely delivered services to the citizens, local networks to clan and ethnic membership provided individuals with social security and ensured social cohesion. Local customs, values, rituals and language were binding elements in shaping individuals' sense of belongingness and self-identification to ethnic groups or more specifically to clan membership among the Guji and Koore communities. This strong sense of membership and security at local level contradicts with what Anderson (1981) envisages the building of 'imagined community' or sense of common identity among populations of a particular territorial entity at a macro level through education, print technology and other forms of communication. For the Guji and Koore people, the state was too far to identify themselves with–far in terms of embracing the subjects and too far for the people to get back what it extracts from their land. The central argument is that it is difficult to ask those who do not feel they belong to something called 'nation state' to pay sacrifices at the expense of the land to which they are strongly connected in economic, cultural and spiritual aspects.

The 1991 political reconfiguration at least brought the state closer to the periphery through decentralization and ethnic based self-government structures. Apart from aspects of self-government along ethno-linguistic lines, the adoption of mother tongue language at regional and local state administrations also reinforced state-society relations along regional lines. Nevertheless, the Ethiopian state remains too far from the ordinary Guji and Koore men and women than their affiliation to their respective ethnic groups. The conceptual distance between the state and society adversely affected how the latter reacted to the state's development projects, owing to the limited level of trust and sense of belongingness. The people consider state development projects as something to be resisted, avoided and seen suspiciously. This reaction has been built upon experiences and memories of state coercive measures such as the 1982 eviction.

With specific reference to the enclosure of former common resources through the government's conservation practices, the imagination of the local people about the state was different across the three regimes and there were considerable differences based upon ethnicity, age and education. Members of both Guji and Koore communities perceived the imperial and military regimes as alien invaders who usurped their local autonomy and customary rights over the land. By contrast, the state became too heterogeneous after the institutionalization of ethnic federalism that made the local people to be inclined towards their ethno-linguistic administrative units, which in turn distances the center from the periphery. Local people's responses to state intervention were also informed by their consciousness of changing political dynamics along ethnicity. As I discussed earlier, ethnic politics in the post-1991 Ethiopia has shaped policy formulations in administration, education, budget allocation, development interventions and other social service provisions in which ethnicity has become the fundamental criteria in these schemes (Abbink 2009). Likewise, government authorities in Oromia and SNNP regional states respond to conservation and tourism development schemes in the national park within the existing political framework–in either pushing for the relocation of the local people or resisting the relocation plans.

While the Nech Sar National Park has become a contested field among actors of different perspectives and interests, ethnic politics remained central in the enclosure, resettlement programs, local resistance and the role of local government authorities in the process. For example, apart from the significance of the territory for their livelihood and other purposes, the Guji were cautious that the resettlement program was designed to remove them from the area and fully incorporate the park into the SNNP regional state. Local government authorities in Oromia region shared the same view and subtly resisted the resettlement of the Guji from this angle. From the part of the SNNP regional state, the Guji were perceived as intruders into two administrative units (Amaro district and Gamo Gofa zone) that belong to the region.

However, it is equally important to discuss about some people who are confronted by loyalty dilemma between their own communities and the state. In this case, it suffices to explain two interrelated cases. The first case refers to the scouts who were recruited from local communities and have been serving in the government institution (the park administration). These people faced the competing visions of their own communities, the state and at times members of other rival ethnic groups. The scouts are mainly from the Koore and Gamo ethnic groups with a few individuals also from the Guji. They are strongly connected to their community through kinship, marriage and social and economic relations. They are also products of the overall social system of the people and share most of the norms, customs, values and cultural practices still practiced within the communities. On the other hand, they had to conform to state bureaucracy, not only for their job security, but also at times owing to the state's indoctrination programs regarding conservation and tourism development. In such cases, the scouts used individual strategies of switching loyalty and different forms of negotiation to remain loyal to both competing actors. They maintain loyalty to both competing actors by not exposing members of their ethnic group and by reporting to the park authority that they failed to capture members of local communities who transgress park rules.

The second case is local government authorities, who like the scouts, had local connections of many dimensions and connected to the state bureaucracy through the political and administrative responsibilities. These individuals are always insecure for their decisions because they fear of hurting either the local people or the government. Therefore, the response of local government authorities to local people's questions of customary right or to government's conservation and development programs in the national park depend on how they negotiate their loyalty.

In drawing the nexus between the political framework, conservation enclosure and local responses, it is essential to briefly outline how grassroots actors were networked with other actors. In the process of accumulating power and establishing networks, local actors used the existing resources (political frameworks, ethnicity, social capital, etc.) to their advantages. In actor networks, the agency of social actors in strategizing their actions, mobilizing resources, collecting information and networking with potential allies is essential to influence external intervention (Long 2001). Long (2001: 182) describes agency as "the capacity [of social actors] to process social experience and to devise ways of coping with problematic situations". In this context, agency involves social relations that require the capacity of social groups to organize and strategize information, and generate/manipulate networks, and channel different discourses, ideas, claims and narratives for effective accomplishment of their interests or deter other actors from certain arenas of common interest. Social groups deploy resources at their disposal and capitalize on their knowledge while establishing networks with other social actors at different scales. In such networks, actors utilize social capital, political frameworks, economic interdependence and religious or cultural issues of connectedness.

The Guji used ethnicity and the new political reconfiguration to their advantage in establishing networks with political elites to counter the park authority. For example, despite their political role in the government, some Oromo political elites specifically among the Guji subtly supported the people at local level by strategizing their actions, informally channeling information from the government to the people, and more importantly by remaining reluctant to implement government plans such as assessment of cost-benefit analysis and relocation of the people. In some cases, they also worked to the benefit of the people, such as the case of increasing the monetary value of properties during the survey and property assessment for the resettlement.

Some important questions should be addressed here regarding the government officials. For example, why were the authorities reluctant in implementing the government resettlement program? Was it a survival politics in this disputed situation or was it just to channel government vision softly or supporting the local resistance in a hidden way? How much space of maneuver do local government authorities have? Moreover, what does that tells us as far as ethnic federalism is concerned? These questions would lead us to the complex political setup in Ethiopia within ethnic politics. Put briefly, ethnic federalism has created circumstances where political elites from a particular ethnic group are sandwiched between ethnic and state loyalty. In such circumstances, these elites are compelled to pragmatically demonstrate their loyalty without diverting their allegiance from the state or the society. I would thus argue that reluctance is part of survival politics to softly channel down the government's policies without compromising local demands.

The other critical issue is whether there has been political space for maneuver within the elites' double positions. Although the broader political setup does not allow the elites to express their deviance from government policies, local government authorities can play their dual positions within limited available spaces. As the previous chapters have also detailed, ethnic federalism has created contested spaces where physical, social, political and cultural spaces have taken ethnic dimensions of contestations. On the other hand, as Migdal (2001) contends, it is not easy to draw clear lines between the state and society, given that the state operates, exists and exercises its power in the society through such intermediaries. Likewise, ethnic federalism has enhanced state intrusion into the society through local ethnic entrepreneurs who serve both the interests of the state and the society depending of circumstances by contextually switching their loyalty.

Guji's agency in winning the support of Oromo elites including the former regional president was through investment in ethnic identity and cultural capital. In terms of cultural capital, the Guji are among a few Oromo groups who still practice the *gadaa* system, which reflects Oromo identity and culture (Hinnant 1977; Baxter 1991). Since 1991, the government used the Guji *abba gadaas* (leaders of the *gadaa* assembly) as an entry route to the Guji populace because for ordinary Guji individuals, the traditional leader is closer and legitimate than the government in Addis Ababa. Cognizant of the place of members of the *gadaa* assembly in the regional politics, the Guji often selected their delegates among these locally respected elders when negotiating with the government. The people were aware of the fact that any humiliation or mistreatment of members of the *gadaa* assembly would risk OPDO's (Oromo People's Democratic Organization–one of the coalition members of EPRDF) constituencies, not only in the Guji territories but also across the entire Oromia. Within such arrangements, local government officials and a few Guji intelligentsias supported the local people particularly by networking with government officials at higher levels. Besides using cultural capital as a resource for winning the support from Oromo political elites, the Guji capitalized on ethnic identity and brought to the stage issues of ethnic-based exclusion from the park. Central to their argument were resettlement plans that exceptionally targeted the Guji to remove them from straddling between two administrative units of the SNNP regional state. Indeed, during the resettlement plans, the government representatives from the SNNP regional state recommended the relocation of the Guji to various sites in Oromia regional state. In this regard, the Guji systematically positioned their contestations within the macro political frameworks where ethnicity has become a major driving force in the country's politics since 1991.

Guji elders were able to establish networks with a few Oromo intellectuals who connected their case to international human rights advocacy organizations such as Survival International and Refugee International. According to a Guji elder who took part in several negotiations with the park authority and other government departments regarding Guji's customary rights, local elders discussed among themselves on strategies how to counterbalance the power positions between their people and those who understand "only paper and pen", according to the elder. This is why the elders began networking with political authorities and intellectuals who had sympathy to their cause. "We traveled to Tore to discuss with our Guji son who was by then district administrator. He showed us some ways to appeal our case to the higher government. He also connected us with another hayyu [intellectual]. The hayyu connected us with foreign organizations that came and studied our problem. They took our case to the world. Later on, the government suspended the resettlement" (informant: Galgalo Edema, March 2012).

However, from a different angle, ethnic federalism brought the Guji closer to the state through local representation along ethnic lines, which in turn enhanced state intrusion into the society. By appointing local authorities to political and administrative positions within ethnic-based administrative units, the government has curbed local resistances and aimed to build legitimacy. Nevertheless, as I discussed earlier, local government authorities played dual roles in connecting the state and society and at times fall into the category that I referred elsewhere as ethnic entrepreneurs (Asebe 2010). In fact, it becomes an exaggeration to imagine the state's heterogeneity to the level that it envisions diversified ends that takes away the historical mission of the state in subjugating the peripheries as if it is no longer a unified end of the state. Based upon the hegemony of the federal state and due to lack of political space for local communities and civil society to challenge hegemonic discourses, the historical trend of state encroachment to the periphery persists to date.

On the part of the Koore, they came to terms with the government's conservation enclosure but, even then, they had to position themselves within the ongoing micro and macro political and economic contestations over the national park that prevailed between different actors. Through local committee, the Koore were able to establish good relationship both with the park authority and district administration in negotiating issues of land rights in the Tsalke Valley. In putting Koore's agenda on board, ethnicity and inter-regional politics played paramount role in both cases. Although the public transcript of government institutions conveyed pro-park messages, the authorities have become reluctant to deter Koore's return to the Tsalke area after over five years of hardships in Abulo and Alfacho resettlement sites. There is also a back-stage argument on the part of government authorities in the SNNP regional state that the Koore were not the major threats to the park owing to their engagement less on pastoralism and hunting. Oromia's reluctance to enforce the resettlement plans upon the Guji angered the SNNP authorities and thus they have considerably reduced their support to the national park authority and even in executing legal procedures on people who were accused of transgressing the park laws, particularly after the withdrawal of the African Parks Foundation and centralization of the park administration.

Aside from their strong partnership with park authorities through scouts who were also from the same community, the Koore pragmatically built strong social capital with the Guji through *ilmofacha* tradition and a limited level of inter-marriage. Despite the rivalry between the two groups over the agricultural farmland, the Koore did not break away from their old social ties with the Guji. Notwithstanding the limited flow of people between the two groups in the form of intermarriage because of long history conflict and other related issues, *ilmofacha* tradition reinforced social ties between the Guji and the Koore, enabling the latter to secure land, access to pasture, possession of livestock and other social privileges from the Guji. Like the culture of *gudifacha* (adoption), which was common among other Oromo groups, *ilmofacha* entailed the flow of Koore families or individuals into Guji at an individual or clan level. Nonetheless, *Ilmofacha* was a one-way practice where only Koore individual members 'change their identity' to the Guji mainly for economic reasons. However, through time, it served as a bridge between competing identities between the two groups. Therefore, although they play the blame game against the Guji, the Koore pragmatically strengthened their ties with the former in their pursuit to gain access to land and pasture in Guji territories.

On the contrary, the struggle over the national park turned into inter-ethnic conflict particularly between the Guji and Koore. In 2010, a conflict initially ignited as a dispute between two individuals from Guji and Koore side eventually escalated in to ethnic conflict claiming the lives of five individuals and destruction of properties and homes. According to my informants from both groups, the conflict erupted when a Guji herder grazed his cattle in a farm of a Koore man that instigated anger from the latter. After disputes between the individuals, the Koore farmer shot the cattle killing one and injuring some others. On the next day, the dispute escalated into ethnic conflict where the antagonists mobilized their members and fought against each. The ethnic dimension stretched also to the court system where five Guji boys were convicted for prison ranging from ten years to life imprisonment under Amaro District court, although it was overturned at the Supreme Court level and the boys were set free in 2011. According to one of the boys who were initially sentenced to many years of imprisonment, they were mistreated in the hands of local militia in the Amaro district where the district court also passed a verdict without fair trail (informant Tariku Chari, March 2011).

8.4 Situating the Contestation in Domination/Resistance Relations

In contestations over social spaces, economic resources, political arenas, cultural representations and symbolic meanings, social actors exercise power that mutually constitutes elements of domination and resistance. Domination and resistance, according to Sharp et al. (2000) are entangled into one another and should be understood as conjointly constituted elements rather than being considered as cause-effect relations. In other words, power constitutes features of domination and resistance, as well as within the same spatial settings. Sharp et al. (2000) rejected the orthodox notions that equated power with domination but treated resistance as an act of response to power/domination. The authors described the two versions of power as dominating power and resisting power stating that:

> Power should not be viewed solely as an attribute of the dominant, expressed as coercion or political control, since it is also present in the ability to resist. This resistance can enable resisters to find common ground in struggle (as in the power to mobilize others), and to become empowered in the very act of resistance. The implication is that power is operative in moments of both domination and resistance, and hence can be assessed in both positive and negative terms (Sharp et al. 2000: 2-3).

A particular social group or institution does not exclusively possess power, although it is not evenly distributed. What is relevant in the context of this research is not only the ubiquity of power and its conjoint possession of domination and resistance but also the fact that dominating power engenders resisting power and at times the latter turns to reveal features of the former.

The Ethiopian state revealed its dominating power through attempts to control and enclose the resources, and thereby coerced the people to accept its notions of conservation built upon the perspectives of nature-culture dualism. This dominating power was located within the state's institutions (the national park authority and other administrative institutions) and was exercised in the form of putting nature under the surveillance, control and mapping by these institutions. The government's practices of displacement and restriction of the people from their customary resources and cultural and spiritual spaces were cases of the state's exercise of dominating power. The exercise of dominating power on the side of the state reproduced resisting power on the part of grassroots actors and in some rare cases within the state itself. Whether the use of coercive force such as the 1982 and 2004, symbolic power inherent in the state institutions due to the historical asymmetrical power relations between the state and society in the country or the continuous threats from the park administration, domination did not succeed in suppressing local resistance though it prompted them to opt to different subtle strategies.

I extend the argument of Sharp et al. (2000), which posits power as inherently in possession of aspects of domination and resistance by looking at the outcomes of such interactions. In contestations over resources, meanings and spaces, neither domination nor resistance prevails as the ultimate result of the interplay between competing actors (Few 2001). Acts of domination and resistance are processes rather than products of the interaction; rather, the interaction generates some sorts of negotiation, compromise or appropriation whereby either resisting power or dominating power is subsumed, dominated or both forms of power come to terms.

In the following paragraphs, I will analyze the interplay between dominating and resisting power and its ultimate outcomes in the case of the Nech Sar National Park. Some practices of domination by the government empowered the Guji people by reinforcing their networks, production of knowledge and accumulation of resources, which in turn enabled them to generate resisting power. For instance, the SNNP regional state's continuous marginalization of the Guji in the Nech Sar area from 1991 to 2008 activated Guji's self-identity along ethnic lines and eventually brought their inclusion into Oromia regional state with their own local administrative unit. By exploiting the national political context of ethnic federalism, the Guji were able to win the support of Oromia regional state in their struggle for claims of entitlement to the land they considered belong to their ancestors. Ethnic identity and sense of boundaries between "us" and "them" grew among the Guji to the extent of viewing state institutions in the SNNP regional state with suspicion.

In April 2011 while I was in Bule Hora town (in Borana zone) to meet a Guji scholar for the purpose of interview, I got the chance to meet three Guji men who came from Arda Guddina village. In a short discussion with the men, I was informed that they traveled for four days on foot to deposit their money in Commercial Bank of Ethiopia. Based upon my awareness of the presence of banks in Arba Minch town, which would be a matter of few hours from Arda Guddina village, I pushed further to know why they chose to travel such long distance to save their money in a bank in Bule Hora, although the banks in both towns belong to the state. One of the men mentioned why they decided to travel to Bule Hora as follows:

> We did not know saving money at all. Our money is our cattle. However, after the park people imposed much restriction on our pasture and water, our cattle are dying of hunger and thirst. Now we began sending our children to school but if these cattle perish, how do we support our children? Some people who knew about life in town advised us to sell some of our cattle and save the money. I think that is good although it cannot be equated to our cattle. The problem we have is now where to put the money. Those in Arba Minch are not good to us. We do not have problem with the people but the government people and the park people do not like our existence close to the park. We do not trust them. However, here are our people. That is why we preferred to come here than giving our money to those in Arba Minch (informant Godana Aga, April 2011).

The elder considered saving money in Arba Minch as keeping it with an enemy. The same was true with the construction of some residential houses that Guji residents of Arda Guddina built in Tore (in Borana Zone), despite possibilities to do so in Arba Minch. In short, alienation polarizes the relationship and eventually generates resistance in different forms. Their sense of deprivation, exclusion and domination by the state that they locally conceptualize through the park authority and related institutions has entailed their strong network with other Oromo institutions, intellectuals and even government officials. Such network eventually empowered the Guji in strategizing resistance against state int3rventions in many dimensions.

The government's rhetoric and discourses of modern nature conservation that devalued indigenous knowledge of resource management and ways of livelihood of the local people carried aspects that reproduced counter narratives, knowledge and claims among the local people (both Guji and Koore). In other words, the struggle over a territory is a contestation over meanings that the main sources of social power and "central to the structuring of the social and of the physical world itself" (Escobar 2008:14). As Escobar (2008) argues, conflicts over ideas are also cultural conflicts that emanate from ontological differences and different ways of understanding the world. In this regard, narratives and critical knowledge of the people were built upon historical reconstruction of settlement, history of harmonious coexistence between the people and wildlife and indigenous knowledge of conservation, such as the case of the seasonal burning of grassland. On the other hand, instances of resource degradation, wildlife migration out of the park and rapid reproduction of parasites resulting from the ban on fire were brought as the weaknesses of the state's conservation practices. The dominating power of the state nurtured the conditions for resisting power to emerge as a response to its repressive and exploitative features. By enhancing the production of critical knowledge and counter narratives on the part of the Koore and Guji local people, the state's exclusionist discourse of conservation has sawn the seed of resistance that gradually empowered particularly the Guji to challenge the park authority in their claim for entitlement to their ancestral land.

In the preceding sections, I have discussed different forms of resistance that the Guji employed, including game hunting being mentioned as one such case. Although hunting was practiced primarily as a response to state intervention, the Guji turned resisting power into dominating power in cases where they have threatened the survival of some animals such as the endemic Swayne's hartebeest. In this regard, Sharp et al. (2000) are right in their argument that dominating power and resisting power do not always remain on the same gamut as opposing forces but based upon the agency of actors from either side in their capability, or lack of it, in mobilizing resources, the status of both aspects of power might be traversed. In this case, although Guji's status in exercising dominating power was a subtle rather than overt manifestation in contrast to the state, the park authority sometimes seems to have found itself on a weaker position.

Another good case in point where dominating power engendered resisting power and later remained on subservient position was regarding questions of access to knowledge positions. In the preparation for resettlement of the Guji in the 1990s and 2004, the government used its own instruments of quantifying resources through the language and technology alien to the Guji. Government technocrats came to the Guji homeland with pen and paper, counted their livestock, estimated their fruit trees, converted it into monetary values, and finally compelled Guji elders to put their finger thumbs on documents they were unable to read and understand. When the decision for the resettlement was enforced, the people were told that they agreed to the proposition of the resettlement to enhance the sustainability of the biodiversity in the park and promote the development programs that the government has designed.

By counting their resources in terms of quantifiable measures, government 'experts' considered Guji's resources as items to be converted to monetary value, which is legible for compensation. This generated two dimensions of resistance on the part of the Guji. On the one hand, the government's reliance on 'quantifiable resources' as instrument of ensuring property rights of the people and the fact that these resources were taken as variables determining the amount of compensation necessitated the Guji to invest in fruit trees and other kinds of plants. As I discussed earlier, the Guji invested in these resources as a license for claiming their entitlement to the land as well as to increase the difficulty for the government in paying compensation for huge amount of plant resources. Secondly, The Guji considered themselves weak in their everyday engagement with the park authority and government institutions owing to lack of 'modern' education. This was partly why they have been contributing their resources for the construction of the primary school in the contested territory, Arda Guddina village since 2008. According to the Guji local residents in the contested territories, government officials deceived them due to their lack of education; however, in order to avoid the differential power relations that stemmed from lack of education, they were committed to educate their children. The government's exercise of dominating power using the people's lack of modern education partly nurtured the people's appropriation of modern education for the sake of resistance.

As power mutually constitutes domination and resistance, the cases in Guji's response to government interventions revealed that acts of dominating power on the part of the state nurtured and reproduced practices of resisting power that prompted the state to partially come to terms with Guji's questions of customary rights over the territory. In this regard, I would argue in line with what Li (1999) calls 'compromising power', that the interaction between the state and local communities should not be placed at polar binaries between domination and resistance. As dominating power nurtures resisting power, the interplay between the two forms of power generates different forms of relationship between the competing actors. The iterative engagement between power of domination and power of resistance often compels actors to negotiate and compromise on programs, ideas, discourses and practices (Li 1999). In this process, local people's agency in deploying different forms of power and negotiation strategies is pivotal in shaping state actions. Such articulations on the part of the local Guji people compelled the state officials to compromise and negotiate the exclusionist conservation practices, with the one major evidence being granting the people portions of the park for settlement, grazing and access to water.

The state also introduced the rhetoric of 'community-based' conservation, although the practical implementation of this perspective remains elusive. Moreover, park scouts became reluctant to bring cattle trespassers and hunters to court. In fact, the people co-opted some scouts through social networks. While discussing about discourses of wildlife conservation trajectories in Africa, Garland (2008) argues that interactions between different forms of knowledge, perspectives and practices produce changes in value systems among competing actors. Thus, according to the author, conservation practices in Africa are the outcome of infusions of competing discourses and forms of knowledge between African and European conceptions of nature-culture relations. In other words, dominating power and resisting power might generate possibilities of understanding the worldviews of other actors from either side, which can be termed as knowledge appropriation.

To summarize, the encounter between the Ethiopian state and the Guji and Koore communities particularly through the national park apparently featured out the traditional trends of center-periphery relations in which the center has been continuously expanding to incorporate the periphery. The analysis of the empirical data from the field reveals that the successive regimes in Ethiopia instrumentally used conservation discourses as part of the strategies to broadcast state power to the periphery. In other words, although environmental issues were part of conservation practices in the country, the process and outcomes of such practices have been hegemonic. In the process, the state used both hard and soft power of domination in the form of enclosure, eviction, resettlement and control over multidimensional spaces of the people. Indeed, the practices of dominating power by the state eventually nurtured resisting power among the grassroots actors.

Nevertheless, it can be concluded that despite asymmetrical power relations between the state and grassroots actors that emanated from the state's monopoly of coercive power, the power of the grassroots actors' in their resistance against state intervention should not be underestimated as they often compelled the state to come to terms with their demands. For example, in the case of the Guji Oromo, the people's capability and knowledge in converting the existing political, cultural and social circumstances into viable resources in building resisting power was good case in point. They positioned their contestations within the macro political dynamics in which ethnicity has played significant role in winning the support from the Oromia regional government, Oromo intellectuals and local political authorities.

Guji's engagement in different forms of resistance and their agency in using resources at their disposal in asserting their claim for customary rights demonstrate that grassroots actors are not passive recipients of external interventions. Rather, they strategize, contest, negotiate or come to terms with different discourses and perspectives that affect them. In other words, it has become difficult for the state to exercise absolute dominating power over the people within the context of growing communication opportunities between grassroots actors and other local or global actors. Mobile phone, radio, schools and ethnic networks were among the major awakening elements between the Guji and Koore in their struggle over the national park. Furthermore, Koore's strategy of switching 'allegiance' to the park authority by establishing local inspection committee and their pragmatic strategy of playing the 'blaming' game against the Guji can also be seen as local strategies of negotiation and coming to terms with state conservation practices.

Moreover, local communities capitalized on their commonalities to mobilize their power against the powerful and dominant actor, namely the state, despite their internal heterogeneities. The analysis brought to the scene the fact that grassroots actors had different levels of agency and knowledge in challenging the state intervention, while their level of influence also depends on their capabilities to maneuver resources at their disposal. The discussion in this chapter goes in line with the arguments by Long (2001) and Peet et al. (2011) who contend that intervention programs by 'powerful' actors do not smoothly translate into actions/practices because they face counter narratives from grassroots actors who define and articulate the same discourse from their own local ways of understandings. This is why the government has not gazetted and demarcated the Nech Sar National Park despite the fact that it was among the pioneer national parks in the country. Local resistance against the exclusionist conservation practices of the state has resulted in suspension of the EU project in the 1990s, suspension of relocation of the Guji in 2004, termination of the APF contract in 2007 and centralization of the national park administration in 2009. Despite continued pressure from the park authority and other related government departments on the local people, the local communities have succeeded in securing access to their customary pastureland, water and farmland in the park.

CHAPTER NINE: CONCLUSIONS: SITUATING CONSERVATION ENCLOSURES BEYOND 'ENVIRONMENTAL DEGRADATION THESIS'

The problems in the politics of environmental conservation in Africa are fundamentally embedded in the nature-culture dualism. This dominant perspective was introduced to the continent as part of the colonial process of annexing and controlling African peoples and their resources. The postcolonial African states adopted and further strengthened this perspective in their ambitions to forge nation-states. Thus, conservation enclosures have reinforced state and non-state actors' control over resource owners through restricting them from having access to their customary resources. However, this led to the struggle on the part of resource owners to regain their rights of ownership, utilization, management and access to cultural and economic spaces. In other words, local communities often resist external interventions that result in the control of resources in the protected areas. This is why I argue that setting aside territories in the form of national parks, reserves and sanctuaries, and the social, economic and environmental implications it perpetrates cannot be explained in isolation from political and economic interests of various actors. Rather, the ontological and epistemological foundations and the political and economic reasons behind these practices should be taken into consideration for a comprehensive understanding of human-environmental interactions. Despite the experimentation of different forms of conservation approaches since the colonial period, the dominant protectionist or 'fortress conservation' approach has been dominantly practiced as a hegemonic form of power, knowledge and control in human-environmental interactions.

Therefore, protected areas management in Africa has had a long history of state violence through eviction, control over customary rights, as well as the imposition of fines and other administrative measures on local communities who transgress state enclosure laws. In this regard, local resource users often become victims of tragedies of enclosure because the territorialization of their resource spaces infringes their social, economic, historical and cultural rights over the territories. Moreover, in most cases, this has led to eviction, displacement and coercive control by the state or non-state actors. From environmental conservation dimension as well, the exclusionist approach has produced sense of hostility and antagonism on the part of local resource users. This can be witnessed from the Nech Sar National Park case, in which local communities associate the park and resources within it with the state that has been dichotomously positioned as something to be avoided or resisted.

In the context of colonial Africa, the genesis of the tragedy of enclosure was rooted in the way colonial system was installed. This means that the colonial system denigrated pre-colonial resource management approaches by expropriating customary property rights of the indigenous peoples, and rather institutionalizing a hegemonic form of knowledge and control. After independence, postcolonial states inherited the colonial approaches of resource conservation and property rights, and molded them within the political and economic settings. Therefore, it has become evident that conservation enclosures in Africa, where the case in Ethiopia is not exceptional, conform partly to the nature of state during the colonial and postcolonial periods because such territorialization of environments was part of the state's exercise of authority over the peoples and their territories.

At a general level, there are two fundamental questions that I addressed in chapter four, and reiterate in this chapter. The first question relates to the motives behind the rhetoric of 'saving the world's rare unspoiled places' advocated by the colonial conservation actors. The second one has to do with the fact that postcolonial African elites modelled their political and economic structures based upon the colonial discourses and practices despite the general expectation that decolonization was supposed to dismantle colonial institutions and policies including the hegemonic forms of nature conservation. It is important to investigate the conservation trajectories from political, economic and cultural dimensions as part of dynamics that evolved from the colonial past without fundamental changes towards a better understanding of why, most African states emulated the colonial institutions and ideas embedded in the colonial systems, beyond biodiversity imperatives. The baseline argument has been that conservation enclosures provided both colonial and postcolonial states and other non-state actors with the possibility of controlling African territories and peoples, thereby exercising their authority as instruments of achieving political, ideological and economic goals.

Another point of investigation that some scholars often overlook is how Ethiopia, despite its history as a country never colonized by external forces, shares in common with other African states regarding the issues of emulating ideas, models and ideologies including development and conservation perspectives. Unless we give due emphasis to the establishment of center-periphery relations in the country, Ethiopia's seemingly exceptional in the colonial history blurs our understanding of internal historical trajectories that brought the birth of the modern state in the late-nineteenth century. Moreover, influences of colonization are not necessarily functions of territorial occupation. Colonial ideas, ideologies, models and practices traveled across geographical boundaries and were adopted, emulated and institutionalized within different political settings. Likewise, due to the internal colonial system that the country experienced, as well as the long established north-south dichotomy between the Abyssinian north and the subjugated south, the hegemonic colonial forms of nature conservation and their practices received favorable political and cultural grounds in Ethiopia. To this end, this research aimed at enabling us to comprehend conservation discourses and practices in Ethiopia by positioning the discussions within broader debates on the introduction of national park ideals to Africa.

Apart from competing ontological and epistemological notions of nature-culture relations within the context of nature conservation, the establishment of protected areas in colonial and postcolonial African states, including Ethiopia, ignores fundamental rights of local communities who lived in and around these territories. I reiterate that by territorializing formerly common access resources such as pasture, water, forests and fishing grounds, the state and non-state conservation and business organizations have excluded indigenous communities from their customary resource areas. As the empirical data from the Nech Sar National Park illuminate, such an imposition of hegemonic conservation discourse and practice has prompted the local communities to respond to the external intervention in different forms. In the process of state-society relations within non-participatory, exclusionary and top-down approaches, both the living conditions of the society, and conservation and economic visions of the government are at stake. While historicizing the establishment of national parks was to situate the debates within the broader center-periphery relations in the country, I also analyzed state intervention and local responses within the context of political reordering under ethnic federalism that took place in 1991.

This chapter focuses on four major themes, drawing some conclusions from the arguments and discussions made in the previous chapters. In the first section, I argue that the establishment of national parks in Africa was for motives beyond biodiversity imperatives. Moreover, it is my position that the popular 'environmental degradation thesis' alone does not enable us to understand the interplay between complex political, economic, cultural and ecological motives, as well as the ontological notions behind the politics of nature conservation. In the second section, I position the conceptualization of the Nech Sar

National Park and contestations over the territory within the center-periphery relations in Ethiopia's historical and political trajectories. In the third section, I argue that the responses of local communities to state intervention go beyond the questions for the commonly applauded community participation rhetoric in modern conservation approaches. In the last section, I consider some theoretical and policy considerations in the general conservation debates in Africa and particularly regarding the state's conservation and development discourses that have jeopardized the status of biodiversity conservation, livelihood conditions of the people and human-environmental interactions in and around the Nech Sar National Park.

9.1 Situating the Establishment of National Parks in Africa beyond the 'Environmental Degradation Thesis'

In this section, I present three interrelated findings that explain why and how the colonial officials introduced the national park ideals to Africa and its existence beyond the colonial period. First, in contrast to the popular discourse that presents the establishment of protected areas in colonial Africa as a response to threats to biodiversity within the broader 'environmental degradation thesis', I strongly argue that national parks were part of hegemonic governance and instruments of state exercise of control over the economic, political and psychological schemes of the subjects. Second, the theoretical and empirical analysis in this work illuminates that the notions of nature-culture dualism and the wilderness discourse have been contingent to political, economic and cultural contexts of different actors inasmuch as they have been embedded in ontological orientations. Third, although the national park ideals were initially exported to the continent through diffusions, trajectories in nature conservation in the continent demonstrate a form of traveling model whereby ideas, perspectives and practices from global North traveled to and interacted with different conceptualizations of nature in the global South, informing some changes in the dominant conservation discourses. For instance, the incorporation of 'community-based' conservation approach was partly a response to resistance from indigenous communities against the dominant exclusionist discourse and practice, and partly owing to the failure of the hegemonic form of knowledge in achieving conservation goals.

The 'environmental degradation thesis' is too simplistic for many reasons. First, resource degradation and biodiversity depletion were not major problems in Africa during the commencement of colonial system in the late nineteenth century. Second, environmental problems that featured during the course of colonial rule were mainly caused by colonial forms of resource exploitation in the form of plantation agriculture, mining, logging, road construction, hunting and fishing. Therefore, as long as the colonial economy and the hegemonic form of knowledge were responsible for environmental problems, excluding indigenous peoples from their resources does not justify the establishment of protected areas for the sake of environmental conservation. Third, the 'environmental degradation thesis' understates the complex political, identity and economic provenances of establishing protected areas by prescribing rules that regulate legitimate and illegitimate practices within the designated territories.

From the above arguments, we can conclude that the establishment of modern forms of protected areas in Africa was closely related to the colonial conquest in Africa that partly constituted attempts to exercise political and economic control over African peoples. Although territorial control did not guarantee the full control over the peoples, it was evident that control over resources was tantamount to exercising power over the peoples because it provided the colonial powers with the advantage of restricting the peoples from access to resources. Moreover, through resettlement in the process of creating human-free spaces for wildlife, for example, colonial states aimed to create collectivized villages, where scattered peasants or pastoralists were brought together. This approach was to enhance easy administration, taxation and labor extraction from the subjects. Based upon the existing perceptions and imaginations among the Europeans about Africa prior to the nineteenth century, which was represented as 'natural', 'wild' and 'unspoiled', European explorers, travelers, missionaries and colonial officers of the late nineteenth century and the early twentieth century framed nature conservation within the broader discourses of preserving the 'Edenic' garden (Adams and McShane 1996). Despite some degree of discrepancies between European imagination of Africa as 'natural' and 'wild' on the one hand, and what these early travelers encountered in the continent on the other hand, subtle forms of imposing 'wilderness' on Africa continued throughout the colonial period. Two overt tenets behind colonial environmental conservation policies regarding protected areas were 'threats' of environmental deterioration and aesthetic or ritual concerns "connected with the mental location of 'Edens' and 'Paradises' within various parts of the tropical landscape" (Grove 1990: 18). However, issues of power, knowledge and representation were implicit in these two 'justifications' for the establishment of protected areas in Africa.

Colonial environmental conservation practices were forms of imperial manifestations because they considered customary property rights, indigenous knowledge of resource conservation and customary rules governing resource utilization among the colonial subjects as being incompatible with and inferior to modern approaches of conservation. MacKenzie (1990: 3) noted that, "conservation practices were deeply embedded in the ideologies of imperial rule, and demonstrate its follies and fallacies, limitations and liabilities, as well as its creative concerns." The ascendency of a scientific conservation approach that denigrated indigenous knowledge of conservation created asymmetrical power relations between the colonial powers and the subjects.

The other aspect of the colonial nature conservation scheme was implicit representation of African territories and its wildlife as threatened by 'irrational' and 'backward' human practices and called for intervention through modern conservation techniques. The ambivalent representations of African territories as 'natural', 'Edenic', 'wild' and 'unspoiled' on the one hand, and as a land whose biodiversity was endangered on the other hand, were complementary to each other in the colonial conservation policies. Moreover, again, these were not innocent representations of Africa out of environmental concerns. Rather, Africa has been depicted as 'traditional' in the West's self-image as 'modern' because the more they demoted Africa to the periphery of civilization; the more it was thought to provide colonial powers the self-esteem and legitimacy to install their hegemonic rule (Mbembe 2001). In chapter four, I argued that policies of nature conservation in Africa were part of colonial control that marked the struggle over issues of identity, knowledge and power.

Indeed, it should not be construed as if environmental concerns were not at all part of the conservation endeavors in Africa. Rather, two underlying arguments can be made here. First, there were apparent cases where colonial powers and postcolonial states used the notions of setting aside land in the form of protected areas because it engendered them with the leverage in determining 'legitimate' and 'illegitimate' ways of human-environment interactions, thereby exercising their authority over the peoples for different ends. Compared to other aspects of colonial rule, colonial ecological interventions involved rather profound influences in terms of controlling the peoples, because they entail the exercise of power over the economic, cultural, political and spiritual spaces of the subjects (Grove 1990). Second, although environmental conservation might be the priority of some (state or non-state) actors, the process infringes customary rights of local communities because whether it was within the context of settler colonialism or extraction colonies, enclosure of customary resource areas restricted resource users from utilizing, managing and owning their resources.

Another important point I discussed in chapter four was related to the motives of the colonial powers in establishing national parks in Africa beyond environmental/ecological reasons. Although national parks were established for different purposes based upon the nature of the colonial system, most protected areas were designated for the purposes of political and economic control and as expressions of identity markers of the colonial authorities. Whether it was in settler colonies or extraction colonies and whether it was in British or French colonial Africa, conservation enclosures entailed the exercise of hegemonic forms of power, knowledge and discourse over the colonial subjects. This is why I argued that nature conservation in colonial Africa, which has also been emulated by postcolonial African states, has had a long history of state violence, enabling the state to control the peoples and their resources. While this has been the political dimension of the story, national parks also had economic purposes. For example, they provided exclusive rights to the colonialists in extracting resources (game, forests, fish and others) and in the form of tourism. Moreover, as markers of identity, particularly in British colonies, national parks served as expressions of nationalism and boundary markers between colonialists and the colonies. This was expressed in two ways. First, setting aside land in the form of national parks was part of 'wilderness' domestication because it brought humans' control over the movement of animals, their behavior and reproduction. Secondly, by creating 'legitimate' and 'illegitimate' practices and by delineating between those who had and those who did not have the right to enter, utilize and manage the protected areas, the system contributed to racial and class boundaries between the colonialists and subjects particularly in areas of settler colonialism.

Furthermore, another important point of contestation in the interaction between different actors with competing views of nature-culture relations is their representation of the territory and the notions behind such a representation. Upon synchronizing the theoretical frameworks with the empirical data from the field, the analysis has shown that the meanings different social actors confer to a particular territory emanate from their cultural, economic and political interests as well as expressions of their ontological positions about nature-culture relations. More specifically, while the colonial powers' representation of Africa as 'wild', 'Edenic' and 'aesthetic' or a land with endangered biodiversity was a discursive instrument of colonial control, the Ethiopian state also emulated similar notions in its depiction of territories in the South as 'empty', 'wild' and 'untamed' resources for political, economic and boundary maintenance between the north and south. By contrast, local communities also produce counter narratives and meanings against the dominant discourse. I argue that while Guji's representation of the Nech Sar area as their ritual place, seasonal pastureland, burial site and ancestral home was based upon historical evidence, the people also use these counter narratives as strategies of resistance against the government's resettlement plans. Therefore, I conclude that the notions such as 'wilderness', 'wild', 'empty' and 'uninhabited' on the one hand, and 'home', 'ancestral land', 'spiritual connectedness', and 'life stream' on the other hand are forms of counter knowledge production in the contestation over environmental resources. Although these conceptualizations have ontological roots, they are also derived from cultural, economic and political interests of competing actors.

Nevertheless, the imposition of national park ideals on Africa and the subsequent establishment of conservation enclosures should not be seen as a smooth diffusion of the dominant notions of nature conservation to Africa by displacing traditional knowledge of environmental conservation. Rather, there were iterative processes of interaction between different forms of knowledge, conceptualization and environmental epistemologies. Subsequently, the dominant colonial conservation approach at times triumphed over the traditional approaches because it was backed by the power of science, capitalism and politics. The combined forces of science, capitalism and politics, which were behind the colonial project at large, reinforced the process of control over multidimensional spaces of African peoples (territories, economies, cultures, knowledge, politics and psychological make-up) under the guise of environmental conservation. In some cases within which the colonial system was more of an extractive rather than settler colonialism, the colonial powers allowed the continuation of traditional approaches of environmental management by the colonial subjects, yet used it to enhance their objectives of resource extraction.

It is also important to highlight the dynamics of conservation discourses in Africa across different historical periods and how such dynamics could be positioned within the interplay between local and global environmental discourses, perspectives and practices. Indeed, I concur with the argument that the analytical distinction between the local and the global obscures local communities' responses to forces beyond the local, and it also leads us to dichotomizing global discourses and local conceptualizations of the environment (Gezon 2006). Conservation discourses, objectives and practices passed through different forms of transformation. During the early colonial period when Africa was considered as 'Edenic garden', conservation was framed as a project to protect the 'wilderness' as part of the last reservoirs of the earth's natural places. During the heydays of colonialism, national parks were considered as exclusive spaces as racial and class identity markers in the case of settler colonies and were also designed as institutions of control over the African resources in both settler and extraction colonies. In the postcolonial periods, these spaces have become economic treasures, symbols of sovereignty and fields of contestation between states, non-state actors and local communities. All across these historical periods, different conservation approaches and philosophies have been experimented in Africa. These ranged from complete protectionist approaches, 'dual regulation' and 'community-based' resource management to the ongoing debates on 'back to the barriers' (Hutton et al. 2005). Despite these seemingly different approaches, protected area management in the continent has been common in curtailing local people's customary rights to their resources.

Initially, the ideas, notions and practices of nature conservation were introduced to Africa during the colonial conquest in the form of diffusion from the American and European countryside ideals. However, as the colonial powers encountered different approaches of understanding the human-environmental relations on the part of the colonial subjects, they were at times forced to adopt different strategies to translate their ideas and notions of conservation into practice. This shows that we can conceptualize national park ideals as traveling models not necessarily as one-way flow of ideas and practices but as a two-way interaction in which both sides influence each other. However, the dynamism in notions of nature conservation can be seen as a continuation of conservation enclosure but with different features characterizing its approaches in different periods. Ontologically, early colonial policies of nature conservation were embedded in the notions of nature-culture dualism that resonated around epistemological positions about the dominance of the modern scientific knowledge over indigenous knowledge in the understandings of nature and the place of humans in it (Grove 1990; Garland 2008). In fact, conservation enclosure involved aspects of establishing boundaries between different actors with asymmetrical power relations. As a result, local communities in and around protected areas resisted this form of conservation prompting the colonial officials at times to negotiate with their demands and at times reinforcing harsh administrative measures against those who struggled for their rights. This is exactly what we see in the Nech Sar National Park. In other words, the exclusionist and hegemonic conservation approach installed in the park nurtured the power of resistance on the part of local communities through which they contested external intervention and at times secured access to their customary resources.

9.2 Understanding the Nech Sar National Park from the Center-Periphery Relations in Ethiopia

Understanding the historical and political contexts behind the establishment of national parks in the country is profoundly important in hinting on the reasons, strategies, responses and characteristics of actors involved in the process of setting aside territories as protected areas. It also sheds some light on the nature of power relations between different social actors, namely those who championed for national parks and those who were affected by the process, thereby enabling us to comprehend complex patterns of park-people relations. Likewise, I strongly argue that historicizing conservation enclosures in Ethiopia by positioning the phenomena within the century-old asymmetrical power relations between the center and peripheries is essential to grasp the current contestations surrounding the Nech Sar National Park. Although the history of modern protected areas conservation in Ethiopia dates back only to the 1960s, the fact that over eighty percent of the national parks in the country were located in the south necessitates the need to historicize the phenomena within the political history of the incorporation of this region into the modern Ethiopian state.

Within this section, two major findings in connection to the establishment of the Nech Sar National Park are presented within the gamut of center-periphery relations. It has been ascertained from the discussions in chapter four that Ethiopia shares much in common with other colonial African states in terms of conservation discourses and practices despite the former's exceptionality in colonial history. Indeed, I contend that these similarities can be understood from two angles. First, it should be clear that institutionalization of colonial ideas, discourses and practices are not necessarily a function of colonial occupation. Rather, they travel across political and geographical boundaries and become adopted into mainstream political, economic and cultural practices. Second, like the Europeans' representation of Africa as 'wild', which was used as justifications in the establishment of protected areas, the long established north-south dichotomy that also depicted the south as 'natural' and 'empty' was central in the internal exportation of the hegemonic conservation discourse and practice to the south. Moreover, despite changes in the political order in the country following the institutionalization of ethnic federalism in 1991, the center-periphery relations was not deconstructed particularly in terms of the fundamental identity markers of the Ethiopian state and its representations of the peoples and territories in the peripheries. Consequently, continued state intervention and local responses can be analyzed within the center-periphery relations.

In chapters four and six, I discussed that the establishment of most national parks in the peripheries (namely, in the regions that were historically incorporated to the Ethiopian empire in the late nineteenth century) was part of the state's efforts to control the territories. I contend that the establishment of eight national parks in the historical peripheries (four in the SNNPR, three in Oromia and one in Gambela regional states) out of the ten major national parks nationwide was not merely an ecological choice. Rather, it was part of the political and economic policies of the state and its representation of these parts of the country as 'natural' in contrast to the depiction of the north as 'historical' in synonymous ways with European representations of Africa during the colonial era. As common elsewhere in other African countries during the colonial period, and largely also thereafter, such representation of a territory entails authoritative imposition of certain values, discourses and narratives of dominant actors over others. Notions of wilderness conservation entail the nullification of the history of local communities by imposing narratives and myths of the dominant actors. The 'modernization' and 'civilization' discourse of the state rooted in the legend of 'great tradition' reverberates in the state representation of the territories in the south as 'natural' and awaiting the hands of 'civilizers', namely the state and its mainstream society to be tamed and converted into economically viable resources.

When we talk about development and conservation projects, such as the Nech Sar National Park, it is important to contextualize them within different forms of power/knowledge and representation between the center and the periphery. For instance, the park authority's negligence of traditional resource management approaches of the local communities is far from a mere administrative restriction; rather, it is part of the broader simplification of indigenous knowledge and instead privileging modernist, expert-based and top-down conservation approaches. It was also part of the continuous approaches of the state in broadcasting or extending its power to the periphery by placing the territories under its control through different forms. For example, when the Omo and Nech Sar National Parks were established in 1966 and 1974, respectively, the government literally dispossessed local communities from their customary land under the name of conservation (Turton 2002). Such state encroachment came amidst the national movement when the suppressed peoples across the country rose up against the feudal regime in demand for land right and the right to self-determination. For instance, the peasant movements such as the Bale Oromo rebellion and Gedeo peasant uprising of the 1960s were direct responses to the state hegemony as part of the century-old expropriation of land from indigenous peasants. Likewise, the reactions against state conservation programs in the national parks were also different dimensions of local movement against similar forms of state encroachment in a different context. Indeed, this is why I argue that cattle trespass, hunting, farm land encroachment, forest and fishing grounds by local communities should not be reduced to a mere struggle for livelihood sustenance. As the cases from both Guji and Koore communities in this work show, such transgression of state law part of delegitimizing state authority, as well as their struggle to reclaim their right over the resources.

However, it is worth mentioning that as the center-periphery relations took a different form following the political reordering in the country in 1991, contestations over the Nech Sar National Park also took more of ethnic and regional dimensions. While it officially introduced ethnic federalism thereby departing from the past in terms of official narratives by recognizing the representation of all ethnic groups in the federal system, the new political reconfiguration has also further marginalized the periphery in terms of economic and political engagements. It is my contewntion that despite the political rhetoric of ethnic federalism, decentralization and local autonomy, the core identity provenances of the state reverberate in state-society relations particularly in the periphery. For instance, the ethnocentric representation of agro-pastoral and pastoral communities in the peripheral regions as 'backward' still surfaces in the policies of the EPRDF government. Because the EPRDF inherited the state from its predecessors, with much of its values and narratives about these peoples unchanged, it also strengthened this views when it assigned advisers from the center to advice or cynically control the political, administrative and economic schemes in these regions, for example. The EPRDF has also maintained a boundary between these regions and other regions by labeling the former as 'backward', 'emerging' and 'developing' regions (Dereje 2013). Although the national park is not located within the regions the government labeled as 'emerging regions', it shares much in common with such representation because the economic practices of the people in both cases as similar–agro-pastoralism.

The federal state intervention in the peripheries was initially through political incorporation but later followed by economic encroachment. This can be seen from the fact that for the first time in the history of national parks in the country, the administration of the parks was initially devolved to regional states and their respective lower administrative units in the 1990s. As part of the broader political project, that grants ethnic groups the right to self-government, the decentralization of the management of the national parks was also considered to boost mass mobilization towards the conservation endeavor, thereby contributing toward economic development through tourism. However, the location of the Nech Sar National Park between Oromia and SNNP regional states and its inclusion into the administrative borders of the latter has created tensions over the territory between the two regional states and between ethnic groups that inhabited parts of the territory. In this regard, unlike its predecessors, the EPRDF government systematically decentralized conflict and reserved the power of mediating between conflicting groups to itself.

The claims of entitlement from both regional states and their respective ethnic groups adjoining the national park were indications of the contemporary political landscape in the country partly characterized by (re)activation of latent forms of inter-group hostility that different actors instigate drawing on perceived or real memories and experiences of historical incidents. The central argument here is that although ethnic federalism was presumably presented as a departure from the past by deconstructing the hegemony of the center through decentralization and local autonomy, it exhibited common features with the previous governments regarding issues of property rights of the local communities. This means that throughout the history of the modern Ethiopian state, land and land-based resources became state property despite differences among the regimes in terms of use rights given to the subjects. In this regard, despite its political rhetoric of empowering local communities by granting them 'autonomy' of self-government, the current government is also engaged in a strong expansionist state project that enabled it to successfully engulf the peripheries for the first time in history. The center-periphery relations remain asymmetrically and hierarchically ordered where most of the values of the hegemonic state continue to shape state-society relations such as promoting settled agriculture as the right path to modernization and development at the expense of pastoral and agro-pastoral ways of life.

From the discussions on situating the establishment of the Nech Sar National Park within the center-periphery relations in the country, the findings can be summarized as follows. Firstly, the persistence of north-south dichotomy in Ethiopia built upon the myth that presents the north as a cradle of 'civilization' and the south as 'backward', entails the representation of the territories in the south as 'natural' to be tamed in analogous trend with the European representation of Africa as 'wild'. As a result, the introduction of the dominant notions of nature-culture dualism in the form of national parks has found favorable cultural and political grounds in Ethiopia where successive regimes in the country internally 'exported' it to the subjugated south as part of their nation-building project and extraction of resources through tourism. Secondly, due to the state's consideration of the conquered lands in the south as its uncontested property, local communities' claims for entitlement to their customary land have been perceived as threats to state sovereignty. This is why I conclude that the state intervention and local responses in the national parks micro-level reflections of broader state-society relations in the country and more specifically indications of the asymmetrical center-periphery relations that sustained across different regimes. Thirdly, despite changes in political ideologies across successive regimes, the north-south dichotomy and state intervention in the periphery as a modernist and developmental agent continues.

9.3 Problematizing Participation

Recent trends in the politics of environmental conservation since the 1980s have introduced the primacy of community participation in the conservation endeavors (Hickey and Mohan 2004). This section highlights three core insights regarding community participation, conservation and customary rights. First, it has been ascertained that the often spoken 'community-based' conservation approach that seemingly recognizes the role of local communities in the management of protected areas does not address the complex power relations, interests and responses of different actors. It rather masks hegemonic environmental management approaches under the disguise of participation. Second, I argue that for participation to address the interest of local communities and internal heterogeneity, it should be established on the fundamental tenets of citizen rights. Third, in the Ethiopia's context, participation has been understood and used by government actors as a strategy of channeling government policies down to the grassroots level, as well as an instrument of silencing local resistance. More specifically in the Nech Sar National Park case, the fundamental questions and points of contestation on the part of most local communities were not questions of getting involved in the government conservation and tourism projects. Rather, they questioned the top-down state intervention that has never recognized their rights over the resources, sacred spaces and ancestral homes.

As discussed in chapters two and four, international institutions such as the World Bank, IMF and IUCN, and donor countries began adopting the so-called community-based resource conservation as their major agenda in the requirement for funding such projects in developing countries. As part of the Structural Adjustment Programs of the 1980s and early-1990s, the World Bank and IMF made economic diversification including ecotourism part of their preconditions for foreign aid to developing countries. Influenced by global dynamics of the 1980s and 1990s such as the International Indigenous Peoples' Movements and the transformations in social sciences with the ascendency of postmodernist thinking, these institutions and other international actors adopted what has been commonly called Community-Based Conservation approach (Hutton et al. 2005). It was after facing local resistance and experiences of failure in the modernist discourses that these institutions seem to have been prompted to reassess their failure in imposing modernization discourses upon societies with different cultures, worldviews and ways of living. It should be acknowledged that this approach has brought at least recognition that conservation efforts without the participation of local resource users would remain difficult to achieve its goals. In some cases, like Southern Africa, the colonially inherited "state-centric 'fortress conservation' gave way to post-colonial or 'new' conservation where local residents in Southern Africa dependent on natural resources locked within national parks were allowed to participate in the management of the parks" (Singh and Van Houtum 2002: 255).

Nevertheless, as discussed in chapter two, the philosophy behind the so-called community-based nature conservation should be scrutinized, because environmental policies under the current neoliberal market economy are apparently shaped by the approach of marketing nature in which environmental resources and institutions that govern them are fundamentally considered from economic dimensions. Neoliberal environmentality reinforces the assumption that the sustainability and success of global markets are fundamentally linked to the welfare, capability and receptiveness of remote communities in the South whose engagement in environmental management and protection in their locale would enhance sustainable flow of global market and consumption patterns (Singh and Van Houtum 2002; Agrawal 2005; Fletcher 2010). In other words, involving local communities into the hierarchically ordered, colonially inherited and epistemologically hegemonic forms of conservation represents a strategy of making environmental subjects.

Community participation in this context engenders localized patterns of control over the peoples and their resources, and further disempowers resource users. Most community-based conservation practices in Africa exhibit nominal participation, whereby a few members of local communities or elites have been brought into projects already instrumentally planned and designed by the state or non-state actors. The state and non-state conservation organizations use these forms of participation for mass mobilization towards development or conservation projects and stifle resistance by co-opting different sections of local communities. As argued, one of the fundamental criticisms against the notion of community-based conservation is its consideration of 'community' as a homogenous group, according to which the involvement of few members is presumed to represent the interest of others. Nowhere can we find a purely homogenous community, because a human society has different interests, behaviors, understandings and reactions to its immediate environment, with such diversity further compounded by differences in gender, age, education, livelihood styles, ethnicity and political affiliations.

Moreover, involving local communities within existing hierarchically structured ways of doing things where the state and other non-state actors occupy central roles in deciding on the rules of the game in the participation rhetoric does not guarantee community empowerment. Under such circumstances of asymmetrical power relations between different actors, powerful actors like the state decide the fundamental questions on who participates, at what level of the project the local peoples participate, what outcome expected from the participation and whose knowledge plays fundamental role in the participation. Therefore, community participation is often a tool to legitimize external interventions.

At this point, it is important to interrogate what community participation entails in the Ethiopian context. In Ethiopia, the Federal Constitution provides a political space for peasants and pastoralists to participate in any development program that affects their community (FDRE Constitution 1995: Article 43.2). By contrast, the developmental state paradigm that the EPRDF government adopted limits community participation in many ways. The former Prime Minister Meles Zenawi who was the architect in introducing this political and economic discourse to Ethiopia once acknowledged that developmental state paradigm might suppress democracy and popular participation for the sake of enabling the state in its capacity to deliver development schemes (Meles 2012). This political and economic paradigm gives uncontested right and power to the state to plan and implement policies, and to mobilize citizens for the same purpose. It is elitist in nature because the perspective considers local communities as subjects to be led by 'developmental' elites whereby the knowledge and views of these elites (state and non-state) are considered as the right path to development. The ambitious Growth and Transformation Plan (GTP) of the government designed for the periods 2010-2015 also envisages economic growth and social development by maintaining and strengthening developmental state approaches (GTP Document 2010). Although it is not my prime objective in this study to assess the contents of the GTP, it is evident that elitist and top-down approaches are visible from the preparation in implementing this ambitious plan.

Local government officials are socialized into the discourses and jargons used at the macro level such as rent-seeking, developmental farmers and developmental state without properly understanding the concepts. Later on, I came to understand during my fieldwork that neither local government authorities nor ordinary members of the communities took part in the discussions launched after the government released the GTP document. Government cadres at different levels often use top-down approaches of inculcating the ambitious development plan into the society and consider it as participation. Put briefly, community 'participation' in Ethiopia is used as part of sensitization of the subjects in the process of implementing government policies, strategies and programs. In other words, community participation is the rhetoric of gaining legitimacy on the one hand, and a technology of establishing local regulatory community on the other. In environmental conservation schemes, the politics of participation can be interpreted as one of the techniques of making environmental subjects.

Regarding the Nech Sar National Park and issues of local communities surrounding the park, the central question to be interrogated is whether participation into the government project was/is their agenda of contestation. Put differently, we should question whether the local communities struggle to get involved in the government project or resist the project itself. This question is very important because most conservation studies on issues of local community participation reduce the struggle over national parks to competition over environmental resources for economic purposes (Igoe 2004). Here, it is difficult to give a straightforward answer to whether or not questions of participation were the agenda of the local communities. There were also differences in interests and reactions towards government interventions owing to internal diversities. However, the baseline argument is that the fundamental question for the majority of the Guji and Koore men, women and youth, has been access and entitlement to the territory for different purposes other than conservation for tourism benefits. This means that most of the local people held the view that rather than participating in the conservation project that has been designed for environmental and economic purposes according to the government perspectives of conservation and development, they would prefer if the government negotiates on devolution of state property rights. By state property rights, I mean the customary right of the people over the land that the state appropriated and then institutionalized the land as state property.

Questions of property right that the state expropriated from the communities through state-centered conservation and development projects in the national park remain critical points of contestation between the local communities and the state, and at times between the local communities. Although most people complain about lack of benefit from the national park over the last forty years, such opinion was mainly an expectation generated from the park authorities' rhetoric of conservation for the benefit of the people. I conclude that rather than a question for devolution of management structure of the national park to the community level in which the local people would participate in the state project, the central demand of the Guji and Koore in and around the park was at least to get partial devolution of property right. This means that their priorities include entitlement to critical resource areas and cultural spaces in the park that would ensure their economic, cultural and spiritual interests. This is where the contradiction between the state visions of conservation and development and the local communities' articulation of the same project within their own contexts of knowledge, values, culture and economic practices emerges.

The discussions in the chapters 3, 6 and 7 show that there were vivid differences in representation of the territory. While the state considered the land as a space to be domesticated, commoditized and marketed through conservation and ecotourism, the local communities viewed it as a territory for livelihood sustenance and as a sacred space for cultural and religious practices. Moreover, the state's normative representation of pastoralism as a backward way of life and its obsession with settled agriculture as a path to modernization and development ignores the fact that pastoral mobility is often a strategy of resource conservation and a response to ecological stress in arid and semi-arid parts of the country (Dereje 2013). Although the government has started large-scale projects that collect pastoral communities into sedentarized villages, the social, cultural and economic consequences of such abrupt shift in ways of life should have been taken into consideration in any policy programs.

As I discussed in the preceding chapters (chapters 4, 5 and 6), the identity of the Ethiopian state has been built upon the cultural, mythical and historical self-representation of the highland, agrarian, Orthodox Christian and Amhara-Tigrean people that constituted the center in the Ethiopian Empire. The state continues to suffer from its Amhara-Tigrean centrism. It should be underlined that the local people had many reasons for resisting the government's sedentarization program. First, there are contradicting notions of pastoralism, as a backward way of life that obstructs development and modernization, as well as a viable livelihood that ensures resource conservation on the part of the state and the pastoral communities, respectively. On a practical level, the people resist the government sedentarization programs due to the experiences and hardships they faced in the past through displacement and relocation. For example, in the case of the Guji Oromo, their experience of state induced displacement and resettlement practices from the past has informed their current attitudes of mistrust to the government. Therefore, given the memory of the people about socio-economic hardships under state induced displacement and resettlement programs and due to the limited recognition of community voices regarding their views on conservation, development and resettlement programs, it would be difficult to expect the local people to compromise their customary resources for the sake of state projects. The state should first build legitimacy and trust before it designs projects that would dismantle the local people from their home. Moreover, for local communities to accept state projects like the national park, the government could have designed alternative options through genuine participatory approaches.

In addition, the issues of devolving the state's administrative and property rights are quite different scenarios, which the government was not ready to negotiate, particularly concerning the property right. While administrative devolution has been nominal in the case of the Nech Sar National Park, aspects of devolving the state's property right that would grant local communities the rights of ownership, utilization and management of parts of the territory seem almost unreachable demand within the contemporary developmental state paradigm and the political context in the country. In fact, questions of property right devolution might bring another critical question in the debate on conservation and customary rights. With an increase in human and livestock population from time to time on the part of the Guji, and continued downslope expansion of the Koore to the contested Tsalke Valley, devolution of property rights would inevitably increase the competition over the resources in the national park. On the other hand, under existing conditions where the government intensifies the enclosure and restriction of local resource users from access to their customary resources, and also given the past experiences and memories of the people under state resettlement programs, the national park poses continuous threat to the survival of the local communities. Moreover, the government's emphasis on enclosure and resettlement without comprehensive plans regarding how the local people would cope with changes in their lifestyle prompts us to pose another serious question. For example, given the inaccessibility of medical services for their livestock, the local people prefer to keep large number of animals to ensure that at least a few will be left in times of uncertainty. Besides the cultural values of livestock, the Guji oppose the government officials' proposal of reducing the number of herd each household would possess owing to uncertainties related to disease, local climate variability and the infertility of animals. Therefore, the Nech Sar National Park will remain to be a serious threat to the local communities and the biodiversity in the enclosure undelss the government opts towards inclusive plans and particularly rights-based approach of environmental conservation and development programs that recognizes the rights of local communities to the land or fair treatment in the process of soliciting alternative options.

This disparity between the state's conceptions of the territory, life styles of local communities and its monopoly of property right over the resources on the one hand, and local communities' counter claims of entitlement, on the other hand, prompted the latter's use of different strategies in reclaiming their rights. I conclude that despite the state's monopoly of coercive power in the management of the national park, local communities, particularly the Guji Oromo, were capable of challenging the state intervention employing different pragmatic strategies. Their agency in networking with government authorities, using everyday forms of resistance and resorting to some innovative approaches such as planting fruit trees and building social services to claim property rights over the land are good examples how local communities persistently challenge external intervention. The Koore, on their part, exploited the park administration's stereotypical representation of the Guji as threats to the park. By playing what I called *blaming game* over the Guji and being strategically allied to the government in its efforts to establishing local regulatory community, the Koore were able to maintain their control over parts of the Tsalke Valley. However, in both cases, the demands of local communities were far beyond nominal participation in the national park project for getting economic benefits through tourism.

Put briefly, the so-called community participation, empowerment and benefit sharing practices have masked the hierarchical power relations and the exercise of control by the state and non-state dominant actors. The government's involvement of few local community members in the administration of the national park shifted the field of contestation from state-society relations to intra-community or inter-group conflicts and sense of mistrust. Accordingly, the local communities began challenging members of the regulatory committee who were assigned as local government agents in its efforts to control the territory. By doing so, the state extended its control over the people and the resources without compromising on property rights regimes it introduced. It can be concluded that the rhetoric of community-based conservation approaches has been used by conservationists, business organizations, state agencies and tourism organizations as pragmatic strategy of stifling local resistance and as covert exercise of hegemonic environmental conservation regime rather than for the purposes of empowering local communities.

9.4 Theoretical and Policy Considerations

Regarding discourses and practices of nature conservation in Africa in general and particularly with reference to the Nech Sar National Park in Ethiopia, this work sheds some light on epistemological, theoretical and policy considerations. While the epistemological and theoretical aspects of the discussion help us in better understanding the complex human-environmental interactions, external interventions and local responses, the policy considerations contribute towards the formulation of balanced environmental conservation policies that would ensure both the sustainability of biodiversity and the wellbeing and customary rights of local communities. In the following sub-sections, I will summarize the themes drawing on discussions and analysis in the preceding chapters.

9.4.1 Epistemological/Theoretical Considerations

The analysis of my empirical data and theoretical approaches hints that environmental conservation discourses and practices advocated by state and non-state conservationist organizations largely ignore the multiplicity of knowledge and ways of knowing human-environmental interactions. For example, the emphasis on the modern scientific knowledge as an ultimate source of knowledge is an epistemological notion rooted in positivist thinking that obscures multiple interpretations of a phenomenon or object of knowledge and ways of knowing it. Colonial and postcolonial states adopted this notion of knowledge production, thus denigrating the voices, conceptualizations and ways of knowing of local communities in the human-environmental interactions. Therefore, social constructionist perspective can be used as an epistemological approach in the study of human-environmental interactions because it gives us the advantage of opening our spectrum of investigation for multiple voices and interpretations.
From theoretical dimensions, debates on nature conservation in Africa have rarely positioned the notions behind national park ideals within the broader conceptual discourses on nature-culture relations. Some scholars asserted that the genesis of the modern forms of nature conservation in the continent was linked to the colonial imposition without or with limited focus on how the interplay between different forms of knowledge shaped the conservation trajectories (Neumann 1998; Brockington et al. 2008). On the other hand, others simply relied on the victimization approach–focusing on the impacts of conservation projects on local people and/or the biodiversity itself (Igoe 2004).

Nevertheless, it is my contention that without relocating the notions behind the establishment of national parks within the historical genesis of wilderness conservation, which was part of the European and American imaginations of nature, debates concerning why people are restricted from access to resources in national parks and why indigenous knowledge of resource management is denigrated remain blurred. Although these debates are not new, I underline the salience of this nature-culture dichotomy in influencing policies and practices under the general framework of 'saving the world's endangered biodiversity' that in turn considers local resource owners as threats to the environment. Moreover, the tendency in many academic works to depict the postcolonial nature conservation discourses and practices as mere continuations of the colonial legacy, on the one hand, and resistance from local communities against such external interventions or impositions, on the other, underplays different social dynamics and value changes that occurred through the interplay between competing forms of knowledge. Likewise, reducing local resistance against state conservation projects to struggle over environmental resources for economic purposes is simplistic, given that most local resistances have been part of the general resistance against hegemonic forms of state discourses of governance, development and conservation interventions. Rather, local communities' contestations and negotiations in terms of external interventions should be positioned within broader state-society relations where both aspects of power (dominating and resisting power) interact through the medium of different actors.

Therefore, there is major scholarly gap in the understanding of contemporary nature conservation in Africa, which is to be conceptualized as a product of contestations, negotiations and appropriations of perspectives and practices within the interplay between competing forms of knowledge. Likewise, the empirical discussions in the preceding chapters illuminate that meanings, narratives and knowledge about nature and humans' relation to nature are contingent to cultural, political and economic contexts that shape the representation of a particular territory. Therefore, it is feasible to conclude that meanings and knowledge about environmental problems are socially constructed; thus, there should be recognition of multiple interpretations of a phenomenon to understand why a certain phenomenon is interpreted differently in different sociocultural contexts.

However, the findings of my research depart from the dualist and mutualist perspectives of nature-culture relations. While the dualist perspective presumes humans and non-humans as disjoint and distinct entities, thereby promoting the isolation of the two for the protection of nature (non-humans), the mutualist notion also posits indigenous peoples as 'close to nature'. The latter understanding of nature-culture conjoint constitution entails the representation of the ways of life and worldviews of non-Western communities as inseparable from non-humans. Rather, the findings show that it was not the inseparability between humans and non-humans per se that fundamentally govern the people's understandings of their surroundings but the myriad relations built upon respect, reciprocity and trust. Particularly among the Guji Oromo, human interactions with its surroundings are governed by the people's eco-cosmology that constitutes and embodies values of respect, reciprocity, trust and partnership. Therefore, it is not because they co0nsider themselves as part of the wildlife or forests or the lakes that the Guji care for the environment; rather it is because of the fundamental values embedded in their belief, culture and worldviews about human place in its environment.

Moreover, by employing political ecology framework, particularly the 'conservation and control thesis', this work has ascertained that the politics of nature conservation advocated by state and non-state business/conservation actors entails the exercise of hegemonic forms of knowledge and governance that control local people's access to their customary resources. In contrast to many works in political ecology that discuss protected areas within the 'environmental degradation thesis' or those that situate local communities as victims of conservation enclosures, this work addresses the problems from a different angle. This work conceptualizes national parks as fields of contestation and negotiation between competing actors where the interplay between dominating power and resisting power shape the approaches of nature conservation in the field. By synchronizing political and economic motives of actors with environmental concerns and at the same time by using an actor-oriented approach where local communities are understood as active agents in influencing the course of interaction, the research contributes to the general debate on the politics of environmental conservation.

9.4.2 Policy Dimensions

Apart from the complex state-society relations in Ethiopia with asymmetrical and north-south dichotomy, the post-1991 political reconfiguration of the country did not give precise policy frameworks concerning how to designate and administer national parks that straddle regional borders like the Nech Sar National Park case. A proclamation was only promulgated to centralize such territories under the federal state in 2007 (FDRE Proclamation No. 541/2007). In other words, the 2007 proclamation itself brought the centralization of national parks rather than opting for more devolution of power and property rights to local administrative units and local communities. The proclamation further rolled back the nominal decentralization regarding resource administration, including national parks.

The proclamation has further restricted community engagement in the administration and ownership of conservation areas, because it proclaimed that local communities are entitled to manage only areas not under the federal and regional states' conservation schemes. It reads, "Wildlife habitats other than the conservation areas referred to in Articles, 4, 5, and 6 [conservation areas to be administered by the federal government, regions and private investors respectively] of this Proclamation may be authorized by the concerned regions to be administered by the local communities" (FDRE Proclamation 541/2007: Article 7). This implies that the proclamation does not give the local communities rights of participation in the management, utilization and ownership of wildlife areas that are designated as national parks, reserves and sanctuaries that are administered by the federal and regional states or private investors. This leads us to the understanding that the are disparities between rhetoric of community participation, empowerment and decentralization advocated by conservation organizations, international institutions such as the World Bank and national states.

At a broader level, one of the major drawbacks of international development interventions in developing countries is the consideration of the state as an utmost legitimate and trustworthy institution to lead intervention projects. Even the neoliberal and postmodernist discourses that advocate for limited state involvement have sustained the power-laden concepts such as empowerment, participation, decentralization, devolution of property rights regimes and capacity building. Implicit in these concepts are issues of power and representation, whereby the state or other non-state actors are presumed to be rightful holders of power, property rights and knowledge, which they would share to local communities who are conversely portrayed as subjects without power, property rights, knowledge and capacity. Regarding empowerment, for example, the concept implies a differential power and knowledge relations between those who 'empower' and those who are presumed to be 'empowered' (Nawaz 2013). Likewise, the neoliberal rhetoric of decentralization, local community participation and capacity building downplay the agency of local communities in development interventions, unless the state or non-state actors activate their capability (Mohan and Stoke 2000).

My central point of argument in this regard is that development and conservation projects operating within the hierarchical power relations without recognizing the knowledge, worldviews, perceptions, internal heterogeneity and agency of local communities replicate the classical top-down and exclusionary practices, even within the umbrella of community participation, empowerment and decentralization. On the contrary, it has now become difficult to easily silence and put local communities under the hegemonic forms of knowledge and power, because there have been evident cases that local communities maneuver resources at their disposal (capital, human, information, technology and network) to change any aspect of interactions with the state or non-state actors to their advantage. As the discussions in the chapters seven and eight have shown, the local Guji Oromo in and around the national park challenged state versions of conservation from different dimensions. Ontologically, they challenged notions of nature-culture dichotomy; practically, they contested to deter or undo the implementation of conservation projects that were alien to them.

At this point, I would like to make two points clear regarding buzzwords such as participation, decentralization and empowerment, with a particular focus on the Nech Sar National Park. Firstly, I reiterate that state-society relations in connection to the national park have been marked by the persistence of hierarchical power relations, hegemonic forms of knowledge and exclusionist resource management approaches. Under the successive regimes since its establishment, the national park has been administered under the classical 'fortress' conservation approach, characterized by fines and fences. 'Expert'-based conservation practices whereby park biologists, rangers and other administrative staff strived to impose environmental knowledge that were alien to local communities, such as banning the use of fire in the park and separation of humans from the wildlife, hint that state projects of environmental conservation reflect hegemonic forms knowledge. Moreover, apart from a short period when the African Parks Foundation was involved in the management of the national park, there was no evidence of the integration of neoliberal and disciplinary environmental governance into the management approaches to genuinely accommodate the interests and demands of local communities, at least within the conservation-ecotourism context. The government recruited a few members of the communities as local regulatory committee, through whom it attempted to control the people. By doing so, the government aimed at inculcating modern conservation discourses and creating environmental subjects through coercive approaches. This was why the local communities considered the national park administration as an alien entity that dispossessed them from their customary resource rights without any visible trickledown effect.

Secondly, although involvement in the management of the national park within the context of state conservation and development visions was not a priority for the majority of the local communities, and particularly for the Guji and Koore, the government fails to engage the local people. It would have been possible to create sense of trust and legitimacy of the state on the part of local communities if genuine and interactive forms of participation had been introduced. Interactive participation would entail the establishment of forum of engagement or dialogue, within which both the state and local communities would debate, contest and negotiate conservation and development discourses, practices and benefits on a balanced level of power relations. Contrary to this, the government opted to involve few local elites within the asymmetrical power relations. By nullifying the history of their settlement in the area, thereby labeling them as illegal settlers, the government established an unbalanced platform for any discussion with local communities. Therefore, the government's failure to engage the people in its project and its history of denigrating the customary rights of the people over the land have reproduced a sense of hostility and mistrust on the part of the local communities.

Similarly, without showing the people any practical benefit from the national park that could be in the form of establishing social services, providing fair employment opportunities and alternative options of livelihood that would ensure the people economic and cultural survival, the government's efforts to convince local communities to sacrifice for the sake of 'national interest' would be naïve. It is difficult to expect the local communities to compromise their interests for the sake of a state project that they consider as an intervention and a threat to their survival for two major reasons. First, given the current circumstance in which local communities have no practical reality on the ground to trust the government's rhetoric of conservation-tourism-development nexus, the government would face difficulties in developing trust on the side of the people. Second, because of their experiences and memories of past state-led evictions and dispossessions, the local people, both Guji and Koore are insecure and uncertain about life in the resettlement sites. Moreover, local involvement in the form of being recruited as park rangers and local regulatory committees does not indicate the complex internal power relations within the communities and their attitudes to the park. Members of local communities might participate in the state project out of fear of state coercion, or alternatively as a survival strategy. Such involvement is rather part of co-opting local elites as a technique of control over the people and their resources and does not build sense of trust on the part of the people.

Another important consideration that most works on protected areas conservation overlook is positioning conservation within multidimensional aspects of rights (economic, social, cultural, citizenship and political rights) of local communities. Above all, a rights-based approach to conservation entails the recognition of local communities' rights to socio-cultural, religious and economic practices and citizenship rights for participation in decisions that affect them. It is far beyond the rhetoric of nominal participation because the rights-based approach to nature conservation constitutes transformative political system that ensures local resource control, decision making and the ownership of resources and their knowledge in the overall planning, management and utilization of environmental resources without infringement upon their fundamental rights. A rights-based approach to conservation and development considers participation not as a provision by the government or any non-governmental actor out of free will, but rather as citizens' right, which would guarantee their engagement in dialogues with other contending actors to ensure the maintenance of their political, economic, cultural and social rights. My point of contention regarding the Nech Sar National Park is that conservation schemes have been practiced in isolation from the rights of the people. Accordingly, the central impasse of protectionist conservation approach is its negligence to local knowledge and livelihood of the local people and detachment of the cultural and spiritual connectedness of the people from different landscapes or territories.

Indeed, the state also has the sovereign right to ensure the wellbeing of its citizens and sustainable management of the environment, albeit not through infringement upon the rights of the citizens.

However, for rights-based approaches to be translated to practice, the political conditions should be responsive to citizens' demands for equal representation and entitlements (Hickey and Mohan 2004). According to Hickey and Mohan (2004: 66), "relocating 'participation' within citizenship analysis situates it in a broader range of sociopolitical practices, or expressions of agency, through which people extend their status and rights as members of particular political communities, thereby increasing their control over socio-economic resources." Citizens' control over socio-economic resources would not be feasible without transformative political processes that grant citizens the right of ownership over environmental resources within a political framework whereby they hold the government accountable through legal "provisions which in various ways make participation in governance a legal right" (Gaventa 2004: 30). A rights-based approach and the ensuing participation are different from the rhetoric of decentralization and community-based participation advocated by states and non-state actors because, as Gaventa (2004: 32) contends, even "democratic decentralization simply opens up space for the empowerment of local elites, not for consideration of the voices and interests of the more marginalized. Obstacles of power, social exclusion, and minimal individual and collective organizational capacity mean that few gains will be made by the poor." Thus, from the discussions in chapters one and two, as well as what Gaventa (2004) says, we can conclude that the politics of environmental conservation should be positioned within genuine transformative political systems, within which citizens are considered as decision makers rather than being perceived as mere recipients of state policies. In the conservation trajectories in the Nech Sar National Park, successive governments have failed to recognize the rights of local communities to the land and thus resorted to the exclusionist approaches of conservation.

The case from the Nech Sar National Park illustrates a glimpse of nominal participation, which can be categorized as co-optation because it neither addresses internal diversities within local communities nor recognizes the agency of the communities in determining the course of their life. This is why the conservation and ecotourism development endeavors of the government have faced fierce resistance and failed to accomplish its missions of biodiversity conservation, as well as ensuring tourism development to the level of the government's expectations. From the preceding discussions, two seemingly contradictory positions have surfaced out regarding environmental conservation and concerns to customary rights of the local people who historically lived in and around the

Nech Sar National Park. For example, despite the local people's narratives of traditional co-existence with the wildlife, different social dynamics among the people and their surroundings show that a call for human-non-human co-existence overlooks the utilitarian and business orientations among the people, which would threaten the biodiversity. This means that because most members of the local communities are socialized to the economic dimensions of human-environmental interaction, the traditional human-wildlife coexistence that was built upon respect, reciprocity and trust as part of their local eco-cosmologies seems infeasible today. On the contrary, the 'modernist' and 'conservationist' call for the complete separation of humans from the protected areas, infringing upon the fundamental human rights of the people, including their right to sacred spaces, economic resources and, more importantly, their ancestral rights over the land.

Reconciling the above contradictory demands and visions remains difficult within the current dichotomized state-society relations where the state imposes its ideologies, policies and practices upon the society. Likewise, under circumstances where the society mistrusts the state and considers its policies and programs as interventions that threaten their lives rather than improving their living conditions, it is also equally difficult to expect a compromise from the local communities. Therefore, to establish trust on the part of the society whereby they would consider state development or conservation projects as something that would benefit them rather than something that dismantles them from their land, there should be a fundamental paradigm shift in state-society relations in the country. The old tradition of considering local societies as passive recipients of state policies and practices should be deconstructed; rather, an approach that recognizes their knowledge, rights and agency should be introduced, although it might not bring an overnight solution to the current problem. The challenge in the current political disposition in Ethiopia is the government's obsession with a developmental state paradigm that largely privileges a strong state at the expense of local community participation. Unless the government comes to terms with the fundamental demands of the local communities–the need for recognition of their historical right over the land and the viability of the territory for their socio-cultural and economic existence–the notion of representing the territory as 'uninhabited' would feed up to the persisting competing views and interests. Such recognition is important because it would put both parties on equal footing in the negotiation, though the government should again work on creating trust so that the people would believe that its projects are genuinely for their benefit too.

Another failure of the government rooted in the broader state's conceptualization of the territory as empty before the establishment of the national park is that it has not been showing the local communities the benefit of the national park through tourism. The only practical benefit the local people (Guji, Koore and Arba Minch town dwellers) got from the national park was during the short periods of the African Parks Foundation (APF) from 2004-2007. During this period, the Koore households who were relocated by the government were provided with limited financial and material supports from the APF. Moreover, the company created temporary job opportunities for nearly one hundred women in Arba Minch town. The Guji on their part recall medical supports they got for their people and livestock from the company. Nevertheless, nothing has been trickled down in the form of material support, social service provisions and financial contributions from the government as incentives to build sense of trust and belongingness on the part of local communities.

9.4.3 Issues for Future Research

As one would expect, this work does not comprehensively address the establishment of national parks in Africa with specific details based upon the nature of colonial policies and the specificities of each national park within the complex ecological, economic, political and social settings. It is with a clear understanding of the differences between the major colonial powers such as Britain and France in their policies of nature conservation, for example, that I have aimed to present a general overview of conservation discourses during the colonial era. Similarly, differences in colonial systems such as extraction and settler colonialism also have had significant contributions in terms of exercising varying approaches of nature conservation. Although both had similarities in excluding native Africans from their customary resources, it would be simplistic to group both forms of colonialism within the same analytical approach in the studying the politics of conservation and local responses. Therefore, for a comprehensive understanding of the nexus between colonial forms of national parks management and the postcolonial approaches adopted from the past, historicizing and contextualizing the phenomena within particular historical and political contexts becomes the agenda of further research.

In the case of Ethiopia, this work has also not covered all national parks in the country. Rather, my approach was purposive and involved taking a single national park and situating it within broader debates on nature-culture relations in Africa, as well as in center-periphery relations in Ethiopia. On the other hand, because national parks in Ethiopia also have specific differences based upon their locations, history of establishment, relations with neighboring communities and changes in approaches of conservation, it is difficult to generalize this work to other cases, except on common issues. For instance, regardless of specific differences, the foundation of conservation approaches of national parks in the country was the dominant protectionist or 'fortress' conservation strategy where local communities were excluded from national parks. However, for a full-fledged understanding of the politics of nature conservation, environmental problems, conservation induced crisis and changes in human-environmental interactions in the country, it becomes very important to undertake the research in each national park and concerning all actors involved. This would help policy makers to formulate conservation policies based upon empirically supported evidence rather than through abstract judgments.

Finally, some comparative values can be drawn from this study. First, this work contributes to the general debate in political ecology of conservation by highlighting the role of diverse actors and institutions in environmental problems, as well as by challenging the traditional approaches that used to put the blame on resource users for environmental problems, and yet ignoring their role in redressing it. That means, regardless of local differences, argument that issues of environmental conservation should be understood within the complex political, economic, ideological and ecological contexts involving actors with competing interests can be taken as be taken as a baseline framework in the study of politicized environments elsewhere. Second, although one might not get similar finding by using the same approach in a different context, this study gives an important framework with regard to local responses to modern conservation practices particularly in developing countries. In this regard, it contributes to the body of knowledge in grassroots communities' resistance as it challenges the binaries of domination and resistance interface. Rather, as power constitutes domination and resistance, it has been evident from the findings of this work that hegemonic knowledge and dominating power that the state exercised nurtured resisting power on the side of local communities that was translating into different forms and strategies of resistance. Third, the core thesis in this work is that modern forms of environmental conservation in the form of protected areas are forms of control over resources and people who own, utilize and inhabit these territories. As a result, the 'conservation and control thesis' provides an important analytical framework in the understanding of how state and non-state actors exercise power in order to control the environment and the people for the sake of achieving economic, political and environmental goals. Fourth, this study on the Nech Sar National Park is about not only state conservation projects and local conceptions of the programs. It is also a mirror to state-society relations in Ethiopia and gives us insights how the Ethiopian state represented its subjects particularly those in the periphery across different regimes. Therefore, the center-periphery approach I used in this study can be used to analyze other issues pertaining to state intervention and local responses in the country.

BIBLIOGRAPHY

Aalen, Lovise (2011): *The Politics of Ethnicity in Ethiopia: Actors, Power and Mobilization under Ethnic Federalism*. Leiden: Brill.

Aalen, Lovise (2008): *Institutionalizing the Politics of Ethnicity: Actors, Power and Mobilization in Southern Ethiopia under Ethnic Federalism.* PhD Thesis, Department of Political Science, Faculty of Social Sciences, University of Oslo, Norway.

Aalen, Lovise and Kjetil Tronvoll (2009): The End of Democracy? Curtailing Political and Civil Rights in Ethiopia. *Review of African Political Economy* 36 (120): 193-207.

Abbink, Jon (2009): Ethiopian Second Republic and the Fragile 'Social Contract'. *Africa Spectrum* 44 (2): 2-38.

Abbink, Jon (1997): Ethnicity and Constitutionalism in Contemporary Ethiopia. *Journal of African Law* 41 (2): 159-174.

Abiyot, Negera (2009): Resettlement and Local Livelihoods in Nechsar National Park, Southern Ethiopia. Unpublished MPhil Thesis, University of Tromsø, Norway.

Adams, Jonathan and Thomas McShane (1996): *The Myth of Wild Africa: Conservation without Illusion*. Los Angeles: University of California Press.

Adams, William and Jon Hutton (2007): People, Parks and Poverty: Political Ecology and Biodiversity Conservation. *Conservation and Society* 5 (2): 147-183.

Adkin, E. Laurie (1998): Ecological Politics in Canada: Elements of a Strategy of Collective Action. Keil, R., David V.J Bell, P. Penz, and L. Fawcett (eds.): In: *Political Ecology: Global and Local*.. London: Routledge, 285-316.

African Parks Foundation Annual Report (2007): http://www.african-parks.org/xMedia/PDF/AnnualReport/APN_AnnualReport_2007.pdf (Accessed on August 20, 2011).

Agrawal, Arun and Kent Redford (2009): Conservation and Displacement: An Overview. *Conservation Society* 7 (1): 1-10.

Agrawal, Arun (2005): *Environmentality: Technologies of Government and the Making of Subjects*. Durham and London: Duke University Press.

Aimé, González (2008): Fragile States and Neoliberalism in Sub-Saharan Africa. FRIDE, Comment, African Studies Group, University of Madrid.

Anderson, Benedict (1991): *Imagined Communities: Reflections on the Origin and Spread of Nationalism*. London: Verso.

Anderson, David and Richard Grove (eds.) (1987): *Conservation in Africa: Peoples, Policies and Practice*. Cambridge: Cambridge University Press.

Appadurai, Arjun (1981): The Past as a Scarce Resource. *Man New Series* 16 (2): 201-219.

Archard, David (1990): Paternalism Defined. *Analysis* 50 (1): 36-42.

Aregawi, Berhe (2009): *A Political History of Tigray Peoples Liberation Front (1975-1991): Revolt, ideology and mobilization in Ethiopia*. New York: Tsehai Publishers.

Arthur, John W. (2002): Pottery Use-Alteration as an Indicator of Socioeconomic Status: An Ethnoarchaeological Study of the Gamo of Ethiopia. *Journal of Archaeological Method and Theory* 9 (4): 331-355.

Debelo, Asebe Regassa and Anne T. Kjørholt (2013): Childhood, Food Culture and Eating Practices among the Guji. In: Abebe, Tatek and Anne Trine Kjørholt (eds.): *Childhood and Local Knowledge in Ethiopia: Rights, Livelihood and Generations*. Tapir Academic Press,173-192

Debelo, Asebe Regassa and Meron Zeleke Eresso (2014, forthcoming): Irrecha: The Traditional Oromo Ritual from Local to Global–the Transnational Dimension of Irrecha. In: Adogame, A. (ed.): *Imagining the Religious 'Other': The Public Face of African New Religious Movements in Diaspora*. Ashgate Publishing Press.

Debelo, Asebe Regassa (2012a): Dynamics of Political Ethnicity and Ethnic Policy in Ethiopia: National Discourse and Lived Reality in the Guji-Gedeo Case. *Journal of Oromo Studies* 19 (1 & 2): 1-36.

Debelo, Asebe Regassa (2012b): Whose Conception of Nature? Conflict between State and Local People over Nech Sar National Park, Ethiopia. *African Journal of Governance and Development* 1 (2): 29-46.

— (2012c): Contesting Views on a Protected Area Conservation and Development in Ethiopia. *Social Sciences* 1 (1): 24–43.

— (2012d): Emerging Ethnic Identities and Inter-ethnic conflict: A Study on Guji-Burji Conflict in South Ethiopia. *Journal of Studies in Ethnicity and Nationalism* 12 (3): 517-533.

— (2011a): Contrast in the Politics of Recognition and Indigenous People's Rights. *AlterNative: An International Journal of Indigenous Peoples* 7(3): 258-271.

— (2011b): Contested Terrain: Conflicts between State and Local Communities over the Management and Utilization of Nech Sar National Park, Southern Ethiopia. *Journal of Sustainable Development in Africa* 3 (5): 49-65.

— (2010): *Ethnicity and Inter-ethnic Relations: The 'Ethiopian Experiment' and the Case of the Guji and Gedeo*. VDM Verlag.

Ash, Timothy Garton (2009): A Century of Civil Resistance: Some Lessons and Questions. In: Roberts, Adam and Timothy Garton Ash (eds.): *Civil Resistance and Power Politics: The Experience of Non-violent Action from Ghandi to the Present*. Oxford: Oxford University Press, 371-390.

Assefa, Fiseha (2013): Ethiopia's Experiment in Accomodating Diversity: A Twenty years balance sheet. *Ethiopian Journal of Federal Studies* 1 (1): 103-154.

Assefa, Fiseha (2006): Theory versus Practice in the Implementation of Ethiopia's Ethnic Federalism. In: Turton,David (ed.): *Ethnic Federalism: The Ethiopian Experience in Comparative Perspective*. Addis Ababa: Addis Ababa University Press,131-364.

Awoke, Amzaye (1985): The Kore of Amaro: A Historical Survey. Unpublished BA Thesis, Addis Ababa University.

Bahru, Zewde (2002): *A History of Modern Ethiopia, 1855-1991*. Addis Ababa: Addis Ababa University Press.

Balsvik, Randi (1985): *Haile Selassie's Students: Intellectual and Social Background to Revolution, 1952-1977*. East Lansing: Michigan State University, African Studies Center.

Bargatzky, Thomas (1994): Introduction. In: Bargatzky, Th. and Rolf Kuschel: *The Invention of Nature*. Berlin: Peter Lang, 9-26.

Barth, Fredrik, ed. (1969): *Ethnic Groups and Boundaries. The Social Organization of Culture Difference*. Oslo: Universitets Forlaget.

Baxter, Paul (1991): Preface. In: Van De Loo, Joseph (ed.): *Guji Oromo Culture in Southern Ethiopia: Religious Capabilities in Rituals and Songs*. Berlin: Deitrich Reimer Verlag, 9-13.

Baye, Yimam (2008): Academic Freedom at Addis Ababa University: An overview of its past and current practices. In: Assefa, Taye (ed.): *Academic Freedom in Ethiopia: Perspectives of Teaching Personnel*. Addis Ababa: Forum for Social Studies, 19-62.

Bayisa, Feye (2011): *Local Livelihoods, Conflicts and Its Impact on The sustainability of Nech Sar National Park in Southern Ethiopia*. Lambert Academic Publishing.

Standing Bear, Chief Luther (1998): Indian Wisdom. In: Callicott J. Baird and Michael P. Nelson (eds.): *The Great New Wilderness Debate.* Athens, Georgia: The University of Georgia Press, 217-230.

Bediru, Ahmed (1986): *The History of Koore*. Ministry of Culture, Addis Ababa: Ethiopia (Unpublished).

Behnke, Roy and Carol Kerven (2013): Counting the Costs: Replacing Pastoralism with irrigated Agriculture in the Awash Valley. In: Catley Andy, Jeremy Lind and Ian Scoones (eds.): *Pastoralism and Development in Africa: Dynamic Change at Margines*. London and New York: Routledge, 57-70..

Berhanu, Gutema (2009): Constitutionalism in the Horn of Africa: Lesson from the new constitution of Ethiopia. *DIIPER Research Series Working Paper* 5.

Berkes, Firkes (2008): *Sacred Ecology*. London: Routledge.

Berreman, Gerald (1962): Behind Many Masks: Ethnography and Impression Management in a Himalayan Village. *The Society for Applied Anthropology Monograph* 4.

Binyam, Sisay M. (2010): *Aspects of Koorete Verb Morphology*. Rüdiger Köppe Verlag.

Blaikie, Piers and Harold Brookfield (1987): *Land Degradation and Society*. London: Routledge and Paul Kegan.

Blaikie, Piers (1985): *The Political economy of Soil Erosion in Developing Countries*. London and New York: Longman.

Blower, J. H. (1967): Report on a visit to the Nechisar area, Lake Chamo. Ethiopian Wildlife Conservation Organisation. Addis Ababa, Ethiopia (unpublished).

Bosak, Keith (2008): Nature, Conflict and Biosphere Conservation in Nanda Devi Biosphere Reserve. *Conservation and Society* 6 (3): 211-24.

Boyce, James (2002): *The Political Economy of the Environment*. UK: Edward Elgar Publishing Ltd.

Braun, W. Bruce (1997): Buried Epistemologies: The Politics of Nature in (Post) Colonial British Columbia. *Annals of the Association of American Geographers* 87 (1): 3-31.

Briggs, Philip (2006). *Ethiopia: The Bradt Travel Guide*. England: The Bradt Travel Guide Lit.

Brockington, Dan, et al. (2008): *Nature Unbound: Conservation, Capitalism and the Future of Protected Areas*. UK and USA: Earthscan.

Brockington, Dan (2002): *Fortress Conservation: The Preservation of the Mkomazi Game Reserve, Tanzania.* London: James Currey.

Bryant, Raymond and Sinéad Bailey (1997): *Third World Political Ecology*. London and New York: Routledge.

Bunce, Michael (1994): *The Countryside Ideal: Anglo-American Images of Landscape*. London and New York: Routledge.

Burawoy, Michael (2009): *The Extended Case Method: Four Countries, Four*

Decades, Four Great Transformations, and One Theoretical Tradition. Los Angeles: University of California Press.

Burawoy, Michael (1991): Reconstructing Social Theories. In: Burawoy, Michael, et al. (eds.): *Ethnography Unbound: Power and Resistance in the Modern Metropolis*. Los Angeles: University of California Press, 8-28.

Callicott, J. Baird and Michael P. Nelson (eds.) (1998): *The Great New Wilderness Debate*. Athens, Georgia: The University of Georgia Press.

Campbell, Ben (2005): Changing Protection Policies and Ethnographies of Environmental Engagement. *Conservation and Society* 3 (2): 280-322.

Castree, Noel (2010): Neoliberalism and the Biophysical Environment: A synthesis and evaluation of the research. *Environment and Society: Advances in Research* 1 (1): 5-45.

Castree, Noel and Bruce Braun (1998): The Construction of Nature and the Nature of Construction: Analytical and Political Tools for Building Survivable Futures. In: Castree, Noel and Bruce Braun (eds.): *Remaking Reality: Nature at the Millenium*. London and New York: Routledge, 2-41.

Chape, Stuart, et al. (2008): *The World's Protected Areas: Status, Values and Prospects in the 21st Century*. Los Angeles: University of California Press.

Chen, Ke (1993): Environmental Affordance: A Theoretical Framework for Incorporating Some Behavioural Considerations in Residential Evaluation. *Forum* 2: 57-64.

Clapham, Christopher (2009): Post War Ethiopia: Trajectories of Crisis. *Review of African Political Economy* 36 (120): 181-192.

Clapham, Christopher (2006a): Ethiopian Development: The Politics of Emulation. *Commonwealth & Comparative Politics* 44 (1): 108–118.

Clapham, Christopher (2006b): Afterword. In: Turton David (ed.): *Ethnic Federalism: The Ethiopian Experience in Comparative Perspective*. Addis Ababa: Addis Ababa University Press, 231-240.

Clapham, Christopher (2002): Controlling Space in Ethiopia. In: James, Wendy et al. (eds.): *Remapping Ethiopia: Socialism and After*. () James Currey Ltd., 9-30.

Cohen, Gideon (2006): The Development of Regional and Local languages in Ethiopia's Federal System. In: Turton, David (ed.): *Ethnic Federalism: The Ethiopian Experience in Comparative Perspectives*. Addis Ababa: Addis Ababa University Press, 165-180.

Colding, Johan and Carl Folke (2001): Social Taboos: "Invisible" Systems of Local Resource Management and Biological Conservation. *Ecological*

Applications 11 (2): 584-600.

Croll, Elisabeth, and David Parkin (eds.) (1992): *Bush Base: Forest Farm–Culture, Environment and Development*. London: Routledge.

Cronon, William (1996): The Trouble with Wilderness; or, Getting Back to the Wrong Nature. In: Cronon, William (ed.): *Uncommon Ground: Rethinking the Human Place in Nature*. New York: W.W. Norton and Company, 7-28.

Crook, Charles and Dean Garratt (2005): The Positivist Paradigm in Contemporary Social Science Research. In: Somekh, Bridget and Chatty Lewin, (eds.): *Research Methods in the Social Sciences*. London: SAGE Publications Ltd., 207-214.

Crotty, Michael (1998): *The Foundations of Social Research: Meaning and Perspective in the Research Process*. New York: SAGE Publications Ltd.

Curtin, Philip D. (1973): *The Image of Africa: British Ideas and Action*, 1780-1850 (Volume 2). Madison, Wisconsin: University of Wisconsin Press.

Daniels, Stephen. (1993): *Fields of Vision: Landscape Imagery and National Identity in England and the United States*. Princeton: Princeton University Press.

Delaney, David (2001): Making Nature/Marking Humans: Law as a Site of (Cultural) Production. *Annals of the Association of American Geographers* 91 (3): 487-503.

Denzin, Normank and Yvonnas Lincoln (eds.) (2005): *The SAGE Handbook of Qualitative Research (3rd edition)*. London and New Delhi: Sage Publications, Inc.

Dereje, Feyisa (2013): Centering the Periphery? The Praxis of Federalism at the Margins of the Ethiopian State. *Ethiopian Journal of Federal Studies* 1 (1):155-192.

Dereje, Feyissa (2011): *Playing Different Games: The Paradox of Anywaa and Nuer Identification Strategies in Gambella Region, Ethiopia*. Berghahn Books.

Dereje, Feyisa (2010): Conflicts within the Ethiopian Federation–A Reappraisal. In: Habtu, Alem (ed.): *Ethiopian Federalism: Principle, Process and Practice*. Addis Ababa: Addis Ababa University Press, 215-225.

Dereje, Feyisa (2006): The Experience of Gambella Regional State. In: Turton, David (ed.): *Ethnic Federalism: The Ethiopian Experience in Comparative Perspective*. Addis Ababa: Addis Ababa University Press, 208-230.

Descola, Philippe (1996): Constructing Natures: Symbolic ecology and social practice. In: Descola, Philippe and Gísli Pálsson (eds.): *Nature and So-*

ciety: Anthropological Perspectives. London: Routledge, 82-102.

Descola, Philippe (1992): Societies of Nature and the Nature of Society. In: Kuper, Adam (ed.): *Conceptualizing Society*. London and New York: Routledge, 107-126.

Descola, Philippe and Gísli Pálsson (1996): Introduction. In: Descola, Philippe and Gísli Pálsson (eds.): *Nature and Society: Anthropological Perspectives*. London: Routledge, 1-22.

Dessalegn, Wana (2008): Local People's Perceptions and Attitudes towards the Management of Nech-Sar National Park, Ethiopia. In: Jefrey, Michael I. et.al. (eds.): *Biodiversity Conservation, Law + Livelihoods*. IUCN Academy of Environmental Law Research Studies, 233-250.

Dessalegn, Wanna (2004): Strategies for Sustainable Management of Biodiversity in the Nech Sar National Park, Southern Ethiopia. A Research Report Submitted to OSSREA, Addis Ababa.

Dhadacha, Gololcha (2006): *The Politico-Legal System of the Guji Society of Ethiopia*. Addis Ababa: DUBAF.

Donham, Donald (2002): Introduction In: Donham, Donald et al. (eds): *Remapping Ethiopia: Socialsim and After*. Oxford: James Currey, 1-8.

Donham, Donald (1986): The Making of an Imperial State. In: Donham, Donald and Wendy James (eds.): *The Southern Marches of Imperial Ethiopia: Essays in History and Social Anthropology*. Cambridge University Press, 3-51.

Dowie, Mark (2009): *Conservation Refugees: The Hundred-year Conflict between Global Conservation and Indigenous Peoples*. Massachusetts: Massachusetts Institute of Technology.

Elliott, Jane (2005): *Using Narrative in Social Research Qualitative and Quantitative Approaches*. London, Thousand Oaks, New Delhi: SAGE Publications.

Escobar, Arturo (2008): *Territories of Difference: Place, Movements, Life, Redes*. Durham: Duke University Press.

Escobar, Arturo (1999): After Nature: Steps to an Antiessentialist Political Ecology. *Current Anthropology* 40 (1): 1-30.

Escobar, Arturo (1996): *Constructing Nature*. New York: Routledge.

— (1995): *Encountering Development: The Making and Unmaking of the Third World*. Princenton, NJ: Princenton University Press.

Falola, Toyin and Emily Brownell (eds.) (2012): *Landscape, Environment and Technology in Colonial and Postcolonial Africa*. London and New York: Routledge.

Fanon, Frantz (1952 [1986]): *Black Skin, White Mask*. UK: Pluto Press.

Federal Democratic Republic of Ethiopia (2010): *Growth and Transformation Plan*. Adis Ababa: Ministry of Finance and Economic Development.

Federal Democratic Republic of Ethiopia (2007): *Proclamation No.541/2007: A Proclamation to Provide for the Development Conservation and Utilization of Wildlife*. Addis Ababa: Federal Negarit Gazeta of the Federal Democratic Republic of Ethiopia.

— (2003): *Sustainable Development and Poverty Reduction Program (SDPRP).* Addis Ababa, Ministry of Finance and Economic Development.

— (1997): *Ethiopia's Wildlife Policy.* Ministry of Agriculture: Addis Ababa, Ethiopia.

— (1995): *The Constitution of the Federal Democratic Republic of Ethiopia*. Addis Ababa: Berhanena Selam Printing Enterprise.

Fekadu, Adugna (2011): Overlapping nationalist projects and contested spaces: The Oromo–Somali Borderlands in Southern Ethiopia. *Journal of Eastern African Studies* 5 (4): 773-787.

Ferguson, James (2994): *The Anti-politics Machine*. Minneapolis: University of Minnesota Press.

Few, Rojer (2001): Researching Actor Power: Analyzing Mechanisms of Interaction in Negotiations over Space. *Area* 34 (1): 29-38.

Fisher, Robert, et.al. (eds.) (2008): *Linking Conservation and Poverty Reduction: Landscapes, People and Power*.Earthscan.

Fletcher, Robert (2010): Neoliberal environmentality: Towards a poststructuralist political ecology of the conservation debate. *Conservation and Society* 8 (3): 171-181.

Ford, Caroline (2012): Imperial Preservation and Landscape Reclamation: National Parks and Natural Reserves in French Colonial Africa. In: Gissibl, Bernhard, Sabine Höhler and Patrick Kupper (eds.): Civilizing Nature: *National Parks in Global Historical Perspective*. New York and Oxford: Berghahn Books, 68-83.

Forsyth, Tim (2003): *Critical Political Ecology: The Politics of Environmental Science*. London and New York: Routledge.

Freeman, Dena (2006): Natural Resource Management in and Around Nech Sar National Park: A Situation Analysis. Unpublished Report.

Freeman, Dena (2002): *Initiating Change in Highland Ethiopia*. Cambridge: Cambridge University Press.

Freudenburg, William, et al. (1995): Beyond the Nature/Society Divide: Learning to think about a Mountain. *Sociological Forum* 10 (3): 361-392.

Galaty, John (2013): Land Grabbing in the Eastern African Rangelands. In: Catley Andy, Jeremy Lind and Ian Scoones (eds.): *Pastoralism and Development in Africa: Dynamic Change at Margines*. London and New York: Routledge, 143-153.

Garland, Elizabeth (2008): The Elephant in the Room: Confronting the Colonial Character of Wildlife Conservation in Africa. *African Studies Review* 51 (3): 51-74.

Gaventa, John (2004): Towards Participatory Governance: Assessing the Transformative Possibilities. In: Hickey, Samuel and Giles Mohan (eds.): *Participation: From Tyranny to Transformation? Exploring New Approaches to Participation in Development*. London and New York: Zed Books, 25-41.

Geertz, Clifford (1973): *The Interpretation of Cultures*. New York: Basic Books.

Getachew, Godana (2007): Do People and their Culture Matter in the Conservation of Natural Resources? A Study of impacts of Conservation Policies in Nech Sar National Park and Yayo Forest in Illubabor Bora Zone. Unpublished MA Thesis. Addis Ababa University, Ethiopia.

Gezon, Lisa (2006): *Global Visions, Local Landscapes: A political ecology of conservation, conflict and control in Northern Madagascar*. New York: Rowman and Little Field.

Gossling, Stefan (2004): *Tourism and Development in the Tropical Islands: A Political Ecology Perspective*. Edward Elgar Publishing Limited.

Grove, H. Richard (1990): Colonial Conservation, Ecological Hegemony and Popular Resistance: Towards a global synthesis. In: MacKenzie, John (ed.): *Imperialism and the Natural World*. Manchester: Manchester University Press, 15-50.

Grove, H. Richard (1987): Early Themes in African Conservation: The Cape in Nineteenth Century. In: Anderson, David and Richard Grove (eds.): *Conservation in Africa: Peoples, Policies and Practice*. Cambridge: Cambridge University Press, 21-39.

Guba, G. Egon (1990): The Alternative Paradigm Dialog. In: Guba, G. Egon (ed.): *The Paradigm Dialog*. Sage Publications Inc., 17-30.

Haila, Yrjo (2012). "Genealogy of Nature Conservation: A Political Perspective. *Nature Conservation* 1: 27-52 (www.pensoft.net/journals/natureconservation. Accessed on 10 September 2012).

Hagmann, Tobias and Mulugeta Alemmaya (2008): Pastoral Conflicts and State-building in the Ethiopian Lowlands. *Africa Spectrum* 43 (1): 19-37.

Hameso, Seyoum (2001): *Ethnicity in Africa: Towards a Positive Approach*.

San Jose, California University.

Hammersley, Martyn and Atkinson, Paul (2007): *Ethnography: Principles in Practice*. Taylor and Francis e-Library.

Harmon, David (1998): Cultural Diversity, Human Subsistence, and the National Park Ideal. In: (Callicott J. Baird and Michael P. Nelson (eds.): *The Great New Wilderness Debate*. Athens, Georgia: The University of Georgia Press, 217-230.

Hardin, Garrett (1968): The Tragedy of the Commons. *Science* 162 (3859): 1243-1248.

Hayward, Richard (1982): Notes on the Koyra language. *Afrika und Ubersee: Sprachen – Kulturen* 65 (2): 211-268.

Herbst, Jeffrey (2000): *States and Power in Africa: Comparative Lessons in Authority and Control*. Princenton, NJ: Princeton University Press.

Hinnant, John (1977): The Gada System of the Guji of Southern Ethiopia. Ph.D. Dissertation, University of Chicago.

Hizkias, Assefa (1996): Ethnic Conflict in the Horn of Africa: Myth and Reality. In: Rupesinghe, Kumar and Valerii Aleksandrovich Tishkov (eds.): *Ethnicity and Power in the Contemporary World*. Tokyo, New York and Paris: United Nations University Press.

Hobsbawm, Eric and Terence Ranger (eds.) (1983 [2012]): *The Invention of Tradition*. Cambridge University Press.

Hornborg, Alf (1996): Ecology as Semiotics: Outlines of a Contextualist Paradigm for Human Ecology. In: Descola, Philippe and Gísli Pálsson (eds.): *Nature and Society: Anthropological Perspectives*. London: Routledge, 45-62.

Horsman, Stuart (2006): The Politics of Toponyms in the Pamir Mountains. *Area* 38 (3): 279-291.

Hussein, Jemma (2002): Competition over Resource and Ethnic Conflict in Federal Ethiopia: the case of recent Guji-Gedeo Conflict. Research report submitted to OSSREA, Addis Ababa.

Hutton, Jon, et al. (2005): Back to the Barriers? Changing Narratives in Biodiversity Conservation. *Forum for Development Studies* 32 (2): 341-370.

Huxley, Julian, et al. (1963): The Conservation of Nature and Natural Resources in Ethiopia. Unpublished Report UNESCO/NS/NR/47. http://unesdoc.unesco.org/images/0013/001398/139892eb.pdf (Accessed on 05 April 2012).

Igoe, Jim (2004): *Conservation and Globalization: A Study of National Parks and Indigenous Communities from East Africa to South Dakota*. Belmont: Wadsworth/Thomson Learning.

Ingold, Tim (2005): Towards a Politics of Dwelling. *Conservation and Society* 3 (2): 501-508.

Ingold, Tim (2000): *The Perception of the Environment*. London: Routledge.

Jennings, Ray. (2000): Participatory Development as New Paradigm: The Transition of Development Professionalism. In: USAID, Community Based Reintegration and Rehabilitation in Post-Conflict Settings. Washington, DC. http://www.usaid.gov/our_work/crosscutting_programs/transition_initiatives/pubs/ptdv1000.pdf (accessed 20 June2010).

Jordan, Fiona and Heather Gibson (2004): Let your Data Do the Talking: researching the solo travel experiences of British and American women. In: Phillimore, Jenny and Lisa Goodson (eds.): *Qualitative Research in Tourism Ontologies, Epistemologies and Methodologies*. London, New York: Routledge, 215-235.

Katz, Cindi (1998): Whose Nature, Whose Culture?: Private Productions of Space and the 'Preservation' of Nature. In: Castree, Noel and Bruce Braun (eds.): *Remaking Reality: Nature at the Millenium*. London and New York: Routledge, 45-62.

Kincheloe, Joe and Peter McLaren (2002): Rethinking Critical Theory and Qualitative Research. In: Zou, Yali and Enrique T. Trueba (eds.): *Ethnography and Schools: Qualitative Approaches to the Study of Education*. Maryland: Rowman and Littlefield Publishers, Inc., 87-138.

Kindner, David (2000): Fabricating Nature: A Critique of Social Construction of Nature. *Environmental Ethics* 22 (4): 339-357.

Legesse, Asmarom (2000): *Oromo Democracy: An Indigenous African Political System*. Lawrenceville, NJ: The Red Sea Press.

Legesse, Asmarom (1973): *Gada: Three Approaches to the Study of African Society*. New York: Red Sea Press.

Levin, Donald (2000): *Greater Ethiopia: the Evolution of Multiethnic Society*. The University of Chicago Press.

Lewis, Herbert (1966): The Origins of the Galla and Somali. *The Journal of African History* 7 (1): 27-46.

Long, Norman (2001): *Development Sociology: Actor Perspective*. Routledge.

Long, Norman (1992): "From Paradigm Lost to Paradigm Regained? The case for an actor-oriented sociology of development. In: (Long Norman and Ann Long (eds.): *Battlefields of Knowledge: The interlocking of theory and practice in social research and development*. London and New York: Routledge, 16-46.

Li, T. Murray (2007): *The Will to Improve: Governmentality, Development, and the Practice of Politics*. Durham: Duke University Press.

Li, T. Murray (1999): Compromising Power: Development, Culture, and Rule in Indonesia. *Cultural Anthropology* 14 (3): 295–322.

MacKenzie, John M. (ed.) (1990): *Imperialism and the Natural World*. Manchester: Manchester University Press.

Mamdani, Mahmood (1996): *Citizen and Subject: Contemporary Africa and the Legacy of Late Colonialism*. Princeton, NJ: Princeton University Press.

Markakis, John (2011): *Ethiopia: The Last Two Frontiers*. London: James Currey.

Markakis, John (1994): Ethnic Conflict and the State in the Horn of Africa. In: Fukui, Katsuyoshi and John Markakis (eds.): *Ethnicity and Conflict in the Horn of Africa*. London: James Currey, 217-37.

— (1981). The military state and Ethiopia's Path to 'Socialism'. *Review of African Political Economy* 8 (21): 7-25.

— (1974). *Ethiopia: Anatomy of a Traditional Polity*. Oxford: Oxford University Press.

Markin, M. et al. (1974): *Development Prospects in the Southern Rift Valley, Ethiopia, Volume 2*. Land Resources Division.

Mazrui, Ali A. (2002): Who Killed Democracy in Africa? Clues of the Past, Concerns of the Future. *Development Policy Management Network Bulletin* IX (1): 15-23.

Mbembe, Joseph-Achille (2001): *On the Postcolony, Vol. 41*. Berkley, Los Angeles: University of California Press.

McClellan, Charles (1988): *State Transformation and National Integration: Gedeo and the Ethiopian Empire, 1895-1935*. East Lansing: Michigan State University.

Mekuria, Bulcha (2011): *Contours of the Emergent & Ancient Oromo Nation: Dilemmas in the Ethiopian Politics of State and Nation-Building*. Cape Town: The Center for Advanced Studies of African Societies Series 84.

Mekuria, Bulcha (1997): The Politics of Linguistic Homogenization in Ethiopia and the Conflict over the Status of 'Afaan Oromoo'. *African Affairs* 96 (384): 325-352.

— (1996): The Survival and Reconstruction of Oromo National Identity. In: Baxter, P.T.W., Ian Hultin and Alessandro Triulzi (eds.): *Being and Becoming Oromo: Historical and Anthropological Enquiries*. Uppsala: Nordic Africa Institute, 48-66.

Meles, Zenawi (2012): States and Markets: Neoliberal limitations and the case for a Developmental State. In: Noman Akbar et al., (eds.): *Good Growth and Governance in Africa: Rethinking Development Strategies*. Oxford: Oxford University Press, 140-174.

Meles, Zenawi (2006): *African Development: Dead Ends and New Beginnings*. Unpublished Monograph.

Merchant, Carolyn (2003): *Reinventing Eden: The Fate of Nature in Western Culture*. Routledge.

Merchant, Carolyn (1996): Reinventing Eden: Western Culture as a Recovery Narrative. In: Cronon, William (ed.): *Uncommon Ground: Rethinking the Human Place in Nature*. New York: W.W. Norton and Company, 132-159.

Merera, Guddina (2003): *Ethiopia: Competing Ethnic Nationalisms and the Quest for Democracy, 1960-2000*. Addis Ababa: Chamber Printing House.

Migdal, Joel S. (2001): *State in Society: Studying how states and societies transform and constitute one another*. Cambridge: Cambridge University Press.

Minute (14/09/2003): Resettling the Guji and Koore Ethnic Groups who illegally Settled in Nech Sar National Park. Community Conference held 11-12 July 2003, Arba Minch (Minute in Amharic, Archive from SNNP regional state, Culture and Tourism Bureau).

Mitchell Laura (2012): Appraising Nature. In: Toyin, Falola and Emily Brownell (eds.): *Landscape, Environment and Technology in Colonial and Postcolonial Africa*. London: Routledge, African Studies, 42-61.

Mohammed, Hassen (1996): The Development of Oromo Nationalism. In: (Baxter, Paul et al. (eds.): *Being and Becoming Oromo: Historical and Anthropological Enquiries*. Uppsala: Nordic Africa Institute, 67-80.

Mohammed, Hassen (1990): *The Oromo of Ethiopia: A History, 1570-1860*. Cambridge: Cambridge University Press.

Mohamed, Salih and John Markakis (1998): Introduction. In: Mohamed, Salih and John Markakis (eds.): *Ethnicity and the State in Eastern Africa*. Uppsala: Nordic Africa Institute, 7-16.

Mohan, Giles and Samuel Hickey (eds.) (2004). Relocating participation within a radical politics of development: critical modernism and citizenship. In: Mohan, Giles and Samuel Hickey (eds.): *Participation–From Tyranny to Transformation: Exploring New Approaches to Participation in Development*. London: Zed Books, 59-74.

Mohan, Giles and Kristian Stokke (2000) Participatory Development and Empowerment: The Dangers of Localism. *Third World Quarterly* 21 (2): 247-268.

Nawaz, Faraha (2013): "Power, Empowerment and Participatory Development: Conceptual Linkages. *Open Journal of Social Science Research* 1 (2):

26-30.

Neumann, Roderick (2001): Disciplining Peasants in Tanzania: From State Violence to Self-Surveillance in Wildlife Conservation. In: Peluso, Nancy L. and Michael J. Watts (eds.): *Violent environments: Social Bonds and Racial Hubris*. Cornell University Press, 305-327.

Neumann, Roderick (1998): *Imposing Wilderness: Struggles over Livelihood and Nature Preservation in Africa*. California: California State University Press.

— (1995): Local Challenges to Global Agendas: Conservation, Economic Liberalization and the Pastoralists' Rights Movement in Tanzania. *Antipode* 274: 363-382.

— (1992): Political Ecology of Wildlife Conservation in the Mt. Meru Area of Northeast Tanzania. *Land Degradation and Rehabilitation* 3: 85-98.

Noman, Akbar et al. (eds.) (2012): *Good Growth and Governance in Africa: Rethinking Development Strategies*. Oxford: Oxford University Press.

Ouderkrik, Wayne (2008): On Wilderness and People: A View from Mt. Marcy. In: Nelson, Michael P. and J. Baird Callicott (eds.): The Wilderness Debate Rages On: Continuing the Great New Wilderness Debate. Athens and London: The University of Georgia Press, 435-460.

Pain, Rachel and Peter Francis (2003): Reflections on Participatory Research. *Area* 35: 46-54.

Pálsson, Gísli (1996): Human-environmental Relations: Orientalism, Paternalism and Communalism. In: Descola, Philippe and Gísli Pálsson (eds.): *Nature and Society: Anthropological Perspectives*. London: Routledge, 63-81.

Paulos, Milkias (2011): *Ethiopia*. Santa Barbara: ABC-CLIO, LLC.

Peet, Richard, et al. (2011): *Global Political Ecology*. London: Routledge.

Peet, Richard and Michael Watts (2004): *Liberation Ecologies: Environment, Development, Social Movements*. London and New York: Routledge.

Peluso, Nancy L. (1993): Coercing Conservation? The Politics of State Resource Control. *Global Environmental Change* 3 (2): 199–217.

Phillimore, Jenny and Lisa Goodson (eds) (2004): *Qualitative Research in Tourism Ontologies, Epistemologies and Methodologies*. London and New York: Routledge.

Proctor, James (1998): The Social Construction of Nature: Relativist Accusations, Pragmatist and Critical Realist Responses. *Annals of the Association of American Geographers* 88 (3): 352-376.

Ramutsindela, Maano (2004): *Parks and People in Postcolonial Societies*. Boston and London: Kluwer.

Reichart-Burikukiye, Christiane (2012): The Railway in Colonial East Africa: Colonial Iconography and African Appropriation of a New Technology In: (Toyin, Falola and Emily Brownell (eds.): *Landscape, Environment and Technology in Colonial and Postcolonial Africa*. London and New York: Routledge, 62-86.

Rich, Jeremy (2012): Chimpanzees in the Colonial Maelstrom: Struggles over Knowledge, Race and Commodities in the Gabonese Primate Trade, c. 1850-1940. In: (Toyin, Falola and Emily Brownell (eds.): *Landscape, Environment and Technology in Colonial and Postcolonial Africa*. London and New York: Routledge, 21-41.

Robben, Antonius C.G.M. (1995): The Politics of Truth and Emotions Among Victims and Perpetrators of Violence. In: (Nordstrom, Carolyn Antonius C.G.M. Robben (eds.): *Fieldwork Under Fire: Contemporary Studies of Violence and Survival*. London, Berkley: University of California Press, 81-104.

Robbins, Paul (2004): *Political Ecology: A Critical Introduction*. Blackwell Publishers.

Saugestad, Sidsel (2005): Improving their Lives: State Policies and San Resistance in Botswana. *Before Farming* 4 (1): 1-11.

Saugestad, Sidsel (2001): Contested Images: 'First Peoples' or 'Marginalized Minorities' in Africa? In: Alan, Barnard and Justin Kenrick (eds.): *Africa's Indigenous Peoples: 'First Peoples' or 'Marginalized Minorities'?* Edinburgh: University of Edinburgh, Centre of African Studies, 299-322.

Schlee, Günther (2002): *Imagined Differences: Hatred and the Construction of Identity*. LIT Verlag.

Scott, David (1995): Colonial Governmentality. *Social Text* 43: 191-220.

Scott, James (2009): *The Art of Not Being Governed: An Anarchist History of Upland Southeast Asia*. New Haven: Yale University Press.

Scott, James (1998): *Seeing Like a State: How Certain Schemes to Improve the Human Condition Have Failed*. New Haven: Yale University Press.

— (1990): *Domination and the Arts of Resistance: Hidden Transcripts*. USA: Yale University Press.

— (1985): *Weapons of the Weak: Everyday Forms of Peasant Resistance*. Yale: Yale University Press.

Sharp, Joanne, et al. (2000): *Entanglements of Power: Geographies of Domination/Resistance*. London: Routledge.

Shibiru, Tedla and Lemma Kifle (1998): Environmental Management in Ethiopia: Have the National Conservation Plans Worked? *Environmental Forum Publications Series* 1, Organization for Social Science Research in Eastern and Southern Africa.

Shkedi, Asher (2005): *Multiple Case Narrative: A Qualitative Approach to Studying Multiple Populations: A Qualitative Approach to Studying Multiple Populations*. Amsterdam: John Benjamins Publishing Co.

Simeon, Mocha (1995): A Brief Description of Cultural Rituals, the Coming of Christianity and Growth of Evangelical Christianity in Amaro District. Unpublished BA Thesis, Mekane Yesus Seminary College, Addis Ababa, Ethiopia.

Simmons, Ian (1993): *Interpreting Nature: Cultural Constructions of the Environment*. London and New York: Routledge

Singh, Jaidev and Henk Van Houtum (2002): Post-colonial nature conservation in Southern Africa: Same emperors, new clothes? *Geo Journal* 58: 253-263.

Slater, Candace (1996): Amazonia as Edenic Narrative. In: Cronon, William (ed.): *Uncommon Ground: Rethinking the Human Place in Nature*. New York: WW Norton & Company, 114-131.

Smith, Donaldson A. (1897) [2008]: *Through Unknown African Countries. The First Expedition form Somaliland to Lake Lamu*. USA: Rarebooksclub.com.

Smith, Stephen, L. (2010): *Practical Tourism Research*. Cambridge: Cambridge University Press.

Solomon, Berhe (2002). *The Burji-Guji Interaction in Hageremariam and Burji Soyema Woredas*. Unpublished B.A. Thesis, Debub University.

Stewart-Harawira, Makere (2005): *The New Imperial Order: Indigenous Responses to Globalization*. New Zealand: Huia Publishers.

Stott, Philip A. and Sian Sullivan (eds.) (2000): *Political Ecology: Science, Myth and Power*. Oxford: Oxford University Press.

Tache, Boku (2013): Rangeland Enclosures in Southern Oromia, Ethiopia: an innovative response or coercion of common property resources? In: Catley Andy, Jeremy Lind and Ian Scoones (eds.): *Pastoralism and Development in Africa: Dynamic Change at Margines*. London and New York: Routledge, 37-46.

Taddesse, Berisso (2009): Changing Alliances of Guji-Oromo and their Neighbours: State Policies and Local Factors. In: Schlee, Günther and Elizabeth E. Watson (eds.): *Identifications and Alliances in Northeast Africa*. London: Berghahn Books, 191-202.

Taddesse, Berisso (2006): The Pride of the Guji Oromo: an Essay on Cultural

Contact and Self-esteem. In: Strecker, Ivo and Jean Lyddall (eds.): *The Perils of Face: Essays on Cultural Contact, Respect and Self-esteem in Southern Ethiopia*. Berlin: Deitrich Reimer Verlag, 207-224.

— (2004). "Planning Resettlement in Ethiopia: the Experience of the Guji Oromo of Nechsar National Park." In: *People, Space and the State: Migration, Resettlement and Displacement in Ethiopia*, (Alula, Pankhurst and Francois P., eds.); 93-101. Addis Ababa: Bole Printing Press.

— (1995). "Agricultural and Rural Development Policies in Ethiopia: A case study of villagization policy among the Guji-Oromo of Jam Jam Awraja", PhD Thesis, Michigan State University, Michigan.

— (1994). Warfare among Guji Oromo of Southern Ethiopia. *Proceedings of the 11th International Conference of Ethiopian Studies*, IES: Addis Ababa.

Taye, Assefa and Gebre-Egziabher Tegegne (eds.) (2007): *Decentralization in Ethiopia*. Addis Ababa: Forum for Social Studies.

Technical Report (1995): Resettling the People Living in Nechisar National Park: Proceedings of a Workshop. National Parks Rehabilitation in Southern Ethiopia Project, EU Project No. 7 ACP ET068, Technical Report No. 4, DHV Consultants.

Tesfaye, Tafesse (2009): The Predicaments of Amhara Migrant-Settlers in East Wollega Zone, Ethiopia. In: *Proceedings of the 16th International Conference of Ethiopian Studies*. Trondheim, 851-865.

Teshale, Tibebu (1995): *The Making of Modern Ethiopia: 1896-1974*. Lawrenceville, New Jersey: Red Sea Press.

Thomas, Karen (2004): The Research Process as a Journey: from positivist traditions into the realms of qualitative inquiry. In: Phillimore, Jenny and Lisa Goodson(eds.): *Qualitative Research in Tourism Ontologies, Epistemologies and Methodologies*. London, New York: Routledge, 197-214.

Thomson, Terry (2005): Ethiopia: Local People Burned Out of Homes to Make Way for National Park, *Refugee International*, 19 April 2005 (http://www.unhcr.org/refworld/pdfid/47a6eebb0.pdf (accessed 12 April 2013).

Turton, David (2011): Wilderness, Wasteland or Home? Three ways of Imagining the Lower Omo Valley. *Journal of Eastern African Studies* 5 (1): 158-176.

Turton, David (2009): The Lower Omo Valley: three stories and one landscape. Paper Presented on Royal Geographical Society Annual Conference on Narrating Landscape and Environment, Manchester.

— (2006): Introduction. In: Turton, David (ed.): *Ethnic Federalism: the Ethiopian Experience in Comparative Perspective*. Addis Ababa: Addis Ababa University Press, 1-31

— (2002): The Mursi and the Elephant Question. In: (Chatty, Dawn and Marcus Colchester (eds.): *Conservation and Mobile Indigenous Peoples: Displacement, Forced Settlement and Development*. New York and Oxford: Berghahn Books, 97-118.

Ullendorff, Edward (1965): *The Ethiopians: An Introduction to Country and People*. Oxford: Oxford University Press.

Van de Loo, Joseph (1991): *Guji Oromo Culture in Southern Ethiopia: Religious Capabilities in Rituals and Songs*. Berlin: Dietrich Reimer Verlag.

Vaughan, Sarah (2003): *Ethnicity and Power in Ethiopia.* PhD Dissertation, University of Edinburgh, Scotland.

Wagner, Claire and Chinedu Okeke (2009): Quantitative or Qualitative: Ontological and Epistemological Choices in Research Methods Curricula. In: *Teaching Research Methods in the Social Sciences*, (Mark, G., Claire, W. and Barbara, K., eds.); 61-70. Farnham: Ashgate Publishing Limited.

Willers, Bill (2001). "The Postmodern Attack on Wilderness." *Natural Area Journal* 21(3): 259-265.

Wilshusen, Peter, et al (2002). "Reinventing a Square Wheel: Critique of a Resurgent "Protection Paradigm'' in International Biodiversity Conservation." *International Biodiversity Conservation, Society & Natural Resources: An International Journal* 15(2): 17-40.

Young, James (2012). Ethiopian Protected Areas a 'Snapshot'." A Reference Guide for Future Strategic Planning and Project Funding (Unpublished).

Zerihun, A. Woldesillasie (2004). "Ethnic Politics and the Question of Participatory Development in Siltie Zone: Some Preliminary Research Hypotheses." *Ethiopian Journal of Social Sciences, Anthropology and History* 1(2): 123-139.

Zeleza, Paul, (2008) "Introduction: the Causes and Costs of War in Africa: From Liberation Struggles to the 'War on Terror'." In: *The Roots of African Conflicts: The Causes and Costs*, (Alfred Nhema and Zeleza, eds.); 1-5. UK: James Currey Ltd.

List of Informants

No.	Name	Age (estimated)	Date of interiew	Context of interview
1.	Girma Tamer	42	April 2005	He was serving as park biologist by the time. I interviewed him while I took my University students to the national park for excursion.
2.	Boru Galgalo	Early 60s	July 18, 2006	Interviewed as part of my Master's thesis fieldwork in a Guji village of Dhoqqee.
3.	Gudo Mamacha	Late 50s	October 13, 2008	The elder was lived in Darba Mamana but I interviewed him in Amaro Kelle town as part of another research project.
4.	Gagasa Adola	Mid 70s	April 2012	This Guji elder was interviewed while heading his cattle to Malkaa around Lake Abaya.
5.	Gacha Atura	70s	Feb. 2012	A Koore returnee from resettlement site.
6.	Gagasa Edema	Early 80s	April 15, 2011	Interviewed in Abulo village (Arba Minch) on his way from hospital.
7.	Former Park Warden	48	April 23, 2012	In Hawassa town through formal appointment.
8.	Shora Mamacha	50s	Feb. 25,2012	A Koore former returnee interviewed in Darba Manana.
9.	Almaz Beyero	51	April18, 2012	Interviewed in Hawassa as former Warden and head of culture and tourism bureau of SNNP regional state.
10.	Ararsa Gumi	Late 70s	May 2012	This Guji elder was interviewed in his wet-season encampment site.
11.	Danbala Badacha	72	May, 2011, March	Took part in negotiations for Guji's request to get land in Gaattira.

No.	Name	Age (estimated)	Date of interiew	Context of interview
			2012	
12.	Waaqoo Dhu-go	65	April 2011	A Guji resident of Abulo village close to Arba Minch.
13.	Galgalo Jilo	Late 40s	April 2011	Resident of Golbo and was interviewed on his way back from Darba market.
14.	Danbala Ar-arsa	16	May 2012	I first met him while he was heading his cattle to pasture at night and interviewed him on next day.
15.	Chari Gamasa	Late 70s	June 2012	Took part in meetings during 2004 resettlement plan of the Guji.
16.	Ayano Guyo	Early 50s	April 2012	Discussed with him in his ficha (farm land) in Golbo.
17.	Galgale Aga	Late 50s	March 2011	A Guji woman from Maddo. I interviewed her at the shore of Lake Abaya while waiting for Jaatara (boat).
18.	Shado Udado	60s	March 2011	A Koore resident of Darba Manana village who had also plots of land in Tsalke.
19.	Gashe Bufa	60s	Feb. 2012	Took part in resettlement discussions of the Koore in 1990s and 2004.
20.	Ayele Machamo	40s	March 2012	Returned from resettlement and resided in Tifate village.
21.	Edema Aga	37	April 2011	
22.	Chari Galgalo	60s	April 2011	
23.	Dhugo Waqayyo	80s	April 2012	Was evicted in 1982 and in 2004 from Ergansa. I met him in Gode where he temporarily settled since 2008.
24.	Warkinesh Ware	60s	Feb. 2012	Evicted from Tsalke in 1982 and settled in Tifate village

No.	Name	Age (estimated)	Date of interiew	Context of interview
				since then.
25.	Former District Administrator (Galana-Abaya District)	54	April 28, 2012	We had long discussion seated under a mango tree in the official's compound in Guangua town.
26.	Halake Chari	38	Feb. 2012	A resident of Golbo. I interviewed him in Darba Manana where he went for business.
27.	Kasahun Ayele	42	April 2011	A government official in Amaro district.
28.	Molla Alemayehu	34	May 2012	A tourism 'expert' in Gamo Gofa Zone.
29.	Former Warden	48	Feb. 2011	Served as Warden under APF.
30.	Ayele Channe	67	May 2011	A retired official of the former regime.
31.	Shibiru Ayele	Late 70s	Feb. 2012	Moved to Tsalke in 1940s with his family but retained plots of land in Darba Manana.
32	Former Administrator of Amaro Especial District	46	May 2012	The former official was 'promoted' to regional level in Hawassa.
33.	Ayele Danche	64	April 2012	The informant worked in Gamo Gofa Province during the previous regime. He knew about the 1982 eviction of the Koore and Guji people.
34.	An Official in Borana Zone	45	March 2012	Worked as deputy administrator of Galana-Abaya district.
35.	Shube Dalacha	Late 40s	May 2012	Was displaced from Belte in 2004 and settled at Arda Guddina.

No.	Name	Age (estimated)	Date of interiew	Context of interview
36.	Former Head, Culture and Tourism Bureau of SNNP Regional State	53	April 2012	A key actor during the APF's take-over of the Nech Sar National Park, served as park Warden and in other different positions.
37.	Edema Galgalo	19	April-June 2012	Field Assistant
38.	Gagasa Adola	Early 80s	April 2012	Narrated his experiences of Guji-state interactions and the evictions.
39.	Gamade Halake	Late 60s	April 2012	A Guji resident of Oddo Miqe. He explained to me circumstances where Christians might take part in rituals.
40.	Aga Shota	70s	March 2011	Among the Guji Qaalluu clan. The elder discussed about food taboos.
41.	Dalge Robale	Late 50s	April 2008	I interviewed the woman in 2008 for a research on Guji children and their food culture.
42.	Edema Danbala	21	May 2012	Guided me from Arba Minch to Guji villages and along the way shared me his childhood experiences about interaction with the park.
43.	Gagasa Edema	69	March 2011	A resident of Guji village in Maddo that adjoins the park on the northeastern direction.
44.	Dokale Shanxara	60s	April 2012	As a non-Christian Guji man, he discussed issues of intergenerational and inter-religious differences on human-environment interactions.
45.	Kasahun Shibiru	59	Feb. 2012	A resident of Darba Manana who has been engaged in

No.	Name	Age (estimated)	Date of interiew	Context of interview
				small scale agri-business in Tsalke.
46.	Ayele Chamo	56	Feb. 2012	
47.	Darba Manana Village Chairman	48	Feb. 2012	The interview was made immediately after the chairman finished a meeting on land dispute between Guji and Koore households in Tsalke.
48.	Hailu Dacha	38	May 2012	A Chairman of Chamo Fishermen Association.
49.	A Senior Park Scout	72	April 2012	The interview was conducted after the scout returned from patrolling duty.
50.	A Park Biologist	32	April 2012	Worked in the park for 3 years.
51.	Tadelech Tibesso	Late 40s	June 2011	Sustained her livelihood on collecting and selling firewood from the park.
52.	Behailu Ayele	44	May 2012	Chairman of "See us Tour Operator" based in Arba Minch.
53.	Laura White	58	May 2012	An American tourist whom I met in Tourist Hotel (Arba Minch).
54.	Gule Mamacha	60s	March 2012	
55.	Galana District official	53	March 2012	Worked on different positions and had information about park-people conflicts.
56.	Head of Scouts	61	April 2012	This official was delegated as park Warden by the time of my interview and shared with me his view about park-people relations.
57.	Shalaqa Admasu	56		A park scout who worked in the park for several years
58.	Arda Guddina	42	May 2012	The official was invited as

No.	Name	Age (estimated)	Date of interiew	Context of interview
	village Chairman			guest speaker on students' school prize ceremony.
59.	Deputy Administrator (Amaro District)	51	March 2011	The official worked as a coordinator of the 2004 Koore resettlement program.
60.	Shadi Macho	65		Discussed about hunting behavior among the Koore.
61.	Kaahun Beyene	53	March 2011	A technocrat in Amaro District Agriculture office (natural resource management department).
62.	Godana Aga	48	April 2011	A Guji man whom I found in Bule Hora town where he went to save his money in a bank.
63.	FGD with 7 Guji school children	15-20	March 2011	The discussion was made in Tore town where the pupils attended the school.
64.	FGD with Guji elders (19 in 3 groups)	Above 60	March 2011, April 2012, May 2012	The elders had knowledge and experiences on many topics related to the research.
65.	FGD with Guji out of school children	16-20	May 2012	The children shared their daily experiences of herding and encounters with scouts.
66.	FGD with 5 Koore women	47-56	Feb. 2012	Mainly discussed about market and trade relations with Guji.

Appendices

Tourist flows to Nech sar National park

Nech Sar National Park

Tourists and Revenue Data

1987-2004

	Year(E.C)	Foreign tourists	Domestic Tourists	Total Tourists	Revenue
	1987	447	600	1047	27185
	1988	713	885	1598	42181
	1989	2241	1081	3322	60036
	1990	1035	999	2034	54029
SNNPR	1991	1201	1286	2487	66712
	1992	1275	1281	2556	68811
	1993	1113	1066	2179	58362
	1994	1163	1017	2180	153255
	1995	1700	1329	3029	222610
	1996	2049	2396	4445	234813
	1997	1703	1716	3419	265,477
APF	1998	5113	2328	7441	639,714
	1999	7267	8528	15795	695,419
SNNP	2000	8112	12854	20966	838,098
	2001	8992	9607	18599	996593
EWCA	2002	10261	8469	18730	1033459
	2003	12660	10049	22709	1331,075

5

Apendix 1: Nech Sar National Park Annual Report 2012 Showing Tourist Flow and Revenue Generated

AGREEMENT

Between

THE SOUTHERN NATIONS, NATIONALITIES AND PEOPLE'S REGIONAL STATE

and

THE GOVERNMENT OF THE FEDERAL DEMOCRATIC REPUBLIC OF ETHIOPIA (MINISTRY OF AGRICULTURE)

and

AFRICAN PARKS (ETHIOPIA) LIMITED

concerning

THE MANAGEMENT OF NECHISAR NATIONAL PARK

Appendix 2: Agreement Letter between APF and Ethiopian Government, page 1

7.10 The Company and the Government will establish a Steering Committee that will advise and report on all non-disruptive scientific monitoring and research regarding the ecology, distribution and population dynamics of game in the Park as well as the condition of the habitat with a view to assessing the Company's compliance with this agreement and any impact on the ecosystem that the Company's activities may have.

7.11 The Company will encourage independent research that is relevant and which may be carried out by scientists attached to universities and other appropriate institutions.

7.12 The Company may negotiate to lease land from the neighbouring communities or local authorities on such terms as are agreed by the Company and the neighbouring communities or local authorities and to manage this land as if it were part of the Park. Government will use its good offices to assist in this process.

7.13 The Company shall be entitled to employ any Ethiopian national as part of its staff complement either on a temporary or permanent basis and will not be restricted to employing only Government staff.

8. **The Government's Responsibilities**

8.1 The Government will support an application by the Company to the relevant Ethiopian revenue authority for duty exemption on all imports of equipment and motor [illegible] use in the management of the Park; for equipment needed to [illegible] the commercial and private lodges; and for any [illegible] concessions [illegible] offered to new investors into Ethiopia.

8.2 The Government will declare through the Negarit Gazetta the park as National Park [illegible] grant the park legal personality with all the attendant rights.

Appendix 2: Agreement letter between APF and Ethiopian Government, page 2

8.3 The Government, in accordance with its intention that the Park shall meet the IUCN criteria for Category II Protected Areas and for other reasons, undertakes to [illegible] all people living within the boundaries of the Park, their livestock and possessions, and to deal with all matters of compensation that may arise.

8.4 The Government agrees to negotiate in good faith on the realignment of the boundaries of the Park if this is considered necessary for improving the ecological integrity and visitor potential of the Park, taking cognisance of existing population pressures and community land rights.

8.5 The Government guarantees to assist with the restocking of the Park by allowing the Company to capture animals in other national parks, game reserves, sanctuaries, controlled hunting areas or other land in Ethiopia at no cost to the Company of the species detailed in Appendix A. The Company will cover only the cost of capture and transportation of such animals.

8.6 The Government reserves the right to collect crocodile eggs and hatchlings for its own use in crocodile ranching. Such collection of eggs or hatchlings will be in accordance with CITES guidelines, and will be at a level that doesn't threaten of the crocodile population in the park.

8.7 The Government guarantees to provide work permits for up to three expatriate staff to be employed by the Company to support park management efforts and permits for the respective spouses or partners of such staff to work in Ethiopia. The Government also guarantees to provide work permits for up to eleven expatriate staff and their spouses or partners to be employed by the commercial and private lodges.

8.8 The Government guarantees to provide any permits and licenses necessary for the lawful carrying out of any and all operations and activities expressly and impliedly allowed for in this agreement in a manner and extent that enables the Company to fulfill the terms and conditions of this agreement at

10

Appendix 2: Agreement letter between APF and Ethiopian Government, page 3

10.6 No variation to the terms of this agreement or the appendices hereto shall be effective unless agreed upon in writing and signed by the authorized representatives of the parties hereto.

10.7 Any notices to be given pursuant to this Agreement shall be in writing and may be given by sending the same by paid registered post, fax, or recorded personal delivery addressed to the party concerned at its address given hereunder or such other address as such party may have notified in writing to the other party for this purpose or by delivering the same to the party concerned at such address any notice so given shall be deemed to have been served:-

(a) If served by registered post: within ten days from the date of posting;

(b) If sent by recorded delivery: on the day on which delivery was acknowledged;

(c) If sent by fax: within two hours of receiving a remittance advise of successful transmission.

For purposes of this clause the addresses for the parties hereto are as follows:_

The Regional Government:	The President The Southern Nations, Nationalities and People's Regional State P.O. Box 18 Awassa Ethiopia
The Federal Government:	A/Minister, Ministry of Agriculture P.O. Box 3735 Addis Ababa Ethiopia
The Company :	African Parks (Ethiopia) Limited P.O Box 3658 Addis Ababa Ethiopia

13

Appendix 2: Agreement letter between APF and Ethiopian Governmnet, page 4

IN WITNESS WHEREOF the undersigned being the duly authorized representatives of the Government and the Company have signed this agreement at Addis Ababa on the day and year shown.

( )	( )	( )
For and on behalf of the Government of the SNNP Region	For and on behalf of the Government of Ethiopia	For and on behalf of the Company
Date: Feb. 17th 2004	Date: Feb-17/2004	Date: Feb 17 /2004

14

Appendix 2: Agreement letter between APF and Ethiopian Government, page 5

21/10 2008 16:17 FAX 251 06 810541 GAMOGOFA-Z.A.O.

001

Att. to W/ro Almaz

1

በደቡብ ብ/ብ/ሕ/ክ/መንግሥት
የጋሞ ጎፋ ዞን አስተዳደር ጽ/ቤት.
Gamo Goffa Zone Office of Chief Administration

ቁጥር 22/ክ/ጽ/3509/23
ቀን 2/10/2000

ለደቡብ ብ/ብ/ሕ/ክ/መ/ርዕሰ መስተዳድር ጽ/ቤት
ለደቡብ ብ/ብ/ሕ/ክ/መ/ሆቴሎችና ፓርኮች ኤጀንሲ.
ሀ ዋ ሣ

ጉዳዩ:- <u>የነጭ ሣር ብሔራዊ ፓርክ አስተዳደርን ይመለከታል</u> ፣

እንደሚታወቀው በአሁኑ ጊዜ ነጭ ሣር ብሔራዊ ፓርክ ተብሎ የሚጠራው ፓርክ ላለፉት 60 ዓመታት 1940ዎቹ ጀምሮ በአካባቢው ባሉት ብሔረሰቦች / በአማሮ ፣ በቡርጂ እና በጋሞ ብሔረሰብ / በጋንታ ካንቻሞ፣ ገንታ ኦቾሎ/ ብሔረሰብ አባላት የአካባቢ እሴቶችና እሣውንቶች ሲጠበቅ ከቆዩ በኋላ የሀገሪቷ ብሔራዊ ፓርክ መሆን አለበት ተብሎ ከተወሰነበት ጊዜ ጀምሮ በእነዚህ ብሔረሰቦች ሕልውናው እየተጠበቀ የነበረ ነው:: ታሪካዊ ሂደቱ ይህ ሆኖ ሳለ በአባያ ሐይቅ አካባቢ ለግጦሽና ውሃ ፍለጋ የሚዘዋወሩ ጉጂ ኦሮሞ ብሔረሰብ አባላት ከብቶቻቸውን ይዘው እዚያ አካባቢ ብቅ ብቅ ማለት ጀምሩ፣ እነኝህ ጉጂ ማኅበረሰብ አባላትም ሆነ በአማሮ ልዩ ወረዳ ያሉ 8 ቀበሌያት የሮ፣ መነና ደርባ ክሊኒክ ጀምሮ በአርባ ምንጭ ዙሪያ ወረዳ ሲተዳደሩ እንደነበርና ማንኛውም መሠረታዊ አገልግሎት ከአርባ ምንጭ ዙሪያ ወረዳ እንደሚጠቀሙ እሙን ነው::

መንግሥት የፓርኩን አስተዳደር ለማጠናከር በወሰደው እርምጃ ፓርኩን አፍርካን ፓርክስ ለተባለው ድርጅት 25 ዓመታት እንዲያስተዳድር ውል በመፈራረም አስረክቧል:: የአፍሪካን ፓርክ በሚያስተዳድሩበት ጊዜ በአካባቢው ግጦሽ ፍለጋ ከመጡት ጉጂ ኦሮሞዎች ጋር በየጊዜው ከሚፈጠረው የሥራ አለመግባባትን ለመቅረፍ ሲል መንግሥት ጉዳዩን አጥንቶ ዘላቂ እልባት እስኪወሰን / ጉጂ ኦሮሞዎችን ወደ ሌላ አካባቢ ወስዶ እስኪያሰፍር/ ድረስ ወደ ዋናው ፓርክስ / Core area / እንዳይገቡ ጊዜያዊ ውል መፈራረሙ ይታወሣል:: የሀገራችን መንግሥት ከአፍሪካ ፓርክስ ጋር በተዋዋለው ውል ውጭ የተፈጸመ ድርድር ስለሆነ ተቀባይነት እንደሌለው የተቀጠረ ነው:: ሆኖም የፓርኩ ችግር ከላይ በተጠቀሰው ሁኔታ መላሽ ባለመግኘቱ የሁለቱ ክልል ርዕሰ መስተዳድሮች እስከ ቦታው በመሄድ አሁን ያለውን ችግር እንዲፈታ በዋናው ፓርኩ

ማስረጃ ቁ. 01 ፋክስ 06-81-05-41 አርባ ምንጭ Arba Minch ስልክ 046-881-05-...
መልስ ሲጽፉልን የኛን ቁጥር ይጥቀሱ
When replying please mention our reference No

Appendix 3: Letter From Gamo Gofa Zone Administration to SNNP Regional State President Office Complaining Guji's presence in Nech Sar National Park

21/10 2008 16:18 FAX 251 06 810541 GAMOGOFA-Z.A.O. 002

.../2

ያሉት ቤቶችም ሆነ ሌሎች ቁሳቁሶች ከፓርኩ ዋናው ክፍል እንዲወጡ ወስነው መምጣታቸው እናሣስውሣለን፡፡

በመሠረቱ የነጭ ሣር ብ/ፓርክ Main area and non main area የለውም በጊዜው የተደረሰው ውል እነኝህ ብርቅዬ የሀገር ሀብት የሆኑ እንስሳትን እንዳይጠፉ የተደረገ ጊዜያዊ ስምምነት ነው፤ አሁን በሮሚያ በኩል እየተካሄደ ያለው አሠራር ከ Main area ውጭ ያለው ቦታ የኦሮሚያ ይዞታ ስለሆነ ማንኛውም መሠረተ ልማት መገንባት ይኖርብናል ሰዎች እዚያው በቋሚነት መኖር አለባቸው የሚል ነው፡፡ ይህንን ሁኔታ ከታሪክ ጋር አገናዝበን ስናየው ኦሮሚያ በነጭ ሣር በኩል ከጋሞ ጎፋ ዞን ጋር ተዋስኖም አያውቅም ድንበርተኛ አይደለም፡፡ ስለዚህ አሁን ኦሮሚያ ክልል ከጉጂ ዞን ጋር ሆኖ እየሄደ ያለው ሁኔታ የአካባቢውን ብሔረሰቦች መብትና/ ሕልውና / የሚነካና በሕዝቦች መካከል ያለውን ሰላማዊ ግንኙነት የሚያበላሽ በመሆኑ ከዚህ በፊት ከአፍሪካ ፓርክስ ጋር የደረሱት ጊዜያው ውል በሌሎች ተቀባይነት ካለገኘ የሚፈርስ መሆኑን የሚጠቁም በመሆኑና ውሉም የሚመለከታቸውንና ጉዳዩ የሚያገባቸውን ክፍሎች ያላሳተፈ፤ በመሆኑ የነጭ ሣር ብሔራዊ ፓርክ ክልል ሕልውና መጠበቅ ከአርባ ምንጭ ከተማ ሕይወት ጋር ቀጥተኛ ተያያዥነት ያለው በመሆኑ አሁን ጉጂ /ሮሚያ / ዞን ቦታውን ለቋሚ ይዞታ ለማድረግ ሰላማዊ እንስሳትን በየቀኑ እየገደለ / በቂ መረጃ አለን/ በዋናው ፓርክ ክልል በተጨማሪ በረቶችን እየሰራ የጀመረው የማስፋፋት ጥያቄ በአስቸኳይ እንዲቆምንልንና አሁን የፓርኩን ክልል ለቋሚ መኖሪያ አካባቢ እያመቻቸ ጤና ጣቢያ ፣ ት/ቤት ውሃ ተቋማት ግንባታ / ያለው ሁኔታ በአስቸኳይ እንዲያቆም እና ለጉጂ ኦሮሞ አርብቶ አደሮች ዘላቂ የመኖሪያ ቦታ ከኦሮሚያ ክልል እንዲመቻችላቸው የክልሉ መንግሥት ከሚመለከታቸው ጋር [illegible] ጥረት እንዲያደርግልን አበክረን ደጋግመን እናሣስባለን፡፡

ከሰላምታ ጋር

S/N/N/P/R/ Government Gamo Goffa Zone Office of Chief Administrator

ዶሣ ማዴ
Dosa Made
ዋና አስተዳዳሪ
Chief Administrator

ግልባጭ፣
ለጋሞ ጎፋ ዞን ፍትህና ፀጥታ መምሪያ
ለጋሞ ጎፋ ዞን ፖሊስ ጽ/ቤት
ለጋሞ ጎፋ [illegible]
ለአርባ ምንጭ ከተማ ሽ/አስተዳደር ጽ/ቤት
ለአርባ ምንጭ ዙሪያ ወረዳ አስተዳደር ጽ/ቤት
አርባ ምንጭ፣ [illegible]

Translation of the above letter:

Letter sent from Gamo Gofa Zone office of Chief Administrator

To: **SNNP Regional State, Office of the President**

To: **SNNP Regional State Hotels and Parks Agency**

Date: June 10, 2008

Subject: **Concerning the Administration of the Nech Sar National Park**

As it is known, the park that is currently known as Nech Sar National Park was managed and protected by local communities, elders and chiefs (Amaro, Burji and Gamo–Ganta-Kanchame and Genta Ocholo) for the last 60 years since 1940s and continued to be protected by these communities even after it was designated as national park. Although the historical process was like this, members of the Guji Oromo who move from place to place in search of water and pasture began to appear around Lake Abaya. The Guji nationality and the 8 villages under Amaro district (Yero, Manana Derba, Klinic villages and others) were administered under Arba Minch zuria woreda [district] and got social services from Arba Minch zuria woreda.

Based on the government's initiative to strengthen the park administration, African Parks Foundation took over the management of the park through 25 years contract. During its administration, the African Parks made temporary agreement with the Guji Oromo (only until the government would relocate them permanently to other places) and the agreement was reached to restrict the Guji people from entering the Core area of the park. However, since the agreement was made irrespective of the general agreement that African Parks made with the government of our country, we consider it as illegitimate. Eventually, since the problem of the park was not resolved, the regional presidents of the two regions visited the area and decided for the removal of houses and properties from the park core area.

Basically, the division of the Nech Sar National Park into Main area and non-main area was only made by the time to protect these endemic animals which are precious national heritages. Nevertheless, we came to understand that the Oromia regional state claims areas out of the main area as part of its administration and has begun establishing different social services and settlements in the areas. When we investigate this from historical perspective, Oromia has never had any boundary with the national park nor with the Gamo Gofa zone. Therefore, this process of Oromia's expansion to the area through Guji zone [Borana zone] violates the rights of other ethnic groups in the area and also threatens peaceful coexistence among different groups. Thus, as the agreement with African Parks would not be considered binding without recognition from other stakeholders, because the agreement did not participate concerned bodies, because the issue of Nech Sar National Park is directly connected to the existence of Arba Minch town, and because that the Guji Oromo continued killing peaceful animals and continued settlements for humans and livestock, we urge that the regional president should intervene in this case without delay because we have evidences that the Guji are continuing to establish healthcare centers, residence houses, school, water, and other social services. We also strongly request that the regional government should deal with Oromia region and resettle the Guji in Oromia region.

With Regards,
[Stamp and signature of the Gamo Gofa Zone administration]
Cc
To: Gamo Gofa Zone Justice Office
To: Gamo Gofa Police Department
To: Arba Minch town Transitional Administration
To: Nech Sar National Park Office
Arba Minch